# SECONDARY ION MASS SPECTROMETRY

# SECONDARY ION MASS SPECTROMETRY

## AN INTRODUCTION TO PRINCIPLES AND PRACTICES

Paul van der Heide

Published by John Wiley & Sons, Inc., Hoboken, New Jersey
Published simultaneously in Canada

For general information on our other products and services or for technical support, please contact our
Customer Care Department within the United States at (800) 762-2974, outside the United States at
(317) 572-3993 or fax (317) 572-4002.

Wiley also publishes its books in a variety of electronic formats. Some content that appears in print
may not be available in electronic formats. For more information about Wiley products, visit our web
site at www.wiley.com.

*Library of Congress Cataloging-in-Publication Data:*

Van der Heide, Paul, 1962– author.
  Secondary ion mass spectrometry : an introduction to principles and practices / Paul van der Heide.
    pages cm
  Includes bibliographical references and index.
  Summary: "This is presented in a concise yet comprehensive manner to those wanting to know
more about the technique in general as opposed to advanced sample specific
procedures/applications"– Provided by publisher.
  ISBN 978-1-118-48048-9 (hardback)
1. Secondary ion mass spectrometry. I. Title.
  QD96.S43V36 2014
  543'.65–dc23
                                                    2014011448

10 9 8 7 6 5 4 3 2 1

# CONTENTS

Development of the techniques Secondary Ion Mass Spectrometry (SIMS) and Secondary Neutral Mass Spectrometry (SNMS) during the last three decades ranks as one of the most important advances in Chemical Physics and Surface Science.

Information derivable from these techniques is of vital importance in the understanding of the atomic, molecular, ionic, solid-state, and electronic processes that occur at the surface and within the bulk of materials. The dynamics of ion and neutral species, as they approach, penetrate, charge-exchange, diffuse, dissociate (molecules and molecular ions), and sputter some of the substrate atoms, molecules, ions, and clusters, provides a wealth of information. Some of this information cannot be obtained by any other means. Although SIMS is now moving out of adolescence into a stage of maturity, there is still much to be learned about the mechanisms and applications of SIMS and SNMS.

This book provides both a pedagogic function and a research need. The pedagogic function is particularly evident in the early chapters. These chapters are written at the level of senior undergraduates or beginning graduate students with backgrounds in Chemistry, Physics, and/or Engineering. Many spectra and illustrative diagrams are included to exemplify the discussions. The research function is particularly evident in Chapter 3. These chapters contain state-of-the-art quantum mechanical and classical treatments as well as new experimental processes that are at the brink of current research.

The approach has been to amalgamate theory and experiment throughout this book. Both classical and quantum models are used as they are both important in understanding the sputtering, ionization, neutralization, quantification, dissociation, implantation, chemical reactions, and cluster formation that are encountered in the sputtering process.

J. Wayne Rabalais

*J W Rabalais*

Distinguished Professor of Chemistry and Physics
Lamar University
P. O. Box 10022
Beaumont, Texas 77710, USA

Secondary Ion Mass Spectrometry (SIMS) is a microanalytical technique used to understand the composition (isotopic, elemental, and/or molecular) of any predefined microvolume from any solid or made to be solid region. This region can include a solid's surface, the interface between two or more chemically distinct solids, and/or any internal volume of the solid. Some examples of fields in which SIMS has been applied (listed in order of application), or is being introduced to, include:

1. The Material Sciences. As an example, SIMS is the technique of choice for defining dopant distributions in the semiconductor industry. SIMS is also applied in the energy, plastics, and automotive industries and so on.
2. The Earth Sciences (Geochemistry, Atmospheric Chemistry, etc.) along with Archeology, Chronology, and Cosmochemistry. Note: SIMS is the benchmark for dating polychronic zircon populations.
3. The Biosciences inclusive of Pathology and Proteomics, along with Metabolomics, Lipidomics, and Pharmacology. This arises from the ability to image elemental and molecular distributions over submicron regions.

In addition, SIMS has experienced extensive growth and sophistication within each of these divergent fields over the past few decades, with its commercialization resulting in the availability of numerous instrument types and geometries with price tags ranging from several hundred thousand US dollars to several million US dollars. As a result, SIMS is now considered the most heavily used of the ion spectrometries for examining submicron scale regions on or within any solid material.

SIMS derives compositional information by directing a focused energetic ion beam at the surface of interest. These ions, referred to as *primary ions*, induce the emission of atoms and molecules from the solid's surface, a small percentage of which exist in the ionized state. The emitted ions, referred to as *secondary ions*, are then collected and passed through a mass spectrometer, hence the name secondary ion mass spectrometry.

The popularity of SIMS stems from its ability to:

1. measure any isotope of any element (H-U) from any solid (conductors through insulators). Note: Liquids can also be examined if frozen
2. examine molecular ions exiting a solid's surface along with their fragmentation patterns, with some novel chemical experiments also possible

3. record many elements/molecules to high sensitivity, high dynamic range, and extensive detection limits (some elements to sub parts per billion levels)

4. define the location of elements/molecules with a spatial resolution of 1 μm or better and depth resolution values reaching or even exceeding 1 nm

5. collect the needed data with relative ease and, in most cases, minimal sample preparation.

Indeed, the full potential of SIMS is yet to be realized. As an example, the recent introduction of large cluster primary ion sources has allowed for the mapping of organic molecules in all Three Dimensions (3D) to levels not possible with any other technique. Likewise, the incorporation of Fourier Transform Ion Cyclotron Resonance (FT-ICR) mass filters has allowed for unprecedented mass resolution values (>100,000) to be reached. These are but a few of the new areas being researched.

The areas being researched are probably best exemplified in the types of manuscripts reported at the International Conference on Secondary Ion Mass Spectrometry, i.e. in 1991, over 60% were devoted to atomic secondary ion emissions, where as 20 years later, over 60% covered molecular emissions. This shift also illustrates the maturity in the use and understanding of atomic secondary ion emissions. Indeed, SIMS when used in this manner is applied more heavily than any other area with most applications seen within the industrial sector. Indeed, the refinement of industry-specific practices (in some cases, these are considered intellectual property) has resulted in this form of SIMS being used more like a metrology-based technique.

That being said, SIMS remains a technique lacking a complete understanding of the physics (fundamentals) leading up to the recorded signal. Indeed, there is no one model that describes all secondary ion emissions from all surfaces.

The inspiration for this book arose when teaching both the fundamental and the practical aspects of SIMS. More precisely, this arose on realizing how the collective works and experiences could be funneled into a book that could further facilitate this transfer of knowledge. The premise used in putting together this book was *easily attainable answers to all of the questions asked over the years*. For example, how is it that SIMS can probe molecular distributions as a function of depth when the sputtering process is known to introduce severe damage to the lattice structure?[1] In adhering to this premise, all sections are prepared such that they can be read independently of each other, all equations are presented using the most commonly used units, and all fundamental aspects are discussed using classical analogies, where possible, over the more correct quantum mechanics descriptions.

---

[1] The possibility of depth profiling organic molecular species, and so mapping such species in all three dimensions, arises from the fact that large cluster ion-induced sputtering can, under optimized conditions, induce the removal of the entire damaged region per sputter cycle (the damaged region resulting from the sputtering event is also minimized) such that the exposed underlying surface is effectively damage free. In addition, many large cluster ions tend to evaporate from the sputtered region during/following the sputtering event.

This book is subdivided into two broad sections following a brief introduction (Chapter 1). These sections cover the following:

1. Fundamental aspects (Chapters 2 and 3). This not only covers the process of sputtering and ionization (those responsible for secondary ion formation/survival) but also presents a brief description of solids, atoms, and molecules as well as the parameters used in describing secondary ion formation/survival.

2. Analytical aspects (Chapters 4 and 5). This covers various instrument types, configurations, and setup conditions required for various specific types of analyses encountered. The more common data acquisition modes and data conversion practices in both Static SIMS and Dynamic SIMS are covered.

A brief compilation of other analytical techniques is then presented in Appendix A along with tables and additional concepts of interest to those in the field of SIMS.

The diverse topics covered are presented as simply and concisely as believed possible with the main emphasis given to attaining a practical understanding of all aspects of SIMS. It is also understood that for many analytical applications, knowledge of the underlying theory is not necessary. Knowledge of both is, however, useful in setting up new analytical protocols, examining new areas of application, and of course, in fundamental studies. This knowledge also provides for the ability to recognize, understand, and even predict the nature of various distortions that may be introduced during SIMS analysis. This, in turn, will allow for a more effective translation of the recorded data into maps representative of the original compositional variations.

It is hoped that this book will find use as an effective stepping stone to the wealth of information presently available on this technique, from the numerous journal articles to some of the more detailed books covering more specific application areas. Some examples of existing tests include Time-of-Flight (TOF)-SIMS (Vickerman and Briggs 2012), Cluster SIMS (Mahoney 2013), and SIMS in the Earth Sciences (Fayek et al. 2009). Moreover, there are the general surface analysis texts, with some examples including those by Riviere and Myhra 2009, Vickerman and Gilmore 2009, and O'Conner et al. 2003.

# ■■■■ ACKNOWLEDGMENTS

Although this book has profited from many people, there are several, in particular, whose names more than deserve mention. First of all, a special thanks goes to Kim van der Heide for the support provided during the many iterations of this project and for being a first draft editor. Secondly, I need to express a deep level of appreciation to Wayne Rabalais for the invaluable tips, support, encouragement, and discussions provided over the years. Finally, I would like to indicate my gratitude to Paul Ronsheim, Michel Schuhmacher, Nathan Havercroft, Scott Bryan, and Christopher Penley for the useful discussions and additional information.

Thank you all.

## LIST OF CONSTANTS

| | | |
|---|---|---|
| Boltzmann's constant | $k_B$ | $1.381 \times 10^{-23}$ J K$^{-1}$ or $8.616 \times 10^{-5}$ eV K$^{-1}$ |
| Elementary charge | $q$ | $1.602 \times 10^{-19}$ C |
| Mass of electron | $m_e$ | $9.109 \times 10^{-31}$ kg |
| Mass of neutron | $m_n$ | $1.675 \times 10^{-27}$ kg |
| Mass of proton | $m_p$ | $1.673 \times 10^{-27}$ kg |
| Planks constant | $h = \hbar.2\pi$ | $6.626 \times 10^{-34}$ J s or $4.136 \times 10^{-15}$ eV s |
| Speed of light | $c$ | $2.98 \times 10^8$ m s$^{-1}$ or $2.98 \times 10^{10}$ cm s$^{-1}$ |

# CHAPTER 1

# Introduction

## 1.1 MATTER AND THE MASS SPECTROMETER

The world we live in is a highly customized environment tailored to maximize our comfort level. This comfort level (which pertains to our well-being, environment, security, transportation, information access, etc.) is acquired through our capacity to fabricate materials that do not exist in nature. This capacity is aided through our ability to understand what has been fabricated, how this interacts with its environment, and how this can be tailored to our needs. This understanding is provided through the act of *analysis.*

Our ability to customize our environment is something that can be said for almost every age the human race has progressed through. Indeed, some eras are associated with the material developed. Examples include the *Bronze Age* (~3300 BCE to ~1200 BCE) and the *Iron Age* (~1200 BCE to ~500 CE). Two more recent examples of this customization include the use of Carbon for creating *plastics* and Silicon for constructing *computer chips* in what is now referred to as the *computer* or *information age.*

Indeed, plastics have become one of the most ubiquitous materials in today's everyday life. Plastics, fabricated from crude oil, are composed of a Carbon backbone formed from $n$ repeating units ($n =$ an integer $>1$) of some *monomer* (a molecule that binds to other molecules or atoms), hence the name *polymer.* For example, Polyethylene is defined as $(C_2H_2)_n$, Polypropylene as $(C_3H_6)_n$, and Polyvinylchloride (PVC) as $(C_2H_3Cl)_n$. Some applications of these three examples are as follows:

1. Polyethylene is used in manufacturing plastic bags, bottles, containers, and so on.
2. Polypropylene is used in packaging/labeling materials, textiles, bottles, and so on.
3. PVC is used in manufacturing specific types of tubing, signs, furniture, and so on.

*Secondary Ion Mass Spectrometry: An Introduction to Principles and Practices*, First Edition.
Paul van der Heide.
© 2014 John Wiley & Sons, Inc. Published 2014 by John Wiley & Sons, Inc.

In addition, there are many more types of plastics and applications.

The element below carbon in the periodic table is that of silicon. When purified from sand (one source), this is the basis of the solid-state semiconductor industry, as we know it today. Indeed, owing to the increasing prevalence of Complementary Metal Oxide Semiconductor (CMOS)-based integrated circuits and the ever-decreasing size of the transistor (over a billion transistors can now be squeezed into a single cm-by-cm-sized chip), more silicon-based transistors have been manufactured than anything else summed over the entire history of mankind.

Interestingly enough, the element below Silicon in the periodic table has played a pivotal role in the continued scaling of transistor dimensions. This stems from the fact that introducing Germanium into substitutional sites within the Silicon lattice induces strain, which, in turn, enhances charge mobilities. Post CMOS-based devices, on the other hand, may be Graphene based. Note: Graphene is an allotrope of Carbon (allotropes are composed of the same elements but have different geometric structures).

Our ability to fabricate a material that exhibits properties specifically tailored to the desired need has arisen from the knowledge attained from the way in which *matter* interacts with each other and its environment. The common definition of *matter* is *anything that has mass and volume* (Barker 1870). According to this definition, *all matter* is composed of atoms irrespective of the phase (solid, liquid, or gas) it exists in (Note: We do not directly interact with the fourth state of matter, otherwise referred to as *plasma*).

The physical properties of *matter* can be defined by the knowledge of the following:

1. the type of atoms present, i.e. which elements are present,
2. the bonds between the atoms,
3. the molecular or crystalline structure.

Atoms are the smallest divisible unit of mass that exists under the conditions we live in. Each element displays a different chemical reactivity (Note: Atoms can only be broken down in high-energy plasmas, energetic sub-atomic particle collisions, etc.). Most atoms, however, do not like to exist as individual entities, rather they prefer to combine with other atoms. Some examples include $N_2$, $O_2$, and $CO_2$ as is present in the air, NaCl in table salt, and more complex combinations present in plastics, semiconductors, and so on.

As first realized in 1909 (Rutherford 1911), atoms are composed of a dense nucleus, which is made up of protons and neutrons around which electrons orbit. The reactivity of an atom is defined by the electrons. The number of electrons in a neutral atom is defined by the number of protons (the number of protons equal the number of electrons in neutral atoms) with the number of protons defining the element (Carbon has six protons). The number of neutrons is generally equal to the number of protons, but it can differ.

Atoms with the same number of protons but with a different number of neutrons are referred to as *isotopes*, with the mass of the specific atom defined by the sum of the protons and neutrons. For example, Carbon 12 (mass equals 12 u) has six protons and six neutrons. Its chemical symbol is $^{12}C$. Carbon 13 (($^{13}C$)), on the other hand, has six protons and seven neutrons. Isotopic mass is covered in Section 2.1.1.1.

Although isotopes of the same element display the same reactivity, their ratio can provide insight into adsorption/diffusion characteristics, past events/environments, and the date at which any such events occurred. The study of the latter is termed *Chronology*. The ability to derive such information stems from the fact that isotope ratios change in a predictable manner over time owing to what are referred to as *fractionation effects*. This ability and the ability to date materials are discussed further in Sections 1.2.3.

The existence of naturally occurring isotopes was first reported by J.J. Thomson in 1913 (Thomson 1913) and later confirmed by F.W. Aston in 1919 using Magnetic Sector-based Mass Spectrometry (Aston 1922).

Magnetic Sector-based Mass Spectrometry separates the isotopes of the elements by passing the monoenergetic beam of ions (atoms that have had an electron removed or added such that it has a charge) through a magnetic field placed normal (perpendicular) to the ion beam's initial direction of travel. This causes the deflection of the beam based on the mass-to-charge ratio ($m/q$) of the ion as illustrated in Figure 1.1. Note: This ratio is also specified as ($m/z$). If all ions have the same charge, as is the case in Figure 1.1, the deflection is then simply dependent on the mass of the ion. As the vast majority of an atom's mass is defined by the protons and neutrons within the respective nuclei, Mass Spectrometry provides a method for separating the isotopes and hence the elements/molecules of different masses. All forms of Mass Spectrometry can thus be viewed as scales for weighing individual atoms or combinations thereof, i.e. molecules.

Mass Spectrometry has become a highly effective technique for elucidating the type and amount of any isotope, element, or molecule present within unknown samples in the gaseous state or made to be gaseous state (originally liquid). Secondary Ion Mass Spectrometry (SIMS) is a variant of Mass Spectrometry that is used in the

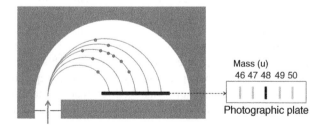

Mass (u)
46 47 48 49 50

Photographic plate

**Figure 1.1**   Cross-sectional image of how the isotopic constituents of a Ti$^+$ ion beam are separated while passing through a Magnetic Sector mass filter. Note: The magnetic field direction is perpendicular to the trajectory plane of the beam, i.e. perpendicular to the page and $q = 1$. The original instruments recorded mass separated images on photographic plates, as is shown.

chemical analysis of solid or made to be solid (frozen) materials. SIMS is discussed further in Section 1.2.

## 1.2 SECONDARY ION MASS SPECTROMETRY

SIMS now represents a fully commercialized technology that is widely used in both industry and academia for defining the isotopic, elemental, or molecular composition over highly localized microscopic regions within the surface and/or near the surface region (just below) of any solid. In specialized cases, frozen liquids can also be examined. As noted in Appendix A.10, an *ion spectrometry* is one that derives their information by recording *ions* as opposed to *electrons* or *photons*.

The popularity of SIMS stems, in part, from:

1. The ability to detect all of the elements within the periodic table (H–U) and combinations thereof, i.e. those that make up molecules. In many cases, this information can be collected quasi-simultaneously from the same surface/volume.
2. The detection limits provided (the ability to detect small concentrations, which in the case of SIMS can extend down to sub parts per billion levels), along with the associated sensitivity (the ability to detect small concentration differences) and the dynamic range (the ability to measure signals over some range, which in SIMS extends to $\sim 10^9$ when using multiple detectors).
3. The ability to map the distribution of any isotope or collection thereof (this ability can be used to replicate elemental or molecular distributions) on or within any solid to spatial and depth resolution values of 1 µm or less (10 nm represents the ultimate physical limit in SIMS) and $\sim 1$ nm, respectively.
4. The minimal sample preparation procedures required before analysis (in most cases, no preparation is needed).

The primary disadvantage commonly associated with SIMS lies in the relative difficulty in quantifying the recorded signal/s. This stems from the often unpredictable variations in the intensity of the recorded signal with substrate type and with the analysis conditions used. As a result, quantification requires that matrix-matched reference materials be analyzed in concert with the sample of interest (methodologies are covered in Section 5.4.3). This is in contrast to, for example, X-ray Photoelectron Spectroscopy (XPS) where quantification can be carried out without the aid of matrix-matched reference materials. Note: The use of reference materials in XPS can, however, enhance the accuracy of the measurement.

The information of interest in SIMS is carried within the ions emitted from the surface of a solid following energetic ion impact (a focused beam of between $\sim 0.1$ and 50 keV is most commonly used) as schematically illustrated in Figure 1.2. The impacting ions are referred to as the *primary ions*, whereas the emitted ions are

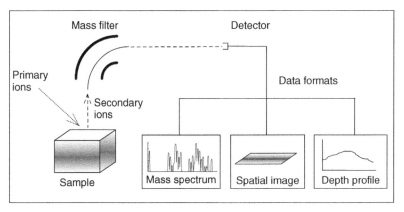

**Figure 1.2**    Pictorial illustration of the main components found in a SIMS instrument along with the common data output noted in Dynamic SIMS. Note: In Static SIMS, the data output is limited to mass spectra and spatial imaging for reasons discussed in the text.

referred to as the *secondary ions*. This figure also reveals the typical data formats available to SIMS as a whole, with these being:

1. mass spectra of all ionized species removed from the solid,
2. spatial distribution mapping of any signal of interest,
3. depth distribution mapping of any signal of interest (this is only available in the Dynamic SIMS mode of operation as is discussed later in this section).

   Combining depth distribution mapping with spatial distribution mapping allows for three-dimensional volumetric images to be generated.
   The mass filter used in commercially available SIMS instruments defines the instrument type, with the respective instruments referred to as

1. magnetic sector SIMS instruments (similar to those used in the original mass spectrometers)
2. quadrupole SIMS instruments (first appeared in the 1970s)
3. Time-of-Flight SIMS instruments (first appeared in the 1980s)

   Note: There are other possible mass filter types, all of which are covered in Section 4.2.3.1.
   In addition, there are two distinct modes in which SIMS analysis can be carried out. These modes, whose capabilities have been realized for sometime (Benninghoven et al. 1987), are otherwise referred to as *Static SIMS* and *Dynamic SIMS*. Static SIMS provides insight into the elemental and/or molecular distributions from the outer undisturbed monolayer, whereas Dynamic SIMS provides the elemental and/or molecular distributions over many atomic layers (in depth). Although the physics underlying signal generation in both modes share many

similarities (see Chapter 3), the analytical requirements along with the data collection procedures differ (see Chapters 4 and 5).

As Static SIMS can be used for probing the chemical composition as well as providing molecular signatures specific to the outermost monolayer of the surface of interest, this has attracted significant interest in fields extending from Materials Sciences to Biosciences (for further details, see Section 1.2.3) whether in academic or industrial settings.

Dynamic SIMS using chemically active atomic or small molecular primary ion beams ($Cs^+$, $O^-$, $O_2^+$) provides the utmost in detection limits in localized elemental analysis. As a result, this is the technique of choice for following dopant distributions as a function of depth in the semiconductor industry (a sector of Materials Sciences) and for mapping isotopic distributions in the Earth Sciences (for further details, see Section 1.2.3). Little in the way of molecular information is accessible under these conditions.

An exception to the above rule (the access to molecular information in Dynamic SIMS) is noted when using large cluster primary ions ($C_{60}^+$, $Ar_n^+$, etc., where $n$ can equal several thousands). This is realized because these ions introduce the possibility of molecular depth profiling and imaging. Cluster ion beams can also be used to examine surface distributions under Static-like conditions. As a result, the use of cluster ion beams in SIMS opens many new fields of application (Mahoney 2013).

Further discussion on the definitions of Dynamic and Static SIMS whether in their conventional forms or otherwise can be found in Sections 4.1.1.1–4.1.1.3.

At this point, it is worth mentioning some acronyms used in this field and its related fields. As the probe beam used in SIMS is often scanned across the solid surface, the term "SIM," which refers to *Scanning Ion Microscope*, has been used. Likewise, the term *"Imaging Mass Spectrometry"* (IMS) has been used to describe image analysis, which is carried out using some form of mass spectrometric technique. To avoid confusion with *Ion Mobility Spectrometry* (also referred to as IMS), the term *"Mass Spectrometry Imaging"* (MSI) is now more commonly used. Note: The term *"SIMS"* should not be confused with that now most commonly relayed in search engines on the World Wide Web. Here, the vast majority of results concern *multiple SIMulationS* as applies to the virtual gaming industry. The SIMS acronym is also used for *Stereoscopic Imaging and Measurement Systems*. This describes a methodology for relaying depth information during imaging.

## 1.2.1 History

The basis of SIMS can be traced back to the beginning of the twentieth century with the first experimental evidence of secondary ions reported by J.J. Thomson in 1910 (Thomson 1910). As quoted *"the secondary rays ... which were on the most part neutral with a small fraction of positively charged particles, were recorded following bombardment of a metal sample by primary kanalstralen* (Canal rays)." These positive ions were noted within a *Crookes tube* when a metal specimen held under vacuum conditions was bombarded by ions. The resulting emissions were then passed through an electric field and recorded on a photographic plate. The term *kanalstralen* was introduced by Eugen Goldstein in 1886 after he noted the production of rays (later understood to be positive ions) through holes of an anode from a low-pressure electrical discharge tube.

The first positive ion mass spectrograph from a gaseous sample was collected by Aston in 1919 after the discovery by Wilhelm Wien in 1898 that magnetic fields are effective in altering the trajectories of ion beams. In light of this discovery, Wilhelm Wien is often considered the forerunner of Mass Spectrometry (for a review, see Downard 2007). The production of secondary ions was first reported by Woodcock and Thompson in 1931 (Woodcock 1931; Thompson 1931). In their experiments, $Li^+$ primary ions were directed at NaF and CaF substrates with the negative secondary ion population recorded. Between 1936 and 1937, Arnott and Milligan, considered the forerunners of SIMS, used a magnetic field to separate the secondary ions on the basis of their $m/q$ ratio (Arnot and Milligan 1936) in much the same manner as that used by Aston.

The first experimental SIMS installation did not, however, appear until 1949 (Herzog and Verlock 1949), with the first dedicated Magnetic Sector-based instruments not constructed until the 1960s (Liebl 1967; Castaing and Slodzian 1962). Note: This timeline is closely associated with the development of the capability to produce the vacuum conditions required for surface analysis and the recognition of the potential of this methodology. The construction of Quadrupole-based SIMS instruments followed (Benninghoven 1971; Wittmaack 1975; Magee et al. 1978), with Time-of-Flight-based SIMS instruments appearing shortly thereafter (Chait 1981). These instrument types have since been fully commercialized (the first being the ims-101 constructed by Liebl in 1963 (Liebl 1967)) to increased levels of sophistication with each new generation. A discussion on instrument types is presented in Chapter 4.

The vast majority of the early work concerned the development of Dynamic SIMS for analyzing both extraterrestrial samples and samples for the semiconductor industry (e.g., see Magee and Honig 1982; Benninghoven et al. 1987; Zalm 1995). Since then, the prospect of examining molecular distributions from an undamaged outer surface has been introduced (Benninghoven 1969), with the true capabilities only fully realized with the development of Time-of-Flight-based SIMS instrumentation (Benninghoven 1994). Most recently, the possibility of carrying out three-dimensional molecular imaging using large cluster ion impact has been demonstrated (for reviews, see Vickerman 2009; Mahoney 2013), thereby opening up new areas of research.

The relatively late development and commercialization of SIMS can be traced to

1. the lack of knowledge concerning the fundamentals of secondary ion generation that, to some degree, still exists,
2. the extreme variability in secondary ion yields even for the same element emerging from different matrices, an observation termed the *matrix effect,*
3. the difficulty in generating the required vacuum (although analysis can be carried out at vacuum as low as of $10^{-6}$ torr, $10^{-10}$ torr, or better is preferable),

4. the complexity of the technique (numerous different instrumental conditions are available even within a specific instrument, with specific conditions required for specific types of analysis),
5. the large amount of data produced, as is particularly evident when using Time-of-Flight SIMS instrumentation.

### 1.2.2 Physical Basis

Conceptually, SIMS can be considered a straightforward and direct technique. In practice, there are many complexities introduced as a result of the various methodologies that can be applied, whether in the static or in the dynamic mode of SIMS. This exists because there are numerous conditions under which SIMS can be carried out. Each condition is optimized to deal with the analysis of a different elemental or molecular species, from different solid matrices. In addition, relating the output to the compositional variations that may occur on or within the solid being examined can be problematic. This stems, in part, from the complexities surrounding secondary ion generation, or more precisely, the *matrix effect*. As the term suggests, the *matrix effect* describes the effect of the matrix on the population of ions emitted. Matrix effects and their associated transient effects are discussed in Section 3.3.3.1.2.

SIMS can be considered a straightforward technique in that an energetic focused ion beam ($\sim$0.1 and 50 keV) is directed at the solid surface of interest. This beam is referred to as the *primary ion beam*, with some of the more common species being $O^-$, $O_2^+$, $Cs^+$, $Ga^+$, $In^+$, $Au_n^+$, $Bi_n^+$, $SF_5^+$, $C_{60}^+$, $C_{84}^+$, $C_{24}H_{12}^+$, $Ne^+$, $Ar_n^+$, $Xe^+$, etc., with $n$ being an integer starting from unity. Some of these species may be multiply charged. This primary ion beam induces the ejection of atoms and/or molecules from the irradiated area. This is referred to as *sputtering*. A schematic example of a sputtering process is shown in Figure 1.3. The small percentage of the sputtered population existing in an ionized state ($+1$ or $-1$, depending on the species, being by far the most common) is then collected, focused, and passed through a mass spectrometer. This ejected ionized population is referred to as the *secondary ion beam*, hence the name SIMS.

SIMS can be considered a direct technique in that the information of interest is derived from the mass of the secondary ions measured. As outlined in Section 2.1.1.1, the mass of an atom is defined by the number of protons and neutrons within the nucleus. Also mentioned in Section 2.1.1.1 is the fact that the number of protons defines the element (no two elements have the same number of protons) and hence its atomic number ($Z$). The number of neutrons is roughly the same as that of the number of protons with most elements exhibiting a distribution in the number of neutrons. These are the isotopes which, terrestrially, follow a *more or less* fixed distribution with the average referred to as the *natural abundance of the isotopes* (listed in Appendix A.2). Note: The term "*more or less*" is used as slight variations are noted according to various physical and/or chemical fractionation processes (Coplen 2002). Molecular mass is simply the sum of the masses of the isotopes of the elements that makes up the molecule. SIMS is thus

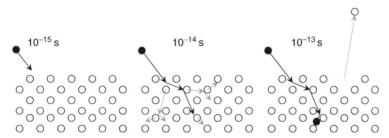

**Figure 1.3** Highly simplified pictorial illustration of the knock-on sputtering process (this is the most common form of sputtering as outlined in Section 3.2.1.1). The solid black ball represents an incoming primary ion, the hollow circles refer to the atoms making up the solid surface, the solid arrows refer to incoming ion trajectories, and the gray arrows refer to substrate atom trajectories. Also illustrated are approximate time scales over which the respective processes take place. As outlined in Section 3.3, some percentage of the sputtered population will exist in the ionized state. It is the ionized population (the secondary ion population) that is recorded in SIMS.

a spectrometry as opposed to a spectroscopy (for the original definition of both, see Appendix A.10) as isotopic mass is a constant, at least under nonrelativistic conditions.

### 1.2.2.1 *Sensitivity and Detection Limits*    Prime attributes associated with SIMS are its detection limits (ability to detect low concentrations) and its concurrent sensitivity (ability to detect concentration differences) to certain elements when analyzed under the appropriate conditions. Indeed, SIMS when operated in the Dynamic mode can routinely provide elemental detection limits extending to parts per billion or even sub parts per billion levels. As an example, Boron in Silicon can be detected to $\sim 0.4$ ppb or $\sim 2 \times 10^{13}$ atoms/cm$^3$ under fully optimized conditions. Note: However, that although SIMS can detect all elements and molecules, not all are detectable to the same levels. This is reflected in the detection limits for Hydrogen through Krypton in Silicon, as provided in Appendix A.5.

An understanding of how such detection limits can be attained over highly localized regions can be realized when considering the following Dynamic SIMS example using a Magnetic Sector instrument.

1. A Boron-implanted Silicon wafer is examined on a magnetic sector-based SIMS instrument using a 4-keV $O_2^+$ primary beam. $O_2^+$ is used as this enhances the $B^+$ ionization yields (see Section 3.3.1) while a Magnetic Sector instrument provides for best sensitivity/detection limits.

2. This primary beam is set to a current that results in an etch rate (sputter rate) of $\sim 1$ nm/s. Typical values range from less than 0.5 to greater than 5 nm/s. A 4-keV $O_2^+$ ion beam is used in this example as this removes $\sim 1$ substrate atom per impinging ion.

3. Fixing the analysis region to $100 \times 100$ μm results in a volume of $\sim 1 \times 10^{-11}$ cm$^3$ being removed per second. Note: As discussed in

Section 5.3.2.3.2, the analyzed region is made much smaller than the sputtered area to remove crater edge effects and thus to improve depth resolution.

4. As the atomic density (volume) of Silicon is $5 \times 10^{15}$ atoms/cm$^3$, the number of atoms within this volume that is removed approximates $1 \times 10^{-11}$ cm$^3 \times 5 \times 10^{22}$ atoms/cm$^3$ or $\sim 5 \times 10^{11}$ atoms/s.

5. SIMS, however, records the emitted ion population. If the emitted atoms experience an ionization efficiency of 1% (typical values range from greater than 10% to less than 0.0001% as discussed in Section 3.3.1), the number of ions approximates as $5 \times 10^9$ ions/s.

6. The likelihood that an ion formed at the sample surface reaches the detector is referred to as the instrument's transmission function (see Section 4.2.3). For the analysis of Boron, this can be maximized to a peak value of around 50% (see Table 4.3).

7. Using the above values, the number of ions reaching the detector per second can be approximated as $1 \times 10^9$ ions/s. As discussed in Section 4.2.3.3, an electron multiplier is capable of detecting every ion impinging on the first dynode (typical background values are less than 0.1 cps).

8. If one of these ions was to have a different mass than that of the elements making up the substrate (Boron has a different mass than Silicon), the detection limit under the above conditions equates to $\sim 1$ in $10^9$ or 1 ppb.

These conditions can be further adjusted to fully optimize the detection efficiency to the specific ions of interest, thereby further improving the respective detection limits. Note, however, that although all elements are detectable, such detection limits are noted only for a limited number of elements. This arises from differences in ionization yields (Step 5) as is discussed further in Section 3.3.2. The concept of volume densities and surface densities are discussed in Section 2.1.1.2.

In the case of static SIMS, a reduced volume is examined. To offset the parallel loss in detection limits/sensitivities associated with the analysis of smaller volumes, the secondary ion transmission functions of Time-of-Flight SIMS instruments (these are best suited to Static SIMS applications) can approach 100%, with all ions simultaneously recorded. As a result, surface detection limits for Iron on a Silicon substrate is in the low $10^8$ atoms/cm$^2$ range, whereas that for Boron on Silicon is in the mid $10^7$ atoms/cm$^2$ range (note the different units used). These detection limits can extend further for the more electropositive/electronegative elements when analyzed under appropriate conditions.

The advantage of increased transmission and simultaneous collection in Time-of-Flight SIMS utilizing pulsed primary ion beam instruments does not, however, translate to Dynamic SIMS studies. This is realized as a sizable portion of the analyzed volume is not recorded during the analysis. This stems from the fact that two ion beams are used, one for sputtering and the other for analysis, with the population arising from the sputter beam not recorded. This is further

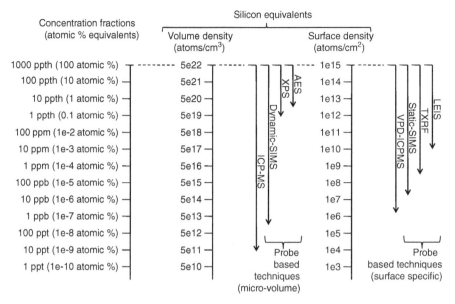

**Figure 1.4**  Concentrations in fractions and percentages. These are translated into volume and surface atomic densities for the Silicon (110) surface. Note: As discussed in Section 2.1.1.2, these densities are a function of the substrate as well as the surface plane. Engineering notation, where e refers to the base 10 (not to be confused with the exponential *e*), is used for the sake of simplicity. Also shown are some additional analytical techniques along with the approximate best possible detection limits noted for specific elements when using fully optimized analysis conditions. Note: ICP-MS, in this case, would either require solid digestion, be coupled with Laser Ablation (LA), or coupled with Vapor Phase Decomposition (VPD) for dissolving surface films.

compounded by limitations in the detection of multiple ions of the same mass within the same pulse.

To put these detection limits into perspective, the volume and the surface atomic densities of Silicon are related to atomic concentrations (in fractions and atomic percent) as shown in Figure 1.4. Note: Different substrates can exhibit different densities (that of Silicon is $4.98 \times 10^{22}$ atoms/cm$^3$). Likewise, different surface planes of a crystalline substrate can exhibit different surface densities. For example, the Silicon (100) plane has a density of $6.78 \times 10^{14}$ atoms/cm$^2$, whereas the (110) plane has a density of $9.59 \times 10^{14}$ atoms/cm$^2$. The (110) plane is used in Figure 1.4 for the sake of simplicity, i.e., is closest to $1 \times 10^{15}$ atoms/cm$^2$. Surface and volume densities are discussed further in Section 2.1.1.2.

Note: Such detection limits cannot be reached when examining large molecular ions because of the lower sputter and ion yields observed. These topics and the elemental emissions are discussed further in Chapter 3.

Also shown in Figure 1.4 are the relative detection limit ranges for SIMS when operated in both its Static and Dynamic modes, as well as the detection limits displayed by some other analytical techniques applied in the analysis of solids or solid surfaces. These include:

1. Auger Electron Spectroscopy (AES),
2. Inductively Coupled Plasma Mass Spectrometry (ICP-MS) whether by itself (digestion required) or coupled with Laser Ablation (LA) or Vapor Phase Decomposition (VPD),
3. Low-Energy Ion Spectroscopy (LEIS),
4. Total reflectance X-Ray Fluorescence (TXRF),
5. XPS (X-ray Photoelectron Spectroscopy)

From this list, only ICP-MS of a digested solution or when combined with LA or VPD, displays, superior detection limits compared to SIMS. It should be noted, however, that ICP-MS and VPD-ICP-MS are not true probe-based techniques, i.e., these are not useful when concentration variations over micro-volume to nano-volume regions are of interest. Furthermore, VPD-ICP-MS is a highly specialized technique developed specifically for the semiconductor industry. Of the micro-probe techniques, SIMS displays the highest elemental detection limits, whether used in the areas of micro-volume analysis/nano-volume analysis (both are forms of Dynamic SIMS) or surface analysis (Static SIMS). Optimal detection limits pertain to those noted from specific element–substrate combinations when examined under optimized conditions.

With the exception of SIMS, the above techniques along with various other related spectrometric and spectroscopic techniques are summarized in Appendix A.10. For completeness sake, various microscopies of interest are covered in Appendix A.11., whereas various diffraction and reflection-based techniques are summarized in Appendix A.12.

### 1.2.3  Application Fields

As SIMS is the solid-state analog of Mass Spectrometry, it has found extensive use in many highly diverse areas within both industry and academia, in which the distribution of isotopes, elements, and/or molecules on or within a specific region of the substrate is of interest. Three areas in which SIMS has found heavy application (listed in the order in which SIMS has been applied) are as follows:

1. Materials Sciences
2. Earth Sciences along with the Environmental Sciences and Cosmology
3. Biosciences

Note: This order does not necessarily relate to the impact SIMS has had in these respective fields. Moreover, of course, there are other fields in which SIMS has been applied, many in *niche* applications.

*The Material Sciences* cover an extremely broad field stemming from many disciplines. As a result, SIMS has been applied in this field more extensively than any other, with all the commercially available instrument types (Magnetic Sector, Quadrupole, and Time-of-Flight SIMS instruments) being applied accordingly. The analysis requirements are also the most diverse because of the range of samples encountered. This is realized as this field examines relationships between the structure of matter at the atomic/molecular scale and the resulting macroscopic properties displayed. For some reviews, see Benninghoven et al. 1987; Wilson et al. 1989; O'Connor et al. 2003; Riviere and Myhra 2009.

Some of the properties include but are not limited to (these are listed in alphabetical order):

1. chemical properties
2. electrical properties
3. mechanical properties
4. magnetic properties
5. optical properties
6. thermal properties

Recall: as mentioned in Section 1.1, the physical properties of matter are defined in one form or another by the elements present (the types of atoms) and how these elements bond with each other.

As an example, adding small amounts (<2%) of Carbon along with other elements to Iron during the forging process enhances alloying and thus the strength/hardness albeit at the cost of ductility. The result, otherwise known as Carbon steel, was heavily used in the development of the sword during the Iron Age and also played a pivotal role in the development of structural materials during the industrial revolution. A more recent example lies in the substitution of trace amounts of Boron, Phosphorus or Arsenic into the Silicon lattice in a highly localized manner. This modifies the local electrical properties, without which the Silicon-based semiconductor industry as we know today would not have developed. Moreover, there are many more examples.

Surface science is a subset of materials science concerned with the macroscopic properties of solid and/or liquid surfaces, and how these properties relate to atomic structure in the surface regions. Note: Most forms of matter present in the solid or liquid phases exhibit a surface layer that is different from that of the underlying material. This difference could be chemical (composition and/or bonding), structural (differences in bond angles, bond lengths, or long range order, i.e. the crystalline structure), or both. How a material is perceived by the outside world thus primarily depends on the form or more precisely the physical properties of the outermost layer, i.e. the surface. These may be in the form of (in alphabetical order)

1. adhesion and wettability
2. adsorption/desorption
3. biocompatibility

4. interfacial electrical properties
5. optical properties in the form of reflectivity
6. reactivity inclusive of corrosion, heterogeneous catalysis, and so on
7. texture
8. wear and tear (also referred to a tribology), and so on.

Note: A solid's or liquid's surface can be defined in several different ways. The more obvious definition is that *a surface represents the outer or topmost boundary of an object*. When getting down to the atomic level, however, the term *"boundary"* loses its definition as the orbits of the bound electrons are highly defuse. An alternative definition would then be that *a surface is the region that dictates how the solid or liquid interacts with its surroundings*. This can also be applied to a solid–solid interface. Applying this definition, a surface or interface can span as little as one atomic layer (0.1–0.3 nm) to many hundreds of atomic layers (100 nm or more) depending on the material, its environment, and the property of interest.

One example of a surface modification lies in the corrosion resistance exhibited by stainless steel. Stainless steel is composed of a range of alloys exhibiting stain resistance, hence its name. Indeed, 316-stainless steel is composed of Iron with 16–18% Chromium, 10–14% Nickel, 2–3% Molybdenum, less than 2% Manganese along with sub 1% levels of Silicon, Phosphorus, and Carbon. The corrosion resistance arises from the segregation of Chromium to the surface whereupon it forms a corrosion resistant oxide ($Cr_2O_3$) of only a few nanometers in thickness. Without this, the bulk material would corrode at a significantly greater rate similar to that noted from pure Iron. Moreover, there are many more examples.

The capabilities of SIMS when applied in this field vary according to the information required, the time constraints, and, of course, the capabilities of the instrumentation used. Indeed, imaging to a spatial resolution of ~1 μm and below is routinely possible on commercially available SIMS instrumentation. This can be pushed to the physical limit of 10 nm imposed by the sputtering process or more precisely the collision cascade (see Section 3.2.1) using highly specialized instrumentation (McPhail et al. 2011; Wirtz et al. 2012).

Likewise, depth resolution values to 1 nm and below are within reach or commercially available instrumentation equipped with Ultra Low Energy (ULE) primary ion impact capabilities (Dowsett et al. 1994). This, however, generally comes at the expense of detection limits. Lastly, detection limits for specific elements can extend down to sub parts per billion levels (see Section 1.2.2.1 and Appendix A.5), albeit at the expense of spatial/depth resolution.

The *Earth Sciences*, *Environmental Sciences*, and *Cosmology* have all long recognized the strengths of SIMS. A prime example of this lies in the fact that NASA funded the instrument built by Liebl (Leibl 1967) for such purposes. Some of the requirements of the application of SIMS in this area include the following:

1. The high mass resolution (see Section 5.2.2.2) at high sensitivity over micron scale regions when isotope measurements for chronology are of interest,

2. Effective sample preparation required to support the small and often insulating but highly stable grains prevalent in these fields. These are covered more in Section 5.1.1.2,

3. Understanding of the Instrument Mass Fractionation (IMF) effect. This results in small but measurable deviations from true isotope ratios as discussed in Section 3.3.1.2.3.

The first requirement has since resulted in the availability of commercial purpose built Magnetic Sector instruments. Examples include the SHRIMP™ series of instruments from the Australian Scientific Institute (ASI™) and the ims-1270/1280™ from Cameca™. These instruments provide the sensitivity to measure mass fractionation effects intrinsic to the samples because of the full transmission at a mass resolution $m/\Delta\, m$ of ~5000 accomplished using large radii electrostatic and magnetic sectors with ultimate mass resolution values extending well past this value. For further discussion on instrumentation, see Chapter 4.

SIMS is primarily applied in these fields for measuring stable and radiogenic isotope ratios from highly localized micron scale volumes as well as for measuring light element distributions not accessible to other techniques (for reviews, see Reed 1989; Fayek 2009). Moreover, there are other applications.

Radiogenic isotope ratios are used in Chronology. Measurement of such ratios is generally carried out using the Ar–Ar, Sm–Nd, or U–Pb radiometric techniques (and there are others) with the latter being the most prevalent. Here, the age of a rock (Zircon being the most popular) can be defined as between 1 million and 4.5 billion years by measuring specific Lead and in some cases Uranium isotope levels (Ireland and Williams 2003). These are of interest since once trapped within the Zircon sample, $^{238}$U decays into $^{206}$Pb with a half life of 4.47 billion years, whereas $^{235}$U decays into $^{207}$Pb with a half life of 704 million years. Zircon is used as it readily incorporates Uranium and Thorium (this decays into $^{208}$Pb), and once crystallized, it is highly stable, thereby forming an effective *time capsule*. As a result of the power of SIMS to derive such information, it has become the technique of choice in this area with such studies being of both academic and industrial interest.

Measurement of stable isotope factions of Hydrogen, Lithium, Boron, Carbon, Oxygen, Sulfur and even Chlorine, and so on are also studied as these elements provide a chemical history of the region of interest. Other areas of interest include trace element detection, diffusion analysis, and surface analysis (Reed 1989). SIMS has also been used to examine, among other things, Arsenic levels in various pyrite grains (Chryssoulis et al. 1989) as this provides evidence for the presence of Gold. Such studies can be carried out on less specialized SIMS instrumentation. Note: Gold and Arsenic are seen in close proximity. Indeed, owing to the increased sensitivity of Canaries to Arsenic, these were used during the early days of Gold mining, hence the term "Canary in a gold mine."

*The Biosciences* present a huge and as yet largely untapped field for SIMS (for some recent reviews, see Boxer et al. 2009; Vickerman and Gilmore 2009;

Passarelli and Winograd 2011; Mahoney 2013). Reasons for this apparent slow adoption lies in the following:

1. The extensive and often highly specific sample preparation procedures required in supporting the often unstable (vacuum sensitive) samples. These are covered further in Section 5.2.3.
2. The challenges in effectively identifying and/or relating many high mass molecular ions (Note: Both low mass and high mass ions are measured in the Biosciences). Two methodologies proving effective in this area are:

   a. cluster ion SIMS discussed in Section 4.1.1.3,
   b. the G-SIMS methodology discussed in Section 5.4.1.1.1.

3. The need for high sensitivity and good detection limits at high mass resolution when constrained to sub micron regions. As will be noted in Chapters 4 and 5, improving one generally comes at the price of the others.

In the case of the third requirement, instrument manufacturers have developed highly specialized Magnetic Sector mass filter-based instruments and time-of-flight mass filter-based instruments. Quadrupole mass filter-based SIMS instruments are ineffective in this area because of their inability to provide the high mass resolution required. Fourier Transform Ion Cyclotron Resonance (FT-ICR) as well as Orbitrap mass filter-based instruments, on the other hand, show significant promise. Mass filters are discussed in Section 4.3.2.1.

An example of a Magnetic Sector-based SIMS instrument that has proved effective in this area is the Cameca nanoSIMS-50. Indeed, this instrument is capable of full transmission at a mass resolution up to $m/\Delta m$ of $\sim 2000$ (higher mass resolution obtained at the expense of transmission). This instrument also incorporates a multi-detection system and normal incidence primary ion beams that can be focused to below 50 nm. Mass resolution, along with definitions, is covered in Section 5.2.2.2.

To date, such instruments have proved useful in probing metals in biological specimens with examples including Sodium, Potassium, Calcium, and Zinc (these are detected as $Na^+$, $K^+$, $Ca^+$, and $Zn^+$). This is generally accomplished by preparing time-sequenced slices of the region of interest. Other high yield ions of interest in biological specimens include isotopes of hydrogen, carbon, nitrogen, oxygen, fluorine, sulfur, and combinations thereof (recorded as: $^1H^-$, $^2D^-$, $^{12}C^-$, $^{13}C^-$, $^{16}O^-$, $^{18}O^-$, $^{19}F^-$, $^{32}S^-$, $^{12}C^2D^-$, $^{12}C^1H_2^-$, $^{12}C^{14}N^-$, $^{13}C^{14}N^-$, $^{12}C^{15}N^-$, $^{13}C^{15}N^-$, etc.), some of which are introduced through isotopic labeling/spiking. An example of these capabilities is illustrated in Figure 1.5 in which the $^{15}N$ uptake resulting from $^{15}N_2$ spiking of *Trichodesmium* trichomes is tracked as a function of depth (Finzi-Hart et al. 2009). Such studies provide the much needed information on metabolic uptake mechanisms. Further discussion on these capabilities along with additional examples can be found in reviews by Lechene et al. 2006 and Orphan and House 2009.

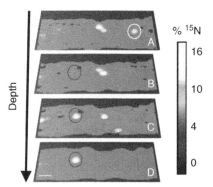

**Figure 1.5**  $^{15}$N spatial distribution maps collected at four different depths (images A–D) on a nanoSIMS 50 instrument. These were recorded from the $^{12}$C$^{15}$N$^{-}$ secondary ions emanating from multiple *Trichodesmium* trichomes spiked with $^{15}$N$_2$. The white and black circles highlight the increased levels noted on two distinct cyanophycin granules. The scale bar (white horizontal line in image D) represents 1 μm. Reprinted with permission from Finzi-Hart et al. (2009) Copyright 2009 National Academy of Sciences.

Time-of-Flight-based SIMS instruments that have proved effective in this area include the Ion-Tof™ TOF-SIMS series, and the Physical Electronics nano-*TOF* and TRIFT™ series. These instruments provide the capability of analyzing higher mass secondary ions than possible in their Magnetic Sector-based counterparts and to retrospectively examine signals not originally thought of interest (in Magnetic Sector-based instruments, all signals of interest must be defined before the experiment). The ability to follow specific molecular species or fragments intrinsic to the sample being examined (as opposed to molecules formed as a result of the collision cascade) also eliminates the need for isotopic labeling (Belu et al. 2003; Fletcher et al. 2011).

Although such instruments have proven effective, limitations resulting in poorer detection limits (due to the poorer molecular ion yields) and some reduction in spatial resolution are noted. Indeed, for the analysis of extremely large molecular ions, techniques such as Matrix-Assisted Laser Ablation and Ionization (MALDI) become more effective. That being said, the introduction of large cluster primary ion beams is opening up new areas not previously accessible to SIMS. This is realized as such beams can allow for three-dimensional mapping of specific molecular species to a spatial resolution of between 1 and 30 μm depending on whether a continuous (former value) or pulsed (latter value) primary ion source is used. Some recent reviews on these capabilities can be found in Vickerman 2009; Passarelli and Winograd 2011; Mahoney 2013.

Lastly, SIMS analysis has been carried out using a FT-ICR mass filter-based instrument. The example reported in the literature (Smith et al. 2011) describes the capability of imaging high mass molecular secondary ions resulting from C$_{60}$$^{+}$ cluster ion impact to a mass resolution better than 100,000 from micron scale regions. Such mass filters also provide greater detection limits for high mass secondary ions, but at the expense of analysis times. FT-ICR-based SIMS is presently, only available in research environments. Further discussion on instrumentation can be found in Chapter 4.

## 1.3  SUMMARY

SIMS is one of the many chemical analysis techniques presently available. Chemical analysis describes the act of elucidating the chemical constituents (elements and/or molecules) making up the material of interest and how these constituents interact with each other (bonding type, etc.). This then affords the possibility of defining the physical properties of the substance of interest.

SIMS can be summarized as a mass spectrometric technique that is applicable to solids (most other mass spectrometric techniques require the sample to be in the liquid or gaseous state). Mass spectrometric techniques differentiate between the different elements and molecules based on their mass-to-charge ratio. To simplify matters, the vast majority exists in the $+1$ or $-1$ charge state. Their mass, in turn, is defined by the number of protons and neutrons making up the element with molecules made up of elements. Note: All matter is made up of the elements with the physical properties of matter thus derivable from the elements present and the way in which they are bound.

SIMS probes a solid by directing a focused energetic ion beam (this is most typically in the 0.1–50 keV range) at the solid's surface of interest. These ions, defined as the *primary ions*, initiate the emission of atoms from the surface, with a small percentage departing in the ionized state. The emitted ions are referred to as the *secondary ions*. These are extracted and focused such that the beam can be directed into a mass spectrometer. Some primary ion beams include, but are not limited to, $O^-$, $O_2^+$, $Cs^+$, $Ga^+$, $In^+$, $Au_n^+$, $Bi_n^+$, $SF_5^+$, $C_{60}^+$, $C_{84}^+$, $C_{24}H_{12}^+$, $Ne^+$, $Ar_n^+$, and $Xe^+$. In some cases, multiply charged ions may also be used with some examples being $Bi_n^{2+}$ and $Bi_n^{3+}$.

The advent/development of SIMS initially hinged on the capability to produce the high vacuum conditions needed. Indeed, it was not until the 1950s that true ultra high vacuum conditions could be reached even though the first secondary ions were recorded in 1910. Note: The first mass spectrometer was not constructed until 1919 with the first experimental SIMS installation not appearing until 1949. Dedicated instruments did not appear until the 1960s. Owing to the popularity of SIMS, this technique has since seen extensive sophistication and diversification, some of which can be traced to the parallel development in computational power.

This sophistication is evident in all forms of SIMS as they exist today. Indeed, SIMS is now routinely used to measure isotopic, elemental, and/or molecular distributions (whether existing at the outermost surface or within a substrate) through the application of highly specific instrumentation. Likewise, SIMS, which has long been able to measure atomic ion emissions, can now do so to unprecedented sensitivity, detection limits, dynamic range, mass resolution, spatial resolution, and depth resolution.

Traditionally, SIMS has been subdivided into the areas otherwise referred to as *Static SIMS* and *Dynamic SIMS*. Static SIMS analyzes the undisturbed population of elements and molecules that are present at the outermost surface of a solid. Dynamic SIMS probes the constituents present at the surface and below through the removal of many layers per analytical cycle. This is carried out using primary ion beams of significantly higher current densities. This form of SIMS also introduces the possibility to derive depth distributions of any element and in some cases, molecular species with relative ease. The ability to derive molecular depth profiles and even three-dimensional images is a recent advent introduced using large cluster primary ion sources on specific substrates.

The popularity of SIMS stems from multiple factors, some of which simply arose as a result of the sophistication driven by both industrial and academic needs. This drive resulted from the initial realization that SIMS was capable of deriving both isotopic ratios and elemental depth distributions over highly localized volumes to extreme sensitivity and detection limits with relative ease. As a result, Dynamic SIMS has, for example, become the technique of choice in such areas as dopant profiling in the semiconductor industry and chronological studies using Zircon in the Earth Sciences. Increased use is also noted in the Biosciences, which is being further facilitated by the development of new methodologies. Indeed, new application fields continue to emerge with the new capabilities being developed.

The full power of SIMS is realized in that it provides insight into past events (when applied in the Earth Sciences inclusive of Cosmology), as well as present-day events (when applied in the Materials and Biosciences), thereby allowing predictions of future events (all sciences).

# PRINCIPLES

# CHAPTER 2

# Properties of Atoms, Ions, Molecules, and Solids

## 2.1 THE ATOM

The *atom* represents the smallest indivisible unit of matter, i.e., it cannot be broken down further unless exposed to an exorbitant amount of energy such as that seen in the sun's core, nuclear reactors, or particle accelerators. The word *atom* comes from the Greek word *atomus*, which means physically uncuttable. This concept is thought to have been first introduced by the pre-Socratic Greek philosopher Democritus around 400 BCE. Sufficient evidence for this concept was not, however, introduced until the early 19th century (post-Socratic era) for which Dalton is primarily accredited.

This chapter provides a cursory introduction to the atom's structure and the properties that result when two or more atoms/ions bind to each other. This includes a discussion on the electronic structure (the distribution of the electrons within an atom, ion, molecule, or solid) in ground or excited states. These are considered important because, as introduced in Section 1.1, the properties of matter are dictated by the type of elements present and how they bond to each other (this can include the long-range lattice structure). Electronic excitations are included because, as covered within Sections 3.2 and 3.3, sputtering can impart a substantial amount of energy into the bound system, which in turn can influence secondary ion yields.

### 2.1.1 Atomic Structure

Atoms are composed of *protons*, *neutrons*, and *electrons*. The protons and neutrons reside in the highly dense *nucleus* around which the electrons orbit. The nucleus is of the order of a few femtometers in diameter with the associated protons and neutrons being ~1.7 fm in diameter. The size of the atom, on the other hand, can range from 62 pm (for Helium) to 520 pm (for Cesium).

The structure of the atom was revealed in 1911 as a result of Ernst Rutherford's interpretation of the 1909 Geiger–Marsden experiment (Rutherford 1911).

*Secondary Ion Mass Spectrometry: An Introduction to Principles and Practices*, First Edition.
Paul van der Heide.
© 2014 John Wiley & Sons, Inc. Published 2014 by John Wiley & Sons, Inc.

Rutherford is also generally credited for the discovery of the proton (Rutherford 1919) even though first recorded by Wilhelm Wien. The neutron was discovered much later by Chadwick (Chadwick 1932). The electron was discovered by J.J. Thompson in 1897 (Thomson 1897). Interestingly, Chadwick was the student of Rutherford and Geiger, whereas Rutherford was a student of J.J. Thompson.

Free atoms prefer to exist in the neutral (zero charge) state. This results when the number of protons equals the number of electrons within the atom. Note: Protons have a unit charge of $+1$, whereas electrons have a unit charge of $-1$ (neutrons have no charge). The number of protons within the nucleus of an atom is described by the atom's *atomic number* ($Z$). This number appears in the periodic table above the elements symbol (see Appendix A.1). The sum of protons and neutrons within the nucleus of a specific atom, or more specifically, that for a specific *isotope* of an element is defined by the *atomic mass number* ($A$), which is also referred to as the *mass number*.

*Ions* are atoms that have gained or lost one or more electrons and thus have gained charge. The number of electrons gained or lost is represented by a superscript number following the elements symbol. As an example, $He^{2+}$ is a Helium atom that has lost its two electrons. Atoms that have not gained or lost electrons are simply represented using the chemical symbol alone, i.e., He for Helium.

*Elements* are atoms that display distinct chemical properties relative to other atoms. These properties are defined by the number and distribution of electrons within the valence shell. The fact that the overall number of electrons bound to a neutral atom equals the number of protons reveals the ordering and systematics in the chemical reactivity of the elements within the periodic table.

*Isotopes* are atoms of the same element with a different number of neutrons. The term isotope, from Greek meaning *same place*, was introduced by Frederick Soddy in 1913 (Soddy 1913). The sum of the protons and neutrons, and thus the identity of the isotope, is generally represented as a superscript number before the elements symbol, i.e., $^1H$ for Protium, $^2H$ for Deuterium, and $^3H$ for Tritium. Note: Protium and Deuterium are stable isotopes, whereas Tritium is not. When the elemental symbol alone is used, i.e. H, a mixture of all of the naturally occurring isotopes is assumed.

The terrestrial distribution of all naturally occurring isotopes (those not produced in reactors or particle accelerators of which tritium is an example) is more or less fixed with the average distribution, which is referred to as the *natural distribution of the isotopes* (Wieser and Coplan 2011; Bergland and Weiser 2011). This list, provided in Appendix A.2, reveals that 99.985% of the natural terrestrial Hydrogen contains no neutrons ($^1H$), whereas 0.015% contains one neutron ($^2H$).

As isotopes of the same element have the same number of electrons, they display the same chemical reactivity. *Isotope fractionation*, also referred to as *mass fractionation*, describes the subtle differences noted in reaction kinetics (speed) displayed by different isotopes of the same element. This is noted as mass ($m$), velocity ($v$), and energy ($E$) are all interrelation through $E = \frac{1}{2}mv^2$. Isotope fractionation can be useful in defining chemical reaction kinetics, secondary ion emission studies (see Section 3.3.1.2.3), and of course chronology (see Section 1.2.3).

***2.1.1.1 Atomic Mass*** The masses of the atoms, or more precisely their isotopes, are all scaled to the mass of the $^{12}$C isotope (1/12th of this equals one *atomic mass unit*). This scaling is a result of much debate between chemists and physicists (De Bievre and Peiser 1992) combined with the fact that there is no bar by which to measure these against. Indeed before 1961, both used slightly different definitions with chemists opting for the definition that one *atomic mass unit* equated to 1/16th of the mass of Oxygen, whereas physicists opted for the definition that one *atomic mass unit* equated to 1/16th of mass of the $^{16}$O isotope. The divergence, albeit small, was removed on the introduction of the *unified atomic mass unit* ($u$), with this equating to 1/12th the mass of the $^{12}$C isotope when set to a mass equal to exactly 12 $u$.

Note: The accepted *SI* unit for mass is the kilogram (kg). *SI* is from French meaning *Le Systéme International d'unités*. However, as the mass of an atom, ion, or molecule is significantly less than the standard *SI* unit, specific mass units needed to be introduced. That accepted by the *SI* is the *Dalton* (Da). More commonly used, however, is the numerically equivalent *unified atomic mass unit* ($u$). Before 1961, the term *atomic mass unit* (*amu*) was also in use. 1 Da equates to $1.6605 \times 10^{-27}$ kg.

Also of note is the fact that the mass of an isotope is not exactly equal to the sum of the masses of the protons and neutrons within the nucleus of the respective atom (electrons are not considered as their masses are a small fraction of the proton/neutron's masses). Rest masses of free protons, neutrons, and electrons are listed along with their mass in *unified atomic mass units* and their charge in units of coulombs in Table 2.1.

As an example, the mass of the $^{4}$He isotope is 4.0026 $u$, whereas the sum of the rest masses of free proton and neutrons (see Table 2.1) is 4.0320 $u$. This loss of mass arises from the fact that energy is required to overcome the electromagnetic repulsion experienced as two or more protons brought in close vicinity to each other. This energy is attained at the expense of mass consistent with the equation $E = mc^2$ where $c$ is the speed of light (this is the basis of how fusion reactors derive energy).

As a result of this electromagnetic repulsion and the fact that all masses are scaled (1 Da equates to 1/12th the mass of the $^{12}$C isotope), isotopes with masses lesser or greater than $^{12}$C display a mass that diverges slightly from the *atomic mass number* ($A$). Recall: $A$ equates to the sum of protons and neutrons within the

**TABLE 2.1 Rest Mass and Charge of Free (Unbound) Ground State Protons, Neutrons, and Electrons.**

| Particle/Symbol | Mass (kg) | Mass (Da) | Charge (C) |
|---|---|---|---|
| Proton/$m_p$ | $1.6726 \times 10^{-27}$ | 1.0073 | $1.6022 \times 10^{-19}$ |
| Neutron/$m_n$ | $1.6749 \times 10^{-27}$ | 1.0087 | 0 |
| Electron/$m_e$ | $9.1094 \times 10^{-31}$ | $5.4858 \times 10^{-4}$ | $-1.6022 \times 10^{-19}$ |

Note: $1.6022 \times 10^{-19}$ C equals one unit ($q$) of charge.

nucleus. This divergence is typically referred to as the *mass deficit* as most of the isotopes (those between $^{12}$C and $^{220}$Rn) display a mass that is slightly less than their *A* value. The masses (*relative isotopic mass*) of the naturally occurring isotopes are listed in Appendix A.2.

Presently accepted definitions of masses (historically preferred by physicists) and weights (historically preferred by chemists) at the atomic scale are as follows:

*Atomic mass* is the mass of an atomic particle, i.e. a specific isotope. When expressed in *unified atomic mass units*, this is called the *relative isotopic mass*. The word *relative* is added to denote the fact that all masses are scaled to that of the $^{12}$C isotope when set to 12 *u*. *Nominal isotope masses* are more commonly used when applying analytical techniques such as Secondary Ion Mass Spectrometry (SIMS) because this significantly simplifies matters without detracting from the information content needed. This represents the number of protons and neutrons within the nucleus, i.e. equal to the *atomic mass number* (*A*). Note: It was the mass spectrograph constructed by Aston in 1919 (the first mass spectrometer from which SIMS evolved as covered in Section 1.2.1) that confirmed the existence of the isotopes, and allowed for the first time, an accurate means of measuring their relative mass (that relative to $^{1}$H, $^{16}$O, or more recently $^{12}$C) and distribution.

*Atomic weight* represents the average weight of a specific element from a specific source, i.e. the weighted average of the *isotopic masses* from a local region. *Relative atomic mass* is numerically identical to the *atomic weight*. The word *relative* again signifies that all masses are referenced to the mass of $^{12}$C. *Standard atomic weights* represent the average terrestrial *relative atomic mass* (or *atomic weight*), i.e. that from many sources/regions collected worldwide. This was introduced to allow the formulation of tables that are globally applicable. Indeed, this is what is used in text books and periodic tables, i.e. see Appendix A.1 with the *atomic weight* presented below the elements symbol as a unitless value. These are also listed in Appendix A.2 in units of g/mol. This unit results from the fact that *atomic weights* can be expressed as the mass of one mole of a specific element. A mole (mol) equates to $6.02214 \times 10^{22}$ (*Avogadro's number*) entities of the item of interest (the entities in this case are atoms). For hydrogen, the *atomic weight* equates to 1.0079 g/mol or just 1.0079.

SIMS, whether used in static or dynamic mode, identifies the types of isotopes, elements, or molecules present on or within a substrate by the *mass to charge ratio* ($m/q$) of the emitted secondary ions. Identification is possible because:

1. Each isotope of each element has a specific mass, and specific molecules are the combinations of specific elements in specific combinations and orientations (methanol is $CH_3OH$, whereas ethanol is $CH_3CH_2OH$). SIMS measures all the isotopes of all the elements.

2. Mass is not affected (this remains an absolute) over the energy range of interest in SIMS (these energies are far from relativistic). As the mass, as defined in SIMS, remains constant, SIMS is defined as a *spectrometry* as opposed to a *spectroscopy* (see Appendix A.10 for a discussion on definitions).

3. The vast majority of secondary ions have a charge of $+1$ or $-1$. In these cases, the ion's mass equates to the recorded mass to charge ratio. For multiple charged ions, the mass is the mass to charge ratio multiplied by the ions charge. Thus, $^{28}Si^{2+}$, which has a nominal mass of 28 u, will appear at 14 $m/q$ (28/2), whereas $^{28}Si^{3+}$ will appear at 9.333 $m/q$ (28/3).

Note: As the vast majority of secondary ions exist in the $+1$ or $-1$ state, the *mass to charge ratio* converts to mass.

**2.1.1.2 Atomic Density** The number of atoms that fits into some spatial or volumetric region is otherwise referred to as the *atomic density*. Only solid state densities are of interest in SIMS.

*Spatial atom densities* $(r_s)$ are important in static SIMS because this allows for the definition of the static limit (1% of a monolayer as discussed further in Section 4.1.1) and because quantification is invariably carried out relative to these spatial units (see Section 5.4.2). Spatial density is derived by accessing the number of atoms in the unit cell plane of interest per unit of area with the latter derived from the unit cell parameter $(a)$. For example, the number of whole Iron atoms in the 110 face of α-Fe (a body-centered cubic structure) is two. From the unit cell dimension of 0.286 nm, the area of the (110) face equates to 0.116 nm$^2$ $(a^2\sqrt{2})$. Thus, the spatial density is two Iron atoms divided by 0.116 nm$^2$, which equates to $1.73 \times 10^{15}$ atoms/cm$^2$. For Silicon, the surface of the (100) and (110) surfaces ranges from $6.78 \times 10^{14}$ to $9.59 \times 10^{14}$ atoms/cm$^3$, respectively.

*Volume atom densities* $(r_v)$ are important in dynamic SIMS (see Section 4.1.2) because, as discussed in Section 5.4.2, concentrations are invariably calculated in such units (atomic % and mass % are less commonly used owing to the confusion that can arise). These can be derived through similar arguments to those used above, i.e. for α-Fe case, the whole number of atoms within a unit cell is two and the volume is 0.0234 nm$^3$ $(a^3)$, which provides a volume atomic density of $8.55 \times 10^{21}$ atoms/cm$^3$. Silicon on the other hand has a volume density of $4.99 \times 10^{22}$ atoms/cm$^3$. In the case of amorphous solids, volume atomic densities can be approximated from elemental mass densities via:

$$r_v \approx \frac{r_m N_A n}{Mr} \qquad (2.1)$$

where $r_m$ is the mass density of the respective elements (listed in Appendix A.2) $N_A$ is avogadros number, $n$ the number of atoms in the empirical formula, that is, the simplest integer ratio of atoms of each element making up the solid, and $Mr$ the molecular weight of the empirical unit.

## 2.2 ELECTRONIC STRUCTURE OF ATOMS AND IONS

The electronic structure describes the energies and spatial location of all electrons bound within an atom, ion, or molecule. Electrons are bound to atoms through the electromagnetic attraction to the respective nuclei (protons have positive charge, electrons have negative charge, and both spin around their own axis). The extent of this attraction, defined by the *electron's binding energy*, is a function of:

1. The number of protons in the nucleus (this defines the element and the number of electrons bound to a neutral atom)
2. The average distance between the electron and their nuclei (defined by the stationary state)
3. The density of electrons around the respective nuclei (influenced by the type of bonding that occurs)
4. Electron–electron interactions (electrons repel each other and shield the nuclei from outer electrons)

### 2.2.1 Stationary States

The electron's binding energy in any element is of a discrete element and level specific value. This arises from the fact that bound electrons can only reside in specific states, otherwise referred to as *stationary states*. The discrete value is also dependent on the charge state of the atom, as well as the bonding if applicable.

Because only a fixed number of electrons can exist in each stationary state (*Pauli exclusion principle*), each bound electron has its own set of *quantum numbers* (these are discussed in more detail in Section 2.2.1.1). Each stationary state is also described by specific notation (*spectroscopic or X-ray notation* as discussed further in Section 2.2.1.2) irrespective of whether they contain electrons or not.

#### *2.2.1.1 Quantum Numbers*   Quantum numbers describe a scheme based on quantum mechanics in which the energy, momenta, spatial distribution, and spin of each bound electron is specified.

The *principle quantum number* ($n$) defines the energy and spatial extent of an electron's orbit around the nucleus. This is an integer number starting from 1 for the most tightly bound electrons (those closest to the nucleus). This number increases in ascending order with the increasing electron binding energy and can contain one or more sub-states depending on the value of $n$. Sub-states are defined by the shape and orientation of the electron's orbit around the nucleus. The *angular momentum quantum number* ($l$) defines the shape of the electron's orbit. This can have integer values from 0 to $n - 1$. For example, electrons in the $n = 3$ state can exist in the $l = 0, l = 1$, and $l = 2$ sub-states. Electrons in $l = 0$ states follow spherical orbits around the nucleus. Electrons in $l > 0$ states follow $l$-dependent non-spherical orbits around the nucleus. The *magnetic quantum number* ($m_l$) defines the orientation of the electron's orbit. This can have integer values that extend from $-l$ to 0 to $+l$. These levels can only contain two electrons, both of which must have opposing spins. The *spin quantum number* ($m_s$) defines the electron's spin (s). These take on values of $+1/2$ and $-1/2$.

The quantum numbers $n$, $l$, $m_l$, and $m_s$ represent the basis of all other quantum numbers. For example, $L$, $M$, or $S$ define the sum of $l$, $m_l$, or $m_s$, respectively. Likewise, $j$ and $J$ are defined via the vectorial addition of $l + m_s$ (also portrayed as $l \pm s$)

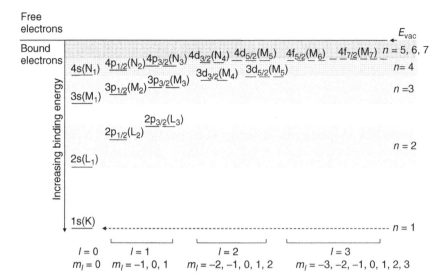

**Figure 2.1** Schematic illustration (not to scale) of the electronic structure within an atom, along with the quantum numbers (bottom and side), spectroscopic notation (above level), and X-ray notation (above level in brackets) used to describe the location of individual electrons in an atom, and the various levels, stationary states, that may or may not be occupied. Reprinted with permission from van der Heide (2012) Copyright 2012 John Wiley and Sons.

and $L + S$ (also portrayed as the sum of $j$), respectively. Thus, both $j$ and $J$ represent the total angular momentum, with $j$ defining that of an individual electron and $J$ that of the entire atom, ion, or molecule, whichever is of interest.

### 2.2.1.2 Spectroscopic and X-ray Notation

Two types of nomenclature are used to describe the various stationary states, irrespective of whether these states are occupied or not. These are *spectroscopic notation* and *X-ray notation*. Both describe the same thing, and both are directly related to the respective electron's quantum numbers if present in these states. These interrelations are illustrated for electrons up to $n = 4$ and $l = 3$ in Figure 2.1. The difference between the two nomenclatures can be traced back to the communities in which these were developed, i.e. spectroscopic notation was developed by chemists, whereas X-ray notation was developed by physicists. Both are used in SIMS. To simplify matters, spectroscopic notation is used throughout the remainder of this text.

*Spectroscopic notation* uses some specific integer followed by some letter to define a specific stationary state. The number used is the same as the principle quantum number of an electron that occupies this level. The letter, on the other hand, relates to the angular quantum number, i.e.:

1. An s orbital (s for sharp) describes the stationary state occupied by electrons with $l = 0$

2. A p orbital (p for *p*rinciple) describes the stationary state occupied by electrons with $l = 1$
3. A d orbital (d for *d*iffuse) describes the stationary state occupied by electrons with $l = 2$
4. An f orbital (f for *f*undamental) describes the stationary state occupied by electrons with $l = 3$

A variation in the energy of a stationary state containing electrons with the same $n$ and $l$ quantum numbers can occur if spin orbit splitting is in effect. The energy levels resulting from this splitting are signified by the quantum number $j$ presented in subscript form following the principle and angular momentum designation. For example, the 2p level has two energetically distinct levels arising from spin orbit splitting that are referred to as the $2p_{1/2}$ and $2p_{3/2}$ levels.

*X-ray notation*, in its simplest form, uses capital letters, $K$, $L$, $M$, $N$, $O$, and so on, to denote the principle quantum number starting with $K$ for electrons with $n = 1$ and proceeding alphabetically to higher principle quantum numbers. As these are all core levels, these are sometimes represented using the letter $C$. Levels within the valence region, i.e. those close to $E_{vac}$, can become indistinguishable in energy (these form bands as discussed in Section 2.2.2.1). As a result, these are often assigned the letter V in place of the capitol letter denoting the principle quantum number.

Sub-levels within the same principle quantum number are designated by an integer value appearing as a subscript following the capital letter. The value of 1 is used for the level containing electrons most tightly bound to the nuclei, and subsequent sub-levels are denoted in ascending order. Thus, the 2s level is represented as $L_1$ and the $2p_{1/2}$ and $2p_{3/2}$ levels are referred to as the $L_2$ and $L_3$ levels, respectively.

*2.2.1.2.1 Stationary State Transition Notation* Spectroscopic notation, X-ray notation, or some derivative thereof is also used to describe transitions between different stationary states. Note: Only electronic transitions, as opposed to vibrational and/or rotational transitions, are discussed in this text.

As an example, the production of a 1s core hole in an aluminum atom (step 1 below) followed by the transition of a $2p_{3/2}$ electron to fill this core hole (step 2) can be described using as the following steps:

Step 1: $1s^2 2s^2 2p^6 3s^2 3p^1 \rightarrow 1s^1 2s^2 2p^6 3s^2 3p^1$

Step 2: $1s^1 2s^2 2p^6 3s^2 3p^1 \rightarrow 1s^2 2s^2 2p^5 3s^2 3p^1$

To simplify matters, only the levels affected may be presented, i.e.

Step 1: $1s^2 2p^6 \rightarrow 1s^1 2p^6$

Step 2: $1s^1 2p^6 \rightarrow 1s^2 2p^5$

If X-ray notation were to be used, step 2 would be represented as a $KL_3$ transition.

As discussed in Section 2.2.2.3.4, such transitions result in Auger electron emission or X-ray emission (fluorescence). In the Auger process, in which the X-ray notation is prevalent, the resulting emission would be referred to as a $KL_3L_2$ electron if the emitted electron came from the $2p_{1/2}$ level. For fluorescence, the emission is referred to as a $KL_3$ X-ray, or if the Siegbahn X-ray convention were to be used (common in the X-ray community), this would be referred to as a $K\alpha_1$ X-ray. In this notation, the Greek letter denotes allowed transitions from stationary states of the next principle quantum number (the order being $\alpha$, $\beta$, $\zeta$), whereas the subscript refers to the relative intensity (1 being the most intense).

A more complex example lies in the charge transfer that occurs on 2p core hole formation in the $Cu^{2+}$ ion when bound within CuO. This process takes the form of electron transfer from the 2p level of an attached $O^{2-}$ ion to the Cu-3d level (Note: Such configurations are only noted during core hole formation). In full spectroscopic notation, this would be represented as:

For $Cu^{2+}$:  $1s^22s^22p^53s^23p^63d^9 \rightarrow 1s^22s^22p^53s^23p^63d^{10}$
For $O^{2-}$ :  $1s^22s^22p^2 \rightarrow 1s^22s^22p^1$

To simplify matters, this can be represented for the ion of interest using only the levels affected. For the $Cu^{2+}$ ion, this takes the form:

$$2p^53d^9 \; L \rightarrow 2p^53d^{10} \; L^{-1}$$

In this case, the ligand is represented by the term L (not to be confused with the italicized $L$ used in X-ray notation for defining a specific stationary state, or the $l$ used to define the angular momentum) and the ligand minus one of its electron by the $L^{-1}$ term.

This can be further simplified using the c and $c^{-1}$ terms to describe the core level to be affected and core hole, respectively, and by dropping the numerals defining the principle quantum number of the stationary state experiencing charge transfer. For $Cu^{2+}$, this would take the form:

$$c^{-1}d^9 \; L \rightarrow c^{-1}d^{10} \; L^{-1}$$

Such transitions describe the satellite structures seen in X-ray photoelectron spectroscopy (XPS) collected from the $Cu^{2+}$ ion present within CuO as discussed in Section 3.3.2.3.2. Spectroscopic notation tends to be preferred in such cases as this more effectively relays the transitions occurring.

### 2.2.1.3 *Ionization Potential and Electron Affinity*    Energy is required to add or remove an electron to a neutral atom or molecule. This stems from the fact that there exists an attraction between negatively charged electrons and positively charge protons.

*Ionization potential* ($I_p$), or more precisely the 1st ionization energy, describes the minimum energy required to remove an electron from a neutral ground state free atom or molecule in the gas phase. The 2nd ionization potential describes the energy required to remove a second electron from an ion of $+1$ charge, i.e. an atom from which one electron has already been removed. Ionization potentials of the elements follow systematic trends that relate to the electronic structure of the atom and the average distance between the electron and its associated nuclei. As with electron binding energies, smaller electron–nuclei distances result in larger ionization potentials because there exists a greater attraction between the two. Ionization potentials of the elements are listed in Appendix A.3.

*Electron affinity* (EA) describes the energy required to add an electron to a neutral ground state free atom or molecule in the gas phase, or in other words the minimum energy needed to form the negatively charged ion of $-1$ charge. Although electron affinities are element/molecule specific, these follow less systematic trends with electronic structure than $I_p$ values. Electron affinities of the elements are listed in Appendix A.3.

Note: The units used vary according to the discipline, i.e. physicists tend to use electron volts (eV), whereas chemists tend to use a molar quantity such as kJ/mol.

### 2.2.2 Bonding and the Resulting Properties of Solids

Atoms bind to each other through the interaction of their outermost electrons. These electrons, otherwise referred to as *valence electrons*, interact through the formation of bonding and anti-bonding states. The resulting states are referred to as molecular orbitals (MOs). Those in unbound atoms are referred to as atomic orbitals (AOs).

Atoms bind as the bonding states formed (the MOs) are more energetically favorable than the states in unbound free atoms (the AOs). An example of this interaction is illustrated schematically in Figure 2.2 for two atoms in close proximity to each other. Also of note is the fact that the energies of the bonding and anti-bonding MO states are a direct function of overlap of the valence electron's wave functions and thus the distance between the respective nuclei.

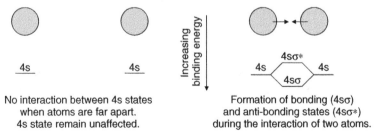

**Figure 2.2** Pictorial illustration of the MOs formed from the 4s states as two As atoms approach each other. The $4\sigma$ bonding and $4\sigma^*$ anti-bonding states (asterisk refers to anti-bonding states) denote the MOs formed.

The bond type formed can fall into one of three different categories, these being:

1. Metallic bonds
2. Covalent bonds
3. Ionic bonds.

*Metallic bonds* describe the situation in which a complete sharing of valence electrons occurs throughout the entire solid formed. As the name suggests, these are typically formed from the bonding of metallic elements such as Iron, Nickel, and Copper. Owing to the extensive delocalization of valence electrons, they display the greatest malleability along with the highest conductivities.

*Covalent bonds* are similar to metallic bonding in that the valence electrons are shared between the bonding atoms. The difference lies in the fact that the bonding electrons are slightly more localized, i.e. only shared among the atoms making up the respective molecule formed. Examples include $H_2$, $O_2$, and $CH_3$. These can also support a charge as in $NH_4^+$, $SO_4^{2-}$, and so on. Organic materials are an example of long-range covalent bonding. These tend to show intermediate properties relative to materials formed through metallic bonding and ionic bonding.

*Ionic bonds* describe the situation where one or more valence electrons have been transferred from one atom to another with the electrons being highly local-ized around the respective ions formed. This can occur between individual atoms or atoms situated within covalently bound molecules. An example of the former is the formation of table salt (NaCl). When this occurs, the sodium atom (Sodium by itself is a reactive metal) transfers one of its electrons to a Chlorine atom (Chlorine by itself is a toxic gas) to form $Na^+Cl^-$. Note: The charge states of the resulting ions are not typically represented. This bonding results in a brittle crystalline struc-ture of low conductivity (because of the large band gap formed as discussed in Section 2.2.2.1). An example of the latter is seen in the ionic bonding that occurs between the charged covalent molecules $NH_4^+$ and $SO_4^{2-}$, which results in the formation of ammonium sulfate or $(NH_4)_2.SO_4$.

Although core level states (the inner electron stationary states) do not partake in bonding, their energies can also be influenced, albeit by a small amount. This primarily arises from variation in the effective nuclear charge felt by the core elec-trons as valence electrons are added or removed from the atom during the course of bond formation.

Variations in electron binding energies also occur when an atom/ion moves from the solid phase into the gas phase. This is a result of the different energy references used, i.e. energies of electrons bound to atoms/ions in the gas phase are referenced to the vacuum level ($E_{vac}$), whereas electrons bound to atoms/ions in the solid phase are referenced to the Fermi edge ($E_F$). The work function ($\phi$) of the solid defines the difference. This represents the minimum energy required to remove an electron from a solid, i.e. is analogous to the ionization potential ($I_p$). These concepts are depicted in Figure 2.3 for free and bound As atoms.

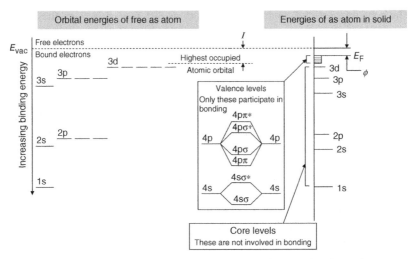

**Figure 2.3**   A schematic diagram illustrating the allowed energy levels (stationary states) for electrons bound to a free Arsenic atom (left), and those for electrons bound to Arsenic when present within the pure elemental solid (right). This bonding results in the formation of MOs as defined by the interaction of the valence electrons (4s and 4p levels). As metallic bonding occurs, pure Arsenic is defined as a metalloid (a nonmetal displaying metal-like properties). Reprinted with permission from van der Heide (2012) Copyright 2012 John Wiley and Sons.

### 2.2.2.1   *Bands and the Density of States*

When valence electrons from two individual atoms interact, bonding and anti-bonding states of lesser and greater energy than the initial unbound states are formed. If the number of interacting atoms increases, so too does the number of possible energy states. When enough atoms interact ($8.55 \times 10^{21}$ iron atoms exist in 1 cm$^3$), this progression results in the formation of bands, where the bonding levels form the valence band and the anti-bonding levels form the conduction band. This progression is illustrated in Figure 2.4. As core level states do not partake in bonding, their energies (core electron binding energies) remain discrete.

In conducting solids (Iron is an example), the valence and conduction bands overlap, hence the high conductivity observed. In semiconductors (Silicon in the solid form is an example) and insulators ($SiO_2$), these bands remain separated by a forbidden region. This energy separation that can be as small as a fraction of an eV to several eV is referred to as the *band gap* ($E_g$). Insulators are those materials displaying band gaps greater than $\sim$4 eV (this translates to a resistivity greater than $10^6$ $\Omega/M$), whereas semiconductors are those materials displaying band gaps less than $\sim$4 eV. Note: The latter are referred to as semiconductors because their properties lie between those of conductors and insulators. The density of states refers to the number of energy levels per unit energy. Note: The density of states in a band is not uniform (pictorial illustrations tend to illustrate the opposite).

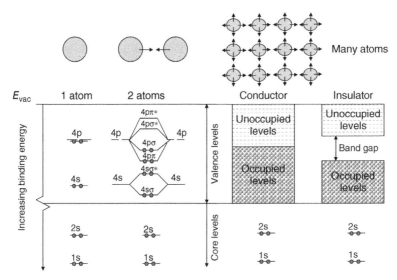

**Figure 2.4** A schematic illustration (not to scale) of the progression in energy states for Germanium (1 atom of Germanium to the left through to many atoms forming the Germanium metalloid to the right). Stationary states are represented using spectroscopic notation with π, π*, σ and σ* representing bonding and anti-bonding orbitals (* for anti-bonding). Note: Solid Germanium is a semiconductor with a band gap of ~0.7 eV at room temperature. For the sake of clarity, the $n = 3$ levels are omitted from this illustration.

### 2.2.2.2  *Work Function*
The minimum energy required to remove an electron from a solid is the energy difference between the vacuum level and the energy of the least strongly bound electron. The minimum energy required to remove an electron from a solid is thus represented as the position (binding energy of electrons) at which a molecular orbital, if present, would exhibit a 50% probability of being filled. This energy is referred to as the solid's *Fermi edge* ($E_F$). According to the *Aufbau principle*, all states below the Fermi edge should be filled and those above the Fermi edge should be empty, at least for atoms in their ground state. Hence, the Fermi edge appears within the band gap of semiconductors and insulators even though this represents a forbidden region.

The position of the Fermi edge thus defines the solid's *work function* ($\phi$). As can be envisaged, the work function is a substrate surface and environment (modified by adsorbates) specific parameter. This generally falls within the range 2–8 eV.

If energy less than the work function is imparted onto an atom making up the solid, this energy will be dissipated within the solid through lattice vibrations. As discussed in Section 2.2.2.4.1, lattice vibrations, referred to as *phonons*, appear as heat. If an amount of energy greater than the work function is imparted, electrons or photons can be emitted from levels below the Fermi edge. Measurement of these emissions reveals the electronic structure of the substrates surface, and hence the elements present. As outlined in Appendix A.10, this is the basis of Auger

Electron Spectroscopy (AES), Energy Dispersive X-ray spectroscopy (EDX), X-ray Photo-electron Spectroscopy (XPS), and X-Ray Fluorescence (XRF).

In essence, the work function is the first ionization potential once modified by the long-range electronic interaction occurring within a solid. The 1st ionization potential and the work function can be related (Wong et al. 2003) through the approximation:

$$\phi \approx \frac{I_p}{1.8} \tag{2.2}$$

**2.2.2.3 Image Field** The long-range electronic interaction in a solid induces an electrostatic field that extends a slight distance from the solid's surface. This is referred to as the image field. If an ion is close to a solid's surface, this image field will modify the ionization potential and electron affinity of the respective ion. The variation in the ionization potential can be crudely approximated for singly charged ions close to metallic surfaces over distances greater than that at which a quasi-molecular orbital (temporary orbital) is formed, as:

$$E_i \approx -\left(I_p - \frac{3.6}{z}\right) \tag{2.3}$$

where $z$ represents the internuclear distances in units of Å (Andrä 1987). This approximation assumes a $1/4z$ dependence from Coulombic interaction.

A similar, but opposite trend, is noted for the affinity level, i.e. the minus sign in Relation 2.3 is replaced by a positive sign and the ionization potential is replaced by the electron affinity. Thus, the affinity level increases in energy relative to the vacuum energy with decreasing distance. The electron affinity represents the minimum energy required to add an electron to a free atom or molecule.

These interrelations are illustrated in Figure 2.5. All energies are with respect to the Fermi edge.

**2.2.2.4 Electronic Excitation** Although each atom has a specific electronic structure, the distribution of electrons can be modified through the deposition of energy through photon, electron, or ion impact. The atom/ion then responds through either:

1. The production of *phonons* (energies are $\sim 0.1$ eV with lifetimes $< \sim 10^{-10}$ s)
2. The production of *excitons* (energies are $\sim 0.1 - 10$ eV with lifetimes $< \sim 10^{-6}$ s)
3. The production of *plasmons* (energies are $\sim 5 - 25$ eV with lifetimes $< \sim 10^{-9}$ s)
4. The production of *core holes* (energies are greater than $\sim 20$ eV with lifetimes $< \sim 10^{-13}$ s)

Which occurs depends on the energy deposited upon the atom/ion and the electronic structure of the atoms that make up the solid. These are discussed in greater detail in the following sections.

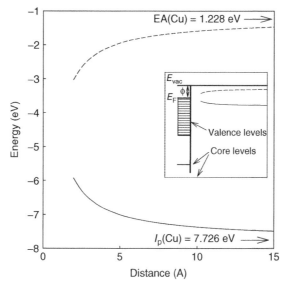

**Figure 2.5**  Variation in ionization and affinity levels of Copper ions as a function of distance from a Copper surface as derived by Relation 2.3. The inset shows a demagnified diagram relating the position of valence and core levels. A work function of 4.65 eV is assumed.

### 2.2.2.4.1 Phonons

Phonons represent collective vibrations (oscillations) in the position of the nuclei of atoms making up a lattice. This, in turn, affects bond distances and hence the energies of the HOMOs/LUMOs. As a result, phonons have the effect of populating states above the Fermi edge/depopulating states below the Fermi edge (the position of the Fermi edge remains unaffected).

The distribution of filled states above the Fermi edge is most commonly defined through the *Boltzmann distribution* as:

$$Fn(E) = \exp\left(\frac{(E - E_F)}{k_B . T}\right) \tag{2.4}$$

The temperature parameter ($T$) is present because such excitations are within the range of energies over which temperature variations are observed.

The entire distribution of filled levels around the $E_F$ is more commonly derived using *Fermi-Dirac statistics*, i.e. via the relation:

$$Fn(E) = \frac{1}{(\exp((E - E_F)/k_B . T) + 1)} \tag{2.5}$$

Solving this reveals how the distribution of filled levels around the Fermi edge spread out as the temperature of the solid is raised. This is illustrated for a substrate with a work function of 5 eV in Figure 2.6.

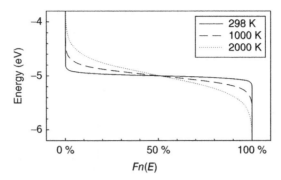

**Figure 2.6** The distribution of filled states around the Fermi edge as a function of temperature as derived via Relation 2.5 for a conductive solid displaying a work function of 5 eV. Note: The work function remains unaffected because, as discussed in Section 2.2.2.2, this is defined as the position at which 50% of the states are filled.

As electrons prefer to reside in the lowest possible energy level, the energy associated with this type of excitation will be released as heat, and in some extreme cases, light (photons). As the photons can have energies in the visible region ($\sim 1.7 - 3$ eV), this explains why extremely hot substances, whether present in the solid, liquid, or gas phase, glow as red hot (these emit photons of $\sim 1.7$ eV) or even white hot (these emit photons at all energies from $\sim 1.7$ to 3 eV).

*2.2.2.4.2 Excitons* Excitons occur in solids in which a band gap exists (semiconductors and insulators). These are formed when a sufficient energy is imparted into the solid so as to induce the movement of an electron situated within a level close to the top of the valence band edge to some level close to the bottom of the conduction band edge. When this occurs, a hole is left behind in the valence band. The electron–hole pairs formed can then move together owing to the Coulombic interaction that exists between the two.

The lifetime of such electron–hole pairs can be significantly longer than those formed in metals because the band gap keeps them separated. As expected, the lifetime of such electron–hole pairs will be a function of the band gap, the crystal structure as this defines the electronic structure, and, of course, the temperature. Lifetimes can be as long as $\sim 10^{-6}$ s, after which the energy is converted into phonons, which is otherwise seen as thermal energy (see Section 2.2.2.4.1).

*2.2.2.4.3 Plasmons* Like phonons and excitons, plasmons represent the energy associated with the collective movement of particles. In this case, the particles are the free electrons present at and around the Fermi edge.

Plasmons are induced through the sudden electrostatic perturbation to the local environment. This can be induced through the passage of fast charged particles (this results in what are referred to as *extrinsic plasmons*) or the formation of a core hole (this results in *intrinsic plasmons*). When this occurs, any free electrons move to account for the sudden charge discrepancy. In doing so, they overshoot their mark and set up *collective oscillations* in the conduction band electrons.

If free electrons oscillate, they prefer to do so at some specific frequency and energy. These are termed the *free electron plasmon frequency* and the *free electron plasmon energy*. Assuming no interband transitions exist close to the plasmon energy (this only occurs for a limited number of elemental solids as described further in the following), the frequency and energy associated with collective oscillations within the bulk of a solid become a function of the density of free electrons in the respective metal. The bulk and surface plasmon energies can be approximated as:

$$E_{\text{pl(bulk)}} = 28.8 \sqrt{\left( \frac{N_v \cdot \rho}{m} \right)} \tag{2.6a}$$

$$E_{\text{pl(surface)}} = \frac{(28.8 \sqrt{(N_v \cdot \rho/m)})}{\sqrt{2}} \tag{2.6b}$$

where $N_v$ is the number of valence electrons in the band that defines the Fermi edge, $\rho$ the mass density in $g/cm^3$, and $m$ the atomic mass in $u$ of the elemental solid. Energy is in units of eV. As the position of the Fermi edge in most metals is defined by the s-p band of electrons in the highest $n$ level, the partial number of valence electrons is limited to these alone. Table 2.2 lists the bulk and surface plasmon energies, the number of valence electrons, the density, and the mass of various elemental solids.

If interband transition energies approach free electron plasmon energies, plasmons will quickly dissipate through the promotion of an electron between the two aforementioned bands. Silver is a prime example in which this occurs, as there exist levels within the valence region, which have an energy separation close to $4\,\text{eV}$ (this represents the transition energy for electrons in the $4d_{3/2}$ and $4d_{5/2}$ levels to the Fermi edge as defined by the 4p-5s band). This transition has the effect of reducing the plasmon free energy from that specified in Relation 2.6 to values

**TABLE 2.2   Bulk Plasmon Energies (Relayed from Relation 2.6(a)) for Select Elemental Solids, along with the Number of Valence Electrons in the Band Defining the Fermi Edge, the Mass Density of the Elemental Solid, and the Mass of the Element.**

| Solid | $E_{\text{pl}}$ (eV) | $N_v$ | $\rho(g.cm^3)$ | $m(u)$ |
|-------|------|------|------|------|
| Na | 5.92 | 1 | 0.97 | 22.99 |
| Mg | 10.87 | 2 | 1.74 | 24.35 |
| Al | 15.77 | 3 | 2.70 | 26.98 |
| Si | 16.59 | 4 | 2.33 | 28.09 |

equal to the interband transition energy. Note: The plasmon free energy for Silver according to Relation 2.6 is 8.97 eV ($N_v$ is taken as equal to one), whereas those observed are closer to 4 eV.

Lastly, plasmons have a finite lifetime ($< \sim 10^{-9}$ s) after which they are quickly converted to thermal energy within the solid.

#### 2.2.2.4.4 Core Holes

Core holes represent the vacancy left behind when an electron is removed from its equilibrium core level stationary state within the associated atom or ion. This can occur via the promotion of the electron to some vacant level of lesser binding energy or the emission of this electron from the atom. The minimum energy required to produce a core hole equates to the electron's binding energy. As core holes represent an extreme form of excitation, these are filled as soon as possible by demoting an electron in a level closer to the Fermi edge into the aforementioned core hole. This typically occurs within $10^{-13}$ s.

The transition energy must, however, be removed from the atom. This is accomplished via the emission of an additional electron or photon of similar energy. If energy is removed through electron emission, the process is referred to as an *Auger* process. For photon emission, it is called *X-ray fluorescence*.

Auger emission is preferred in the lower Z elements, whereas fluorescence is preferred in higher Z elements. This is illustrated in Figure 2.7 for 1s (K level) core hole formation. The same general trend is noted for core holes formed in other

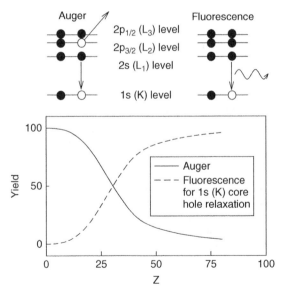

**Figure 2.7** Schematic example of core hole relaxation via a $KL_2L_3$ Auger process (top left) and $K\alpha_1$ X-ray fluorescence (top right), which can occur following 1s core hole formation. Note: $KL_3L_2$, and $KL_3L_3$ transitions can also occur. The probability (percentage yield) for Auger versus fluorescence as a function of $Z$ is shown in the accompanying plot.

levels with the crossover moved to higher Z elements. For historical reason, X-ray notation is predominantly used in describing such transitions.

In the case of fluorescence, Siegbahn notion is most commonly used, again for historical reasons. The different notations are described in Sections 2.2.1.2 and 2.2.1.2.1.

Core holes can be produced in both bound and unbound atoms through the deposition of an amount of energy equal to or greater than the electrons binding energy. This energy can extend from tens to tens of thousands of eV depending on the level and the element. This energy can be imparted through photon, electron, or atomic collisions. In the case of photon and electron interaction, energy transfer is relatively straight forward, i.e. simply transferred from one to the other, albeit less efficient in the case of electron interaction. Atomic collisions represent the least efficient method, owing primarily to the dynamics necessary.

The generally accepted mechanism via which core holes are produced during atomic collisions is described within the *electron promotion model* introduced by Fano, Lichten, and coworkers (Fano and Lichten 1965; Lichten 1967; Barat and Lichten 1972). This describes the formation/modification of MOs from the respective AOs of the two colliding atoms as they approach each other from an infinite distance to distances substantially less than equilibrium bond distances. This modification can be understood as the energy of an MO formed as defined by the interatomic distance (see Section 2.2.2). Correlation diagrams relay the modification of MO energies for all levels of interest as a function of interatomic (internuclear) distance.

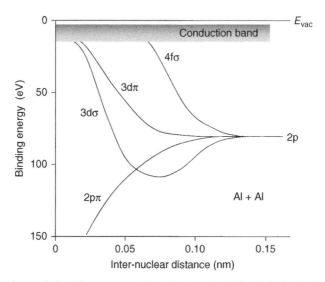

**Figure 2.8**    A correlation diagram approximating how the MOs of the 2p levels of two Aluminum atoms form and are modified in energy as the respective Aluminum nuclei approach each other. This is extrapolated to zero interatomic distance such that they form the nuclei of Iron. Of note is the fact that all but the $2p\pi$ MOs cross into the conduction band.

An example of a correlation diagram constructed from the collision of two Aluminum atoms is shown in Figure 2.8. This shows the modification in the core level MOs formed from the 2p levels as the two Aluminum atoms approach each other to zero interatomic distance (for the sake of clarity, the other levels are not shown). Zero distance is used as this allows for the extrapolation of the MO energies to those of Iron (the nucleus that would be formed). The understanding of how the MOs are modified allows for the prediction of which core holes are most likely formed during such collisions, as well as the minimum energy and distance required (Figure 2.8).

As an example, the fact that the 4fσ level (one of the levels that arises from the 2p core level on MO formation) crosses the conduction band when the two Aluminum nuclei are within 0.7 Å of each other (this is almost half the equilibrium interatomic distance) suggests that a 2p electron can be left behind in the conduction band as the two atoms recede from the collision. This would then leave a 2p core hole (see Joyes 1969a, 1969b), which would then de-excite through either Auger electron emission or fluorescence.

## 2.3   SUMMARY

SIMS measures the elemental, isotopic, and molecular distributions on and/or within a solid. Elemental and molecular elucidation is arrived at by essentially weighing each of the respective elemental/molecular secondary ions. A solid is composed of many atoms/ions and/or molecules, with the solid's properties, inclusive of density, defined by the bonding between the respective atoms.

Bonding of the elements stems from the interaction of their electrons, and hence, the electronic structure of the respective neutral ground state atoms. The three types of bonds that can be formed (ionic bond, covalent bonds, and metallic bonds) dictate the properties to be exhibited by the respective solid formed. The energy to which the bands are occupied in a solid is defined as the work function of the solid.

The electronic structure of the atoms is specific to each element in the periodic table. This describes the distribution in energy and space of stationary states (levels in which electrons can reside). These states are described using *spectroscopic* or *X-ray notation* (these both describe the same thing), whereas transitions can be described via one of the above two notations, or the *Siegbahn notation* (for historical reasons, this depends on the outcome). Electrons in these stationary states are described by a specific set of *quantum numbers*, with no two electrons from the same atom being allowed to share the same quantum numbers. Free electrons are not ascribed a set of quantum numbers.

The minimum energy required to remove an electron from an atom, i.e. to form a positively charged ion, is referred to as the 1st *ionization potential*. The minimum energy required to add an electron to an atom, i.e. to form a negatively charged ion, is referred to as the *electron affinity*. These parameters change as atoms form solids, i.e. the ionization energy becomes the *work function*.

The excitations that can occur within a solid's lattice on removing an electron, atom, or ion from the solid's surface include (listed in order of increasing energy) *phonons, excitons, plasmons,* and the production of *core holes.* Indeed, all of the above forms of excitations can exist during the production of secondary ions. Indeed, core hole relaxation is stipulated as one of the routes via which multiply charged ions are formed. Core hole relaxation can also result in the production of any of the former excitations depending on the lattice in question and the energies involved.

# CHAPTER 3

# Sputtering and Ion Formation

## 3.1 THE FUNDAMENTALS OF SIMS

As with many analytical techniques, an understanding of the fundamental concepts underlying the generation of a recorded signal is not necessary in order to carry out an effective routine analysis. Indeed, the analytical information needed is most often acquired through comparisons of *known good* versus *known bad* samples, or comparisons of *unknown* versus *known* samples (the latter can be in the form of reference materials) using procedures developed specifically for the analysis at hand.

There are, however, situations in which some degree of knowledge of the processes can help. Examples of this fall into two specific areas, these being:

1. Understanding and setting up optimal analytical conditions for new types of analyses. Note: The intensity of various types of secondary ions, which include the atomic and molecular species of many types that can exist as positive or negative ions, are dependent on the following:

    a. The type of atomic, molecular, or cluster primary ion beam used (more common examples used in Secondary Ion Mass Spectrometry (SIMS) include $O^-$, $O_2^+$, $Cs^+$, $Ga^+$, $In^+$, $Au_n^+$, $Bi_n^+$, $SF_5^+$, $C_{60}^+$, $C_{84}^+$, $C_{24}H_{12}^+$, $Ne^+$, $Ar_n^+$, and $Xe^+$, with $n$ being an integer starting from unity, and in the case of $Ar_n^+$ extending to several thousand. In some cases, multiply charged ions are used, with some examples including, but not limited to, $Bi_n^{2+}$, $Bi_n^{3+}$, $C_{60}^{2+}$, and $C_{60}^{3+}$).
    Note: The term cluster ion is commonly applied to $Ar_n^+$, $C_{60}^+$, $C_{84}^+$, $C_{24}H_{12}^+$, and so on, even though ions from large molecules such as $C_{60}$ are in fact molecular ions.

    b. The primary ion impact energy applied (this is typically within 0.1–50 keV range).

    c. The type (atomic or molecular) and polarity of the secondary ions measured.

    d. The composition of the surface examined (this is referred to as the *matrix*).

*Secondary Ion Mass Spectrometry: An Introduction to Principles and Practices*, First Edition.
Paul van der Heide.
© 2014 John Wiley & Sons, Inc. Published 2014 by John Wiley & Sons, Inc.

e. The instrument geometry, i.e. primary ion incidence angle, mass filter type, and so on.

f. Other analysis conditions inclusive of vacuum, $O_2$ leak, and ion optical parameters.

2. Recognizing distortions to the data that are introduced through the act of analysis. These distortions can be induced by the instrument and/or sample with the latter referred to as the *matrix effect*. Note: Such distortions have been referred to as SIMS *artifacts* even though not strictly correct.

That being said, there remains debate surrounding the details of secondary ion formation/survival, which in itself, more than justifies the ongoing research being carried out in this area.

The aim of this chapter is to illustrate the generally accepted state of knowledge concerning the fundamentals of secondary ion emission. This includes a discussion in various models explaining variations in the recorded signal, whether from the sputtering process or from the ion formation/survival process.

### 3.1.1   Secondary Ion Generation

In SIMS, the signal of interest (the secondary ion population) arises from the impact of energetic ions (primary ions) on the substrate's surface. The signal of interest is *most commonly* accepted to arise from a two-step process akin to that depicted in Figure 3.1. Note: The term "*most commonly*" is used as there exist/s other modes by which secondary ions can be generated.

The processes shown in Figure 3.1, albeit highly simplified, describe the ejection of particles from the surface of interest. This can include atomic and/or molecular species depending on the initial conditions used. This is then followed by some form of electron transfer (charge transfer) to or from the ejected population. It is the latter that results in the formation and/or survival of secondary ions. The secondary

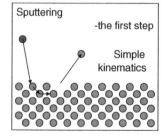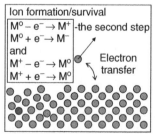

**Figure 3.1**   Highly simplified illustration of a two-step process in which atomic secondary ion emission proceeds via knock-on sputtering resulting from keV primary ion impact (left box) followed by secondary ion formation/survival (right box) from the sputtered population. M in the right box can represent any element with the superscript referring to its associated charge and e⁻ referring to an electron. The ejection of molecular ions appears to follow similar albeit more complicated routes.

ion formation/survival process is generally accepted to occur when the interatomic distance between the nuclei of atoms/ions at the outermost surface of the solid in question and the departing atom/ion is less than ~1 nm, i.e. slightly greater than typical bond distances.

The ejection of particles from a solid's surface is termed *sputtering*. In this text, this will comprise any process inclusive of ion-induced and electron-induced desorption resulting from the initial ion impact. The ejected population most *generally* consists of neutral atoms/molecules, not measurable by SIMS, with only a small fraction leaving as ions. The term "*generally*" is used as highly electropositive or electronegative elements can form a sizable fraction of the ionized population, i.e. approaching or greater than 10%. Although sputtering is more clearly understood than the formation/survival of secondary ions, there is still active research being carried out, particularly with respect to the ejection of large *intact* molecular ions, which are not expected within the *linear cascade* model (this billiard ball-style concept is the traditionally accepted view in atomic sputtering). The various modes of sputtering are discussed further in Section 3.2.

The formation/survival of secondary ions is less well understood. This is made particularly evident in the lack of understanding surrounding the *matrix effect*. The matrix effect describes the significant and often unpredictable variations in secondary ion yields (the number of ions produced as a fraction of the sputtered population) as the composition and/or structure of the surface from which the ions emanate varies. As the secondary ion yield variations can span five orders of magnitude or more, quantification is often difficult. The significant secondary ion intensity variations noted during the initial stages of sputtering with reactive ions such as $O_2^+$ or $Cs^+$ are referred to as the *transient effect*. These effects result from the chemical modification of the substrate as a result of some fraction of the primary ions becoming implanted. Consequently, transient effects can be thought of as a specific form of the matrix effect. The secondary ion formation/survival process is covered further in Section 3.3.

## 3.2 SPUTTERING

Sputtering describes the removal of atoms (or ions as noted in ionic crystals as opposed to those formed as a result of the sputtering process) and/or molecules from a solid's surface following the deposition of energy onto the aforementioned solid's surface. This energy can be deposited by directing the energetic ions, electrons, or photons at the solid of interest. The energy required to induce sputtering must be greater than the bond energy holding the respective atom, ion, or molecule to the solid's surface (see Table 3.1). Sputtering threshold values are in the 15–40 eV range (Malherbe 1994).

The *surface binding energy* defines the energy required to break a bond between an atom (or ion) situated at the outer surface of a solid. In an unperturbed system (one not experiencing significant electronic excitation), this can be approximated from the energy spectra of the sputtered atoms resulting from the linear cascade mechanism as twice the recorded peak energy, or theoretically as the *surface*

**TABLE 3.1    Surface Binding Energies and Cohesive Energies of Various Elements Present on Their Elemental Substrates.**

| Element | Substrate | Binding Energy (eV) | Cohesive Energy (eV) |
|---------|-----------|---------------------|----------------------|
| Al | Al | 4.6 | 3.39 |
| Si | Si | 4.7 | 4.63 |
| Ti | Ti | 4.6 | 4.58 |
| Cr | Cr | 4.2 | 4.10 |
| Fe | Fe | 5.0 | 4.28 |
| Ni | Ni | 5.4 | 4.44 |
| Cu | Cu | 3.5 | 3.49 |
| Ag | Ag | 3.1 | 2.95 |

*potential barrier*. The latter, however, requires solving many body interaction potentials particularly in multicomponent solids (this requires the use of quantum mechanical calculations, which is considered outside the scope of this text). The surface binding energy of elemental solids can also be roughly approximated from the solid's *cohesive energy, heat of vaporization, sublimation energy*, or some multiple of. Surface binding energy and cohesive energy values for specific elements from elemental targets are listed in Table 3.1. The surface binding energy is one of the parameters that control the sputter yield/sputter rate.

The *sputter yield* defines the absolute number of atoms, ions, or molecules removed per impinging primary ion. As discussed in Section 3.2.2, this *absolute* value is differentiated from the *analytical* value to avoid confusion. The absolute definition is used throughout this text unless otherwise specified.

The *sputter rate* defines the sample removal rate resulting from sputtering. This can vary from less than a monolayer during the entire analysis to many monolayers per analytical cycle, with the rate depending heavily on the type of the primary ions and the conditions used, as this defines the energy transferred. How the sputter rate is determined is covered in Section 5.3.1.

Note: Sputtering can also modify the underlying substrate (one that is below the region being sputtered). This modification can range from recoil implantation, cascade mixing, diffusion, segregation, amorphization, re-crystallization, surface roughening, and so on, all of which are discussed further in Section 3.2.3.

### 3.2.1    Sputtering by Ion Impact

Although first noted in 1852 (Groves 1852), the first ion beam-induced sputtering experiment was not reported until 1902 (Goldstien 1902). The following sections describe the salient features and present understanding of the sputtering process as it applies to SIMS. For more in-depth reviews, see Behrisch and Wittmaack 1991; Rabalais 1994; Gnaser 1999; Garrison and Postawa 2008.

Sputtering relies on the fact that energy is conserved (first law of thermodynamics). Thus, by directing an energetic ion beam at a solid's surface (this can be an ion of any element or molecule of sufficient mass and energy), it will interact with the aforementioned solid through the deposition of energy. On depositing all or some fraction of its energy, the incoming ion will then scatter off the surface, evaporate from the surface, fragment, or come to rest within the surface, i.e. become implanted. Energy can be transferred via either a combination of *Elastic collisions* or *Inelastic collisions*.

*Elastic collisions* are those in which energy is deposited from the incoming ion via momentum transfer onto atoms and/or ions that make up the solid. This process results in a change in the trajectory of the incoming ion. *Inelastic collisions* are those in which energy is deposited by the incoming ion via electronic excitation of the atoms and/or ions that make up the solid. This process does not result in any change in the trajectory of the incoming ion.

There also exist several forms of sputtering. These forms can be delineated based on the type of energy transfer process that induces atom, ion, or molecular emission. These are:

1. kinetic sputtering
2. kinetically assisted potential sputtering
3. potential sputtering

Kinetic sputtering describes the removal of surface-bound atoms, ions, or molecules, which occurs purely through the momentum transfer (elastic collisions). This form includes *knock-on sputtering* and *recoil sputtering*.

Kinetic-assisted potential sputtering describes the removal of surface-bound atoms, ions, or molecules, which occurs through various modes of excitation induced as a consequence of the momentum imparted by the incoming ion.

Potential sputtering describes the removal of surface-bound atoms, ions, or molecules, which occurs purely through various excitation modes (inelastic collisions) initiated by the charge imparted by the incoming ion.

As electronic excitation is not considered within kinetic sputtering, the collisions can be likened to an atomic-scale billiard ball game that is initiated on primary ion impact. The valence electron shells of the atoms/ions involved would thus represent the billiard ball's surfaces. The *linear cascade model*, which describes the most prevalent form of ion-induced sputtering, at least that from atomic ions and ions comprising small molecules (common examples used in SIMS include $O^-$, $O_2^+$, $Cs^+$, $Ar^+$, $Xe^+$, and $Ga^+$), assumes a specific form of kinetic sputtering in which a *full isotropic collision cascade* is produced close to the surface. This is one form of *knock-on sputtering*.

A *collision cascade*, as the name suggests, describes a sequence of many collisions involving many atoms within the solid. Each collision is, however, considered separate from each other. As a result, these can be considered a linear sequence of collisions. Ejection of surface-bound atoms, ions, and/or molecules will ensue if, at any time during the collision cascade, there exists sufficient energy transfer (momentum) in the direction away from the solid such that the atom, ion, or molecule can escape the solid surface. Note: The molecules observed may be representative of molecules previously existing on the surface or newly formed molecules generated during the sputtering process. The $MCs^+$ secondary ions, where M is some substrate element, and combinations thereof noted when using $Cs^+$ primary ions represent one such example.

When a full isotropic collision cascade is not formed, the sputtering process becomes more anisotropic. This is noted in *recoil sputtering*, which is another form of *knock-on sputtering*. As fewer collisions occur in this form of sputtering, deviations from the trends implied by the *linear cascade model* are noted. The *linear cascade model* is covered in Section 3.2.1.1.

Cooperative motion of many atoms within a solid can also result during kinetic sputtering when collision events resulting from a single ion impact overlap. This is in contrast to the sputtering described by the *linear cascade model*, which assumes a linear sequence of individual nonoverlapping collisions. Cooperative motion can be observed under large cluster ion impact. Common examples used in SIMS include $C_{60}^+$, $C_{84}^+$, $C_{24}H_{12}^+$, and $Ar_n^+$, with $n$ being an integer extending to several thousand. Under optimized conditions, cooperative motion within specific substrates can result in sputter yields that are significantly greater than that expected from a linear collision cascade.

Energetic large cluster ion impact results in a greater multitude of processes relative to atomic and small molecular ion impact, with some examples including:

1. surface sputtering, radiation damage, crater formation, and introduction of buried clusters,
2. surface melting, alloy formation, plastic deformation, and soft landing of clusters. Note: Surface cleaning and smoothing result from soft landing,
3. cluster implantation, cluster reflection (scattering), cluster splatting, and cluster fragmentation.

To understand the reasons for this, it is worth considering the process in detail. When a sufficiently energetic cluster ion approaches a solid surface, those atoms at the leading edge of the cluster can slow down before those atoms at the trailing edge. As a result, all atoms making up the cluster may effectively *pile up* at the surface within a very short time period. This can increase the pressure/density of atoms/ions in the impacted region and/or send a shockwave through the lattice structure of the substrate impacted. As shockwaves are typically noted only under high energy impacts, i.e. greater than 100 keV (outside the range used in SIMS), these will not be discussed further. The pressure built up in the impacted region can then be released through deformation of the impacted area. This can take the form of the bulging of material around the rim of the crater formed. Indeed, craters of the dimensions of the cluster are often noted ($\sim$7 nm for $Ar_n^+$ clusters on Graphene (Mochiji 2011)). The prevalence of a specific process has been found to depend on several parameters with the following deemed the most important:

1. cluster size ($N_{Cl}$),
2. cluster charge (typically +1 charged clusters are used in SIMS),

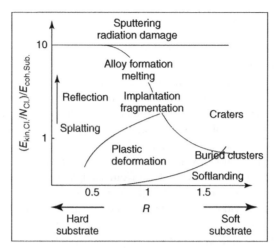

**Figure 3.2**   Cluster-surface collision process space diagram illustrating the dependence of the Cluster-surface collision-induced process on the cohesive energies of the system ($R = (E_{(Coh)}^{Cl})/(E_{(Coh)}^{Surf})$) and reduced cluster ion energy ($E_{kin}^{Cl}/N_{Cl}/E_{(Coh)}^{Cl}$). Reprinted with permission from Harbich (2000) Copyright 2000 from Springer-Verlag.

3. cluster impact energy ($E_{kin}^{Cl}$),
4. cluster impact angle (with respect to the sample normal),
5. the cohesive energies of the cluster ($E_{(Coh)}^{Cl}$) and surface ($E_{(Coh)}^{Surf}$) and the cohesive energies of the bonds formed between the cluster ion atoms and atoms at the substrate's surface,
6. cluster and substrate temperature.

These have been used to generate the process space diagram as illustrated in Figure 3.2 (from Hsieh et al. 1992 once modified by Harbich 2000). This diagram includes many cluster ions not commonly used in SIMS, i.e. $Al_n^+$, $Cu_n^+$, $Ag_n^+$, and so on.

For the small molecular ions (examples commonly used in SIMS include $Au_n^+$, $Bi_n^+$, and $SF_5^+$), energy transfer can also proceed through various modes of electronic excitation. As a result, they are best described as forms of *kinetically assisted potential sputtering*. The various modes of electronic excitation are often discussed in the form of localized elevated temperatures that quickly dissipate.

Lastly, sputtering can occur through *potential processes* alone. This has only been observed on specific nonmetallic substrates under inert gas ion impact with the sputter yield increasing with the primary ion charge. This charge dependence is not observed or predicted during kinetic sputtering.

Models describing all forms of sputtering other than that described by the linear cascade model are covered in Section 3.2.1.2.

**3.2.1.1  *Linear Cascade Model***   Sputtering resulting from elastic collisions (*knock-on sputtering*) is the most well understood of all the other forms of sputtering. The most common form of knock-on sputtering has been modeled by what is referred to as the *linear cascade model* (Sigmund, 1969). This model applies to energetic atomic and small molecular ion projectiles ($O^-$, $O_2^+$, $Ar^+$, $Xe^+$, $Ga^+$, $Cs^+$, etc.), which generate full isotropic collision cascades.

Knock-on sputtering proceeds via a sequence of individual elastic collisions (momentum transfer) occurring between atoms and/or ions as they come into close proximity to each other. How close they approach depends on the energies involved. At and below 100 keV, the distance of closest approach $r_{ca}$ can be defined via the Coulombic potential as:

$$V(r_{ca}) \sim \left( \frac{Z_i . Z_a . q^2}{E} \right) . \Phi . \left( \frac{r}{a} \right) \tag{3.1}$$

where $Z_i$ and $Z_a$ are the atomic numbers of the collision partners (the number of protons and neutrons, which defines their mass), $q$ the unit charge, $\Phi(r_{ca})$ the screening function defining the Thomas–Fermi potential (approximated using *Moliere* or *Zeigler* potentials), and $a$ the *Firsov* screening factor. The $(Z_i . Z_a . q^2 / E)$ term describes the Coulombic repulsion between the two nuclei, whereas the $\Phi(r_{ca}/a)$ term accounts for the screening induced by the electronic clouds surrounding the respective nuclei.

As an example, solving Relation 3.1 for 10-eV $Ar^+$ ion impact on an Al surface reveals that momentum transfer occurs as the valence shells of the incoming ion and surface atom just start to interact. As a result, electronic screening can be said to dominate, and the colliding partners can be envisaged as two solid balls whose dimensions are close to the radii of the respective interacting atoms/ions. It is due to this, and the fact that electronic excitation resulting from ion impact is an inefficient energy transfer process, that kinetic sputtering is often compared to a billiard ball game.

Similar calculations reveal that 1-MeV $He^+$ ions pass right through the electronic cloud of an Al atom and approach to within $2 \times 10^{-3}$ Å of the Al nucleus (the atomic radii of Al is 1.82 Å). The transparency of the solid to $He^+$ and $He^{++}$ ions thus explains the negligible sputter yields resulting on energetic $He^+$ irradiation. Indeed, these conditions are customarily used in the associated technique called Rutherford Back Scattering (RBS) where interaction of the nuclei is needed and where minimal sputtering is desirable.

When an ion of sufficient mass and energy impacts a solid surface, momentum is transferred to the atoms/ions making up the solid and lost from the impacting ion which then becomes *scattered*. The initial interacting atom/ion is referred to as the *recoil*. If treated individually, the change in energy experienced by an ion of mass $m_1$ and energy $E_0$ on colliding with an atom of mass $m_2$ that is initially at rest can then be described in the laboratory frame using *Newtonian mechanics* as:

$$\frac{E_1}{E_0} = \left\{ \frac{\cos \phi_1 \pm \left[ (m_2/m_1)^2 - \sin^2 \phi_1 \right]^{0.5}}{[1 + (m_2/m_1)]} \right\} \tag{3.2}$$

where $\phi_1$ represents the change in trajectory experienced by the incoming ion (the scattering angle) and $\phi_2$ the angle of trajectory experienced by the collision partner with respect to the direction of the incoming ion (the recoil angle). $E_1$ represents the energy of the incoming ion following the collision. Note: Although quantum mechanics is more correct, Newtonian mechanics more than suffices, i.e. provide the same results with a fraction of the calculation complexity.

Using this reasoning, the energy picked up by the atom, $E_2$, initially at rest can be expressed as:

$$\frac{E_2}{E_0} = \cos^2\phi_2 \left\{ \frac{[4\,(m_2/m_1)]}{[1 + (m_2/m_1)]^2} \right\} \tag{3.3}$$

This binary elastic collision along with the parameters used to describe the energy transfer occurring in this event (those used in Relations 3.2 and 3.3) are shown in Figure 3.3.

The scattered ion and recoil atom/ions can then impart further momentum onto other atoms/ions within the solid, and so forth, with the energy eventually dissipated in all three spatial dimensions (isotropic).

Each collision event can be considered separate from all other events (those leading up to or following the respective collision) as the interaction time for each collision ($\sim 10^{-15}$ s) is less than that of any lattice vibrations induced as a result of the respective collisions ($\sim 10^{-12}$ s). Also of note is the fact that full collision cascades terminate within $\sim 10^{-12}$ s. These time scales are revealed through simulations as are discussed in Section 3.2.1.3.

Simulations of the sputtering process also reveal that:

1. in excess of 90% of the sputtered population comes from the outermost atomic layer,
2. the energies of the emitted population peaks at around half the surface binding energy,
3. the angular distribution of the sputtered population peaks toward the sample normal.

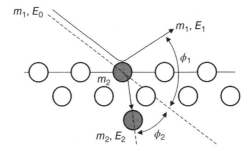

**Figure 3.3** Parameters of interest in binary atomic collisions.

These trends, which are fully supported by empirical data, confer with the supposition that the sputtered population arises from a sequence of many collisions (the collision cascade) with each collision acting independently of all other collisions (linear collision cascade) (Eckstein 1989, Rabalais 1994, Gnaser 1999).

Collision cascades require some minimum energy dependent on the primary ion and the substrate composition. Projectiles of sufficient mass are also required to ensure that the energy is dissipated within the substrate sufficiently close to the surface. As revealed through Relation 3.1, this explains why energetic Helium ions, for example, are not used in SIMS but are used in both Helium Ion Microscopy (HIM) and RBS in which sputtering is not desired.

The energy and angular distributions resulting from a full isotropic collision cascade as described by the linear cascade model have been modeled using the *Sigmund–Thompson relation* (Thompson 1968; Sigmund 1969). This predicts the sputtered neutral energy distribution to scale as:

$$N(E)dE = \frac{E}{(E + E_b)^{3-2x}} \tag{3.4}$$

where $N(E)dE$ is the yield (relative number) of sputtered atoms falling within the energy range $dE$, $E_b$ the surface binding energy, and $x$ a numerical value less than 0.15 characterizing the screened Coulombic interaction potential represented in Relation 3.1. This is usually taken as a fitting parameter. This relation also assumes that the sputtered population follows a cosine distribution centered along the surface normal.

Because the Sigmund–Thompson relation is derived from a *linearized Boltzmann transport* equation (a statistical approach), it can only approximate macroscopic properties arising from an isotropic collision cascade in nonstructured solids, i.e. it does not predict the presence of increased emissions occurring along specific lattice directions from crystalline solids (these are sometimes referred to as *Wehner spots* (Wehner 1955), temperature effects, and so on). In addition, the simplest form of this relation (Relation 3.4) does not account for the finite energy of the incoming ion (an infinite energy is assumed). As a result, deviations at higher emission energies are noted (see Section 3.2.1.3).

The point of emission is easily understood on the basis that surface binding energies are lower than bulk lattice binding energies (4.7 versus 15 eV for Si) and the fact that atoms dislodged deeper within the lattice must have sufficient energy to pass through the respective overlayer/s. It should, however, be realized that kinetically assisted or pure potential sputtering discussed in Section 3.2.1.2 can also cause emissions from the outermost surface.

Also of note is the fact that this form of sputtering is insensitive to the charge of the incoming primary ion beam. This is in contrast to the strong charge sensitivity noted for potential sputtering (discussed in Section 3.2.1.2).

In summary, the *linear cascade model* applies most effectively when using medium to heavy mass primary atomic or small molecular ions (i.e. $O^-$, $O_2^+$, $Ar^+$, $Xe^+$, $Ga^+$, $Cs^+$, etc.) in the low-to-medium keV energy range (within 0.1–50 keV). These energies are used as sputter yields for such ions peak at between 10 and

100 keV depending on the ion. Above ~100 keV, sputter yields decrease as energy is deposited too deep within the substrate to allow for the effective removal of surface atoms. At lower energies, a sharp drop-off in sputter yields is noted with sputtering thresholds noted between 15 and 40 eV dependent on the impacting ion and the substrate (Malherbe 1994). Under such conditions, a full collision cascade is not formed.

Although the *linear cascade model* is extremely effective in describing atomic and certain small molecular emissions, this success does not transfer to:

1. sputtering at energies insufficient to induce a full isotropic collision cascade,
2. sputtering induced by energetic cluster ion impact on:
   a. dense substrates
   b. substrates with low cohesive energies.
3. sputtering of large intact molecular ions, particularly organic molecules on metal substrates, and
4. sputtering of nonmetallic substrates by low energy singly and multiply charged inert gas ions.

Sputtering under conditions that are insufficient to induce a full isotropic collision cascade can still result in measurable sputter yields. These occur also under the same processes described earlier, but with far fewer collisions involved. As a result, a more energetic sputtered population will be noted that displays a more anisotropic angular distribution. This is sometimes referred to as *recoil sputtering*.

As for the remainder, additional mechanisms/models have been suggested. These are based around various inelastic modes of energy transfer, i.e. forms of potential sputtering with some being kinetically assisted. These are covered in Section 3.2.1.2.

***3.2.1.2 Other Sputtering Models***   There are other forms of ion-induced sputtering that, until recently, the SIMS community had paid less attention to. This is, however, changing with the realization that some of these forms of sputtering can induce significant sputter yields from specific types of materials, i.e. organic substrates (see Section 3.2.2). As a result, these are not as fully understood as those resulting from knock-on sputtering.

The various forms of sputtering not described within the context of knock-on sputtering include the following:

1. Energetic large cluster ion impact on organic substrates, with examples of primary ion species including, but not limited to $C_{60}^+$, $C_{84}^+$, $C_{24}H_{12}^+$, and $Ar_n^+$ with $n$ equal to several thousand
2. Energetic dense atomic and small molecular ion impact on organic and/or dense matrices, with examples of primary ion species including, but not limited to $In^+$, $Au_n^+$, $Bi_n^+$, and $SF_5^+$.
3. Low-energy singly and multiply charged inert gas ion impact on nonmetallic matrices such as LiF, NaCl, $Al_2O_3$, and $SiO_2$.

The inability of knock-on mechanisms, inclusive of the linear cascade model, to effectively predict sputter yields in the cases described earlier arises from the fact that such mechanisms describe sputter yields as arising from many individual momentum transfer processes occurring in a linear sequence. However, as outlined in Section 3.2, ejection of atoms/ions or molecules from a solid surface can also occur through:

1. potential sputtering
2. kinetically assisted potential sputtering
3. sputtering through cooperative motion of many adjacent atoms.

As introduced in Section 3.2.1, potential sputtering, whether kinetically assisted or not, results from inelastic energy transfer processes, with electron–phonon inter-actions playing a part. Cooperative motion describes a kinetic process in which a single primary ion impact induces the movement of a collective body of atoms within the solid. All of the above result in sputter yields (these are covered in Section 3.2.2) that are greater than that expected based on the linear cascade model (see Section 3.2.1.1).

The *cooperative motion* mechanism is stipulated to arise from a concerted effect arising from the impact of many atoms within an incoming cluster ion on the respective surface. This, then, induces a meso-scale motion of many substrate atoms/ions within the surface. Under the appropriate conditions (large *reduced cluster ion energy* and large $R$ value in Figure 3.2), this can then act to lift off weakly bound substrate-based molecules, fragments, and atoms in what has been suggested to be a two-step process.

This two-step process is as follows. Firstly, a *fluid-like* motion is generated within the region where the ion initial impacts. This induces the formation of an expanding energized region. Expansion of the energized region then induces the emissions from the surrounding surface regions. Following this, an *effusive-type* motion then follows. This describes the interaction of weakly bound substrate fragments and intact molecules along the walls of the crater formed in the step described earlier. This insight is relayed through Molecular Dynamics simulations (Brene et al. 2011) as discussed in Section 3.2.1.3.

This form of sputtering has attracted significant interest within the SIMS community as of late, as this not only explains the higher than expected atomic and molecular sputter yields but also opens up the possibility for molecular depth profiling and three-dimensional imaging (Gillen and Roberson 1998; Vickerman 2009; Winograd and Garrison 2010). This possibility arises from the fact that a greater fraction of the damage produced on sputtering is removed in each sputtering event as a result of the greater sputter yields noted. Indeed, complete removal of the damage can be noted under optimized conditions (examples include 10 keV or greater $C_{60}^+$, $C_{84}^+$, $C_{24}H_{12}^+$, and $Ar_n^+$) on specific substrates (these tend to be organic based). Cluster ion SIMS, as this mode of operation is referred to as, is covered further in Section 4.1.1.3.

*Kinetically assisted potential sputtering* can take several forms depending on the primary ions, the conditions used, and the matrix examined. For dense atomic and the small molecular ion impact ($In_n^+$, $Bi_n^+$, $Au_n^+$, $SF_5^+$, etc.), these generally tend to assume the presence of overlapping collision events within the lattice that occur as a result of the same initial collision event (the linear cascade model assumes individual events). This overlap ensues when momentum transfer is constrained within a more localized volume and/or when multiple atoms from the same impacting ion strike the same region.

When overlapping collisions occur on dense substrates (large *reduced cluster ion energy* and small $R$ value in Figure 3.2), additional energy loss can proceed via electronic excitation, with the various modes covered in Section 2.2.2.3. Because electronic excitation results in heat, such processes have been equated to the formation of a localized *heat spike* with *thermal evaporation* (Sigmund 1974; Sigmund 1981) and *gas flow models* (Urbassek and Michl 1987; Smartsec et al. 2005) applied. The heat spike analogy is drawn from the fact that translating the kinetic energies of the atoms involved into temperature units reveals values that extend to $10^4$ K. This energy is then argued to dissipate through either or both *Vaporization* of substrate atoms/ions (this results in emissions of atoms/ions and molecular fragments from the centrally affected region) and/or *electron–phonon interactions* (these result in substrate density fluctuations, which move outward from the vaporized region over a period of $10^{-10}$ to $10^{-12}$ s).

If electron–phonon interactions proceed along the substrate's surface or approach the surface from beneath, these interactions can induce further emissions from the periphery of the vaporized region. Owing to the lower energies involved, these emissions can be in the form of complete intact molecular species. This spike-based mechanism differentiates itself from cooperative motion by the fact that the former describes a superposition of events by individual atoms/ions making up the lattice, i.e. overlapping collisions within a single collision cascade, as opposed to a cooperative motion of many adjacent atoms.

Much of the insight into the fundamentals of sputtering resulting from molecular/cluster ion impact has been drawn from Molecular Dynamics simulations (see Section 3.2.1.3). These simulations *tend* to reveal the formation of a central transient under-dense region surrounded by an outwardly moving over-dense halo with preferential ejection of atoms/ions and fragmented molecules from the central under-dense region and ejection of intact molecules from the halo edges as depicted in Figure 3.4 for small molecular ion impact. Large cluster ion impact, on the other hand, *tends* to result in the removal of surface molecules through what appears to be a trampoline-like effect. Note: The word *tend* is used numerous times in this paragraph as the matrix can have a strong impact on the process/s active and hence the outcome.

*Potential sputtering* does not require the assistance of knock-on effects. In other words, no momentum transfer is required. As a result, potential sputtering can occur even without the primary ion directly colliding with the solid's surface. It only needs to be in close enough proximity to allow for electron transfer. This is realized as any charged particle in close proximity to a solid surface will undergo some form

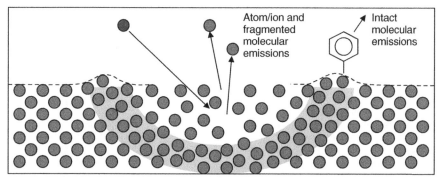

**Figure 3.4**   Highly simplified pictorial illustration of sputtering believed to occur during kinetically assisted potential sputtering induced on small molecular ion impact. This can also describe the first step in cooperative sputtering from large cluster ion impact. The gray region represents an outward moving dense halo.

of electronic interaction with surface-bound atoms (charge transfer processes are covered in Section 3.2). As a result, all primary ions are readily neutralized even before coming into physical contact with the surface (this also occurs in other forms of sputtering; however, it is not noted because of the excessively small sputter yields that generally result).

As potential sputtering results from the neutralization of the primary ion, yields scale with the energy required to produce the primary ion (the ionization potential, as discussed in Section 2.2.1.3) and hence the charge state (greater yields are noted from multiply charged ions). As neutralization occurs through populating the outermost valence levels of the incoming primary ion, a hollow atom is formed (one in which unfilled subvalence levels exist). De-excitation processes will result in further energy release. This will occur via Auger or fluorescence processes (these are described in Section 2.2.2.4.4). In addition, numerous substrate-based de-excitation processes can be induced.

Potential sputtering thus serves to explain why sputtering can be noted, albeit at extremely low yields, below the kinetic sputtering threshold (the kinetic sputtering threshold is between 15 and 40 eV depending on the system) that scales with the primary ion charge (Malherbe 1994).

A pictorial illustration of the steps believed to lead up to this form of sputtering is shown in Figure 3.5.

As can be seen, this form of sputtering is generally accepted to proceed via:

1. Neutralization of the incoming ion before it strikes the surface (occurs with $10^{-13}$ s). This causes electronic excitation of the aforementioned ion in the form of core holes.
2. De-excitation of the neutralized hollow primary ion via an Auger or fluorescence process (the Auger process is shown in Figure 2.6). In the case of $Ne^+$ primary ions, this results in the Ne $KL_2L_3$ Auger electron emissions with an example as shown in Figure 3.6(a).

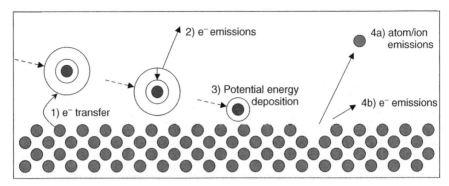

**Figure 3.5** Highly simplified pictorial illustration of the four-step mechanism believed to describe potential sputtering. This multistep process described in the text is initiated through electronic excitations induced as an ion approaches a solid surface.

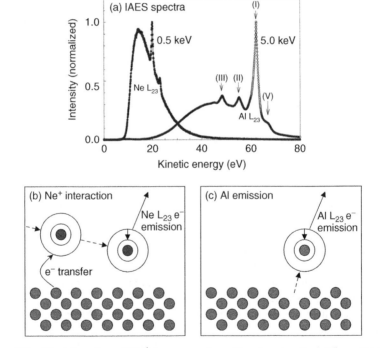

**Figure 3.6** The interaction of $Ne^+$ ions on a clean Aluminum surface (a) as revealed by ion-induced Auger electron spectra (b) with the potential electron emissions from the incoming $Ne^+$ ions illustrated pictorially, and (c) with the kinetic electron emissions from sputtered Aluminum atoms illustrated pictorially. Note: The spectra resulting from 0.5 and 5.0 keV $Ne^+$ impact are normalized and overlaid. These reveal peaks that are ascribed to the Neon $KL_2L_3$ and Aluminum $KL_2L_3$ Auger electron emissions (these are referred to as Ne $L_{23}$ and Al $L_{23}$ emissions in the respective plots). The spectra were collected on a Phi 5700 XPS instrument operated in the Ion-induced Auger Electron Spectroscopy (IAES) mode. (Authors unpublished work replicating that of Valerie 1993, with the same definitions used).

3. Electronic excitation of the surface through the production of plasmons, excitons, and/or phonons depending on the substrate (these excitations are discussed in Section 2.2.2.3).

4. Sputtering or desorption of the surface-bound atoms and/or intact molecules along with additional electron emissions from the substrate as a result of the electronic excitation induced.

The exact process by which sputtering/desorption (steps 3 and 4 as mentioned earlier) occurs is, however, still an area under study with several substrate-specific models having been presented. These models, which can also be kinetically assisted, are as follows:

1. *Coulomb explosion model.* This model developed to explain emissions from insulating surfaces ($SiO_2$) asserts that neutralization of the incoming ion along with subsequent electron emissions results in electron depletion over a localized near surface region. The resulting Coulombic repulsion between ion cores then results in atom/ion ejection from this region.

2. *Intense ultra-fast electronic excitation model.* This model developed to explain emissions from semi-conducting and insulating surfaces (Si, GaAs, $SiO_2$, etc.) asserts that the bonds holding the respective atoms/ions to the surface are destabilized as a result of the promotion of many electrons from their bonding to anti-bonding states on ion neutralization/impact, that is, valence to conduction band transitions.

3. *Defect-mediated sputtering model.* This model developed to explain emissions from insulating surfaces ($SiO_2$, LiF, NaCl, etc.) asserts that the defects are introduced in the form of self-trapped excitons and/or holes in response to the valence band excitation induced on ion neutralization/impact. Electron–phonon coupling then results in desorption.

Except for the interaction of highly charged primary ions, the present consensus tends to favor the defect-mediated sputtering model (Rabalais 1994). For highly charged ions, the intense ultra-fast excitation model appears likely (Aumayr and Winter 1994, 2003). As the probability of inelastic energy loss depends on the overlap of the respective orbital wave functions, simulations of a pure potential sputtering process requires quantum mechanics. As mentioned in Section 3.2.1.3, this is considered outside the scope of this text.

Evidence of both kinetic and potential sputtering can be observed in the spectra arising from the low-energy electron emissions collected at large angles relative to the surface normal during low-energy sputtering of various metal surfaces. Note: Such angles are required in order to observe the respective emissions. Examples of the resulting spectra are shown in Figure 3.6(a). These were produced as a result of 0.5 and 5.0 keV $Ne^+$-induced sputtering of an Aluminum surface. For the sake of clarity, the spectra are overlaid with their peak intensities normalized to unity (in reality, the 5.0 keV-induced emissions are of far greater intensity such that they would otherwise drown out the 0.5 keV emissions). Evidence for potential sputtering is noted in the form of the Neon $KL_2L_3$ Auger peaks (the X-ray notation used is described in Section 2.2.1.2). These are specified as being due to the interaction of $Ne^+$ ions before impact, as schematically illustrated in Figure 3.6(b). Evidence of a kinetic sputtering process, on the other hand, is noted in the form of the Aluminum $KL_2L_3$ Auger peaks observed under 5.0 keV $Ne^+$ impact. These peaks are specified as arising from the de-excitation of sputtered Aluminum atoms. The excitation stems from elastic collisions between Aluminum atoms within the lattice before the ejection of the excited atoms/ions, as illustrated in Figure 3.6(c). The elevated broad background upon which the respective Aluminum $KL_2L_3$ peaks reside are from Auger emissions from Aluminum atoms involved in the collision cascade within the substrate. These suffer nondiscrete energy losses, hence the board peak observed. Further details on these processes as well as on the experimental aspects associated with measuring these processes can be found elsewhere (Valerie 1993).

### 3.2.1.3 Simulations
Effective simulations are possible only if the event being simulated is well understood. As this is the case for knock-on sputtering, inclusive of the linear cascade model, there exist a number of simulation packages, with some freely available. As this same level of understanding is not present for the other forms of sputtering (those described in Section 3.2.1.2), particularly when inelastic effects come into play, these must be dealt with using more complex and highly specific many-body approaches. In the following sections, an overview of the various simulation packages available is presented along with some examples of the various capabilities (for more in depth reviews, see Smith 2005; Behrisch and Eckstein 2007).

The primary reason why various simulations of sputtering resulting from momentum transfer, as described within the *linear cascade model*, are successful lies in the fact that the collision events can be treated using classical arguments, that is, Newtonian mechanics as opposed to quantum mechanics. Owing to the insignificant wavelength of ions ($\sim 10^{-5}$ Å), quantum mechanics is not needed. In addition, calculations are simplified for isotropic linear cascades as these represent a linear sequence of independent collision events that occur over time scales much shorter than lattice vibrations.

As Newtonian mechanics can be used to describe ion–solid interactions, scattering as well as complete collision cascades can be simulated to a good degree of accuracy with relative ease. This includes defining scattering yields, the distribution of implanted ions as well as recoils with respect to their original location, the sputtered population, sputtering yields, any lattice damage, electronic excitation, and so on. The approaches developed can be subdivided according to whether:

1. trajectories are numerically derived for atoms/ions as they traverse through some predefined lattice structure. This microscopic approach can be further subdivided according to whether:

    a. the individual collisions are treated as separate events. These use the *Binary Collision Approximation* (BCA) on/in a crystal structure in the form of the *MARLOWE* code,

b. multiple collisions with all atoms within the lattice modeled are considered together. These are referred to as *Molecular Dynamics* (MD) type simulations with the SPUT93, MDRANGE and KALYPSO codes being examples.

2. trajectories are statistically derived for atoms/ions within solids from a sequence of collisions. This macroscopic approach can be further subdivided according to whether:

a. the location of atoms in space and time are derived via a random number generator. These use the Monte Carlo method with the *TRansport of Ions in Matter* (TRIM) code being the most common. For following sputtering processes, the closely related code of *Stopping and Range of Ions in Matter* (SRIM) is the most appropriate.

b. the location of atoms satisfies space and time invariance, i.e. exist in an infinite homogeneous array. These use the *Boltzmann Transport Equation*, with the *KUBBIC* code being an example.

Simulations in which trajectories are numerically derived are more useful when information of ion trajectories and their effects in crystalline solids are required (sputter yields can vary with crystal orientation, etc.) or when specific information from particular collision events are sought after. Molecular Dynamics, which is the most time intensive, excels in that it does not require surface binding energies (all other approaches do), and it accounts for many body effects, i.e. collisions between more than two atoms/ions (this is not accounted for in other approaches). Thus, only the Molecular Dynamics approach extends down to the eV range thereby providing some insight into kinetically assisted potential sputtering processes, i.e. those resulting in the theorized *heat or thermal spikes* (see Section 3.2.1.2).

All other approaches assume that the trajectories of an energetic ion as well as those of its recoils can be modeled as a sequence of separate binary collisions (elastic component) separated by straight-line segments in which the atom/ion experiences a continuous electronic energy loss (inelastic component). Trajectories are followed until the energy of the atom/ion and/or recoils of interest drop below some predefined value (of the order of 10 eV) at which point the incoming ion and/or recoils come to a halt. Those using statistics require the least computational power and, as a result, are the most effective in providing statistically significant values for energetic ions and recoils in amorphous and polycrystalline solids. Good statistics are required as significant fluctuations in individual scattering events are evident. Furthermore, sputtering includes many billions of collision cascades.

Examples of the use of two commonly used packages for modeling knock-on sputtering as applies to energetic atomic ion projectiles are shown in Figures 3.7–3.10.

In Figure 3.7 are shown the scattered ion yields (the number of ions scattered off the surface per the number of incident ions) of $Cs^+$ ions impinging on an Si surface as a function of energy (250–2.5 KeV) and incident angle (40–80° with respect to the sample normal). As scattering from a solid's surface results from individual collisions events, the MARLOWE code proves to be highly effective.

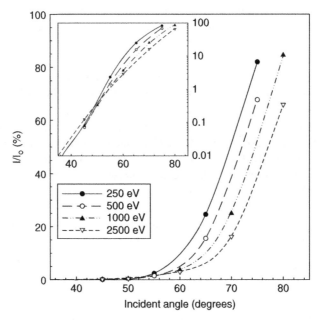

**Figure 3.7**   Scattered $Cs^+$ ion fractions derived by MARLOWE from a Silicon wafer as a function of incoming ion energy and angle of incidence (with respect to surface normal). In the inset is the same data with the scattered fraction plotted on a log scale (ordinate). Both Mollier and Ziegler potentials yield similar results (those using Mollier potentials are shown).

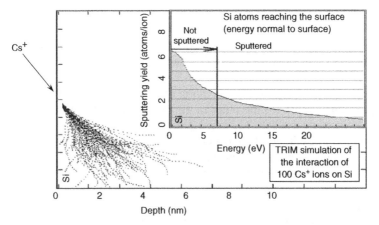

**Figure 3.8**   SRIM simulations of the trajectories of 100, 1 keV $Cs^+$ ions incident at an angle of $60°$ with respect to the surface normal in Silicon. In the inset is shown the energy distribution of Silicon recoils relative to the surface normal toward the vacuum. Note, only those with energy greater than the surface binding energy (4.7 eV for Silicon substrates) can escape. The energy of the sputtered population then decreases by this amount.

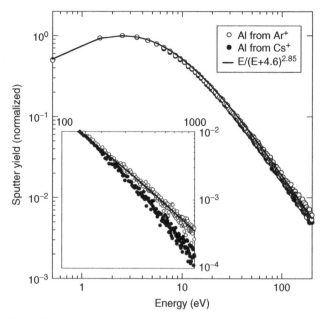

**Figure 3.9** Normalized energy distributions of sputtered Aluminum atoms resulting from 17.5 KeV Ar$^+$ and Cs$^+$ impact on an Aluminum surface at an impact angle of 20° as defined by MARLOWE (symbols) overlaid that predicted by the Sigmund–Thompson relation as defined by Relation 3.4 (lines). Reprinted with permission from van der Heide and Karpusov (1998) Copyright 1998 John Wiley and Sons.

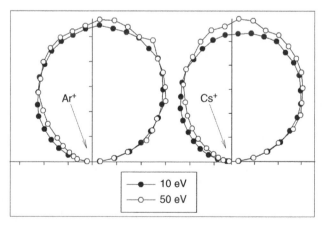

**Figure 3.10** Angular distribution derived by MARLOWE of sputtered 10 eV and 50 eV Aluminum atoms from Aluminum under 17.5 KeV Ar$^+$ and Cs$^+$ impact at 20° with respect to the surface normal. Reprinted with permission from van der Heide and Karpusov (1998) Copyright 1998 John Wiley and Sons.

In Figure 3.8 are shown the trajectories of 100 1 keV $Cs^+$ ions incident on Silicon at an angle of 60° with respect to the sample normal. In this case, the SRIM code is the more effective approach as it readily provides distributions of ions, recoils, any resulting damage of the lattice, and sputter yields (the number of atoms removed per incoming ion). In the inset of Figure 3.8 is shown the distribution of energies of recoils in the direction of the vacuum versus the number of recoils as derived by SRIM. Only those atoms at or close to the surface with energies greater than the surface binding energy are able to depart the solid. This represents the sputtered population. Also evident is the fact that sputtered atoms can emanate 5–10 nm from the point of impact (this defines the physical limit of the spatial resolution in SIMS).

The effectiveness of SRIM primarily arises from the fact that a large number of calculations can be carried out in a relatively short time period, thereby providing good statistics. For example, SRIM can easily compute the outcome of $10^6$ ions or more with ease (only 100 are shown in Figure 3.8 for reasons of clarity). SRIM also proves particularly effective on Silicon as Silicon is an example of a crystalline solid that quickly becomes amorphous on energetic ion impact (the formation of a 2-nm amorphous layer is noted on a single crystal Silicon when sputtered by 1-keV $Cs^+$ ions as discussed in Section 3.2.3.2).

Although energy distributions of sputtered atoms can be approximated using the SRIM approach, MARLOWE proves to be more effective when single crystals are of interest. Polycrystalline solids can also be studied using MARLOWE by randomly altering the crystal axis. An example is shown in Figure 3.9 for Aluminum atoms sputtered normal to the surface from 17.5 KeV $Ar^+$ and $Cs^+$ impact on polycrystalline Aluminum at an angle of 20° with respect to the surface normal. The energy represents the kinetic energy of the sputtered population. These compare well with the trends implied by the Sigmund–Thompson relation (Relation 3.4). Note, however, that deviations are observed above 100 eV particularly for $Cs^+$ impact. This is due to the fact that the Sigmund–Thompson relation does not account for the finite energy of the primary ion.

In Figure 3.10 is shown the angular distribution of the sputtered Aluminum atomic population resulting from 17.5 KeV $Ar^+$ and $Cs^+$ impact on a polycrystalline Aluminum surface at an impact angle of 20° as a function of the kinetic energy. Again, the MARLOWE code proves to be effective, i.e. the resulting cosine-to-cosine-squared distribution slightly offset from the surface normal is observed. As indicated in Section 3.2.1.1, the Sigmund–Thompson theory predicts a cosine distribution because of the isotropic nature of the event.

Examples of the results derived from Molecular Dynamics simulations of large cluster ion impact on organic substrates are shown in Figures 3.11–3.14. Note: Molecular Dynamics simulations are presently the most effective option for following the kinematics (momentum transfer) resulting from such impacts on organic matrices. This is realized as, in many of cases, overlapping collision sequences or even the collective or cooperative motion of many atoms within the substrate are noted as resulting from the same primary ion impact event (see Section 3.2.1.2).

The example shown in Figure 3.11 described the trajectories (shown by the arrows with the length proportional to their emission energy) of sputtered intact

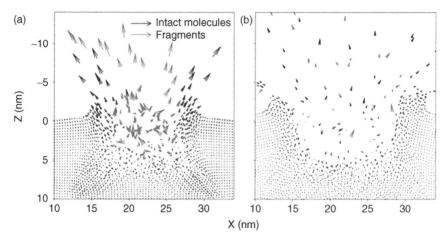

**Figure 3.11**    Time snapshots from Molecular Dynamics computer simulations of the emission of coarse grain benzene molecules (black) and fragments (grey) represented as a vector plot at (a) 2 ps and (b) 8 ps after $C_{60}^{+}$ impact on a 400-nm-thick amorphous Benzo-[a]pyrene film ($C_{20}H_{12}$) held at 85 K. Reprinted with permission from Brene et al. (2011) Copyright 2011 American Chemical Society.

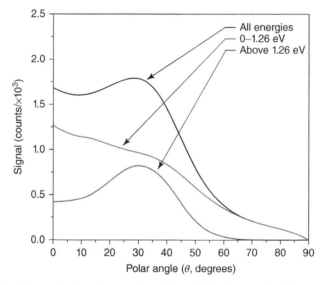

**Figure 3.12**    Polar angle distributions from Figure 3.11. Reprinted with permission from Brene et al. (2011) Copyright 2011 American Chemical Society.

and fragmented molecules arising from 2 ps and 8 ps after 20 keV $C_{60}^{+}$ impact. These times were chosen as they relay sputtering believed to occur via fluid-like and effusive-type motions initiated as a result of $C_{60}^{+}$ impact (Brene et al. 2011). The importance of these motions is discussed in Section 3.2.1.2. Such simulations

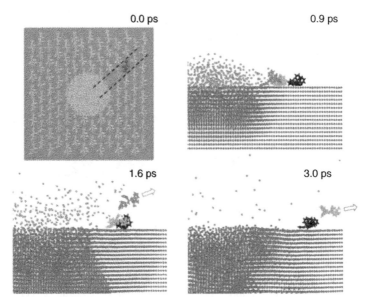

**Figure 3.13** Time evolution of ejected low-energy polystyrene tetramer molecules induced by 15 keV $Ar_{2953}^+$ cluster impact at normal incidence. Only ejected molecules together with their nearest neighbors located within a slice 1.5 nm wide centered at the point of impact are shown. Reprinted with permission from Rzeznik et al. (2008) Copyright 2008 Wiley Periodicals Inc.

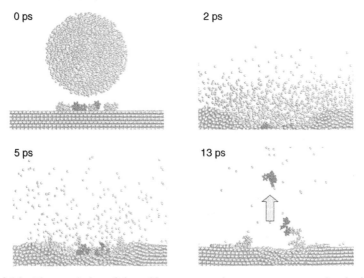

**Figure 3.14** Time evolution of ejected low-energy polystyrene tetramer molecules induced by 15 keV $Ar_{2953}^+$ cluster impact at normal incidence. Only ejected molecules together with their nearest neighbors located within a slice 1.5 nm wide centered at the point of impact are shown. Reprinted with permission from Rzeznik et al. (2008) Copyright 2008 Wiley Periodicals Inc.

also show the incoming $C_{60}^+$ ion to disintegrate into its constituent Carbon atoms on releasing its kinetic energy within the first 100 fs.

Analysis of the trajectories of intact molecular emissions with energies less than 1.26 eV and greater than 1.26 eV reveals quite different polar angle distributions as illustrated in Figure 3.12. In short, those at higher emission energies show an off normal emission tendency as would be expected from an outwardly expanding energized region, i.e. that resulting from the $C_{60}^+$ initiated fluid line motion within the substrate. Those with lower emission energies, however, display a more isotropic angular distribution peaking close to the surface normal consistent with an effusive type motion. The overall distribution shown in Figure 3.12 represents the sum, and hence the off normal distribution (Brene et al. 2011).

Molecular Dynamics simulations also reveal situations in which intact organic molecules can be emitted without crater formation, i.e. through *soft landing*. This has been shown on metal surfaces bearing organic constituents when examined under the appropriate conditions, i.e. large cluster impact at sufficiently low energies per atom (Rzeznik et al. 2008). Such processes are shown in Molecular Dynamics simulations of 15 keV $Ar_{2953}^+$ impact at normal incidence on Silver substrates bearing a polystyrene tetramer film in Figures 3.13 and 3.14.

Of interest is the fact that intact molecules can be emitted from the central or edge locations of the impacted region. In the aforementioned case, a washing mechanism is proposed in which the collective movement of substrate atoms induces the emission of intact molecules at large angles relative to the surface normal, whereas in the latter case, a compression mechanism is noted where collective motion of substrate atoms now induces intact molecular emissions at angles close to the sample normal analogous to that of a trampoline (Rzeznik et al. 2008; Garrison and Postawa 2008).

These are but a few of the scenarios being opened up by the implementation of classically based simulations. Indeed, the effectiveness of the theory used also allows for the irradiation-induced diffusion, segregation, and ripple topography growth apparent on and within various single and multicomponent substrates to be modeled (Liedke, et al. 2013). Note: Although quantum mechanics would provide the more correct theoretical description, sputtering simulations based around such an approach would be far more expensive computationally with little benefit added.

### 3.2.2 Sputter Rates and Sputter Yields

As covered in Section 3.2, sputtering describes the removal of atoms (ions being intrinsic to ionic substrates as opposed to those formed on sputtering) and/or molecules from a solid's surface through the deposition of energy imparted by ions, electrons, or photons directed at the aforementioned solid's surface. In the case of SIMS, energy deposition results from the impact of a highly focused energetic ion beam, referred to as the *primary ion beam*. For the sake of clarity, single-component substrates of amorphous or polycrystalline structure will only be discussed unless otherwise specified (multicomponent systems introduce additional effects as will be discussed in Section 3.2.3.4)

The rate at which material is removed from the substrate's surface is referred to as *etch rate* or the *sputter rate*, with the latter more commonly used in SIMS. Commonly used units include Å/s, nm/s, and nm/min. In some cases, the primary ion current is also incorporated, i.e., Å/nA.s. Sputter rates can range from some negative value (only noted when a sizable fraction of the primary ion beam is implanted into substrates, thereby causing swelling) to values greater than 100 nm/min.

Sputter rates defined using units of Å/s, nm/s, or nm/min, i.e. those not incorporating the primary ion current, depend on:

1. The dynamics (mechanism) of energy transfer which in itself depends on the masses of the colliding species, as well as the impact energy, angle, etc.
2. The material examined, inclusive of long-range lattice structure (amorphous versus crystallinity), the crystal orientation, the temperature of the sample, etc.
3. The instrumental conditions used, i.e. the area over which the beam is scanned (stationary beams are also used in SIMS)
4. The rate at which the incoming ions are striking the surface. Note: This dependence drops out when primary ion current units are incorporated into sputter rate units, i.e. as Å/nA.s

Sputter rates are defined through:

1. surface profilometry (stylus or optical) measurements of the resulting crater,
2. analysis of samples bearing discrete layers at known depths, i.e. markers, interfaces, and so on,
3. micro-balance methods, i.e. weighing the sample before and after sputtering.

As sputter rates are more commonly applied in analytical circles, further aspects on sputter rates and potential sputter rate issues noted in depth profiling are covered in Section 5.3.2.4. Details on methods for measuring sputter rates are discussed in Section 5.4.1.

Closely related to the sputter rate is the rate of removal of a specific atom/intrinsic ion or molecule from the substrate of interest. This is referred to as the *absolute sputter yield*. The term "absolute" is used to differentiate this from the *useful sputter yield*. As absolute values are of interest in theoretical evaluations, sputter yields will refer to absolute values henceforth unless otherwise defined.

*Absolute sputter yield* values refer to the number of secondary ion species removed per incoming primary ion. In the case of atomic emissions, these become atoms removed/incident primary ion or atoms/ion. For molecular emissions from cluster ion impact, these become molecules/cluster. Note: 1 nA equates to $6.28 \times 10^9$ ions passing some point in space per second.

*Useful sputter yield* values refer to the number of secondary ion species detected per incoming primary ion. As the number of species detected depends on the transmission of the instrument used, these will be less than absolute values. These tend to be used in analytical application discussions.

Sputter yields depend on the dynamics (mechanism) of energy transfer and the material examined, inclusive of long-range lattice structure (amorphous versus crystallinity), the crystal orientation, and the temperature of the sample. The energy transfer depends on the masses of the colliding species as well as on the impact energy and angle, and so on. Indeed, values resulting from pure isotropic linear cascades (a specific form of knock-on sputtering as described in Section 3.2.1.1) can be simulated for many materials. The same cannot be said for sputtering resulting from other processes as covered in Section 3.2.1.2.

Sputter yields for pertinent species are most commonly defined using some form of post-ionization technique. These effectively operate by ionizing the plume of neutral species shortly after their departure from the surface from which they are sputtered, with all ions then passed through a mass spectrometer, most commonly a time-of-flight mass spectrometer.

### 3.2.2.1 Sputter Yield Dependence on Primary Ion Conditions

Because there exist a large array of primary ion types (common examples used in SIMS include $O^-$, $O_2^+$, $Cs^+$, $Ga^+$, $In^+$, $Au_n^+$, $Bi_n^+$, $SF_5^+$, $C_{60}^+$, $C_{84}^+$, $C_{24}H_{12}^+$, and $Ar_n^+$ with $n$ being an integer starting from unity and some being multiply charged) along with several sputtering mechanisms (kinetic sputtering, kinetically assisted potential sputtering, and pure potential-based sputtering mechanisms, as covered in Section 3.2.1), the preceding sections concentrated on the more *general* trends observed when using conditions consistent with those most typically used during routine SIMS analysis, i.e. primary ion impact energies within the 0.1–50 keV range with incidence angles ranging from 0 to 80° relative to the surface normal.

*Atomic sputtering yields* resulting from atomic ion or small molecular ion impact on literally every substrate type have long been known to display dependences on:

1. the mass of the colliding partners,
2. the energy dissipated on primary ion impact, and
3. the angle of the primary ion impact relative to the surface normal.

These dependencies, however, *tend to* be substrate specific, hence the inclusion of a separate section discussing substrate-induced variations (see Section 3.2.2.2.1). Note: The term *tend* is used as substrates comprising elements displaying similar Coulombic repulsion (see Relation 3.1) with similar densities and similar surface binding energies can display similar sputtering trends.

Assuming a specific substrate, the parameters of *mass* and *energy* are interrelated through *momentum* (momentum equates to the product of mass times velocity, with the velocity related to energy as the product of half of the mass times the velocity squared). As a result, sputter yield dependences on these are discussed together in Section 3.2.2.1.1. These, in essence, define the sputtering dynamics.

The dependencies on sputtering angle are discussed separately in Section 3.2.2.1.2. This is done as the angle defines the subsurface region affected by the collision dynamics. In other words, although the dynamics may be the same at different incidence angles, different sputtering rate/yield variations are noted, which are then simply a matter of how close to the surface the collisional processes take place.

*Molecular sputtering yields* resulting from atomic or molecular/cluster ion impact (small or large) display a much greater diversity of trends. At this point, it is worth noting the distinction between intact molecular species directly emitted from the surface of interest and those formed as a result of the sputtering process itself. The latter makes up the vast majority of molecular emissions noted under Dynamic SIMS conditions when using all but large cluster primary ions. The former, on the other hand, is seen under Static SIMS conditions and Dynamic SIMS when large cluster primary ions are used. As these can provide significantly more information on molecular distributions on and/or within the substrate, these are discussed henceforth.

The emissions of intact molecular species from a specific substrate as a result of atomic or small molecular ion impact tend to dissipate quickly, with a direct correlation noted with the primary ion current density. This disappearance results from accumulation in damage imparted by the respective primary ion on the substrate being analyzed. This damage is thus responsible for what is referred to as the *static limit*, and hence the need to perform Static SIMS under primary ion dose values less than 1% of the equivalent molecular density, assuming complete coverage, on the surface.

The fact that it is difficult to measure intact molecular sputter yields under Static SIMS conditions, particularly from covalent organic materials, the loss in such signals (secondary ion signal) as the primary ion dose is increased above this limit is typically reflected in what is referred to as the *disappearance cross section*. This relays the loss in signal, as measured from a substrate bearing a monolayer of the molecule of interest, as being due to the removal of the molecular species of interest from the probed regions and damage introduced as a result of the sputtering process to the surrounding regions. This is commonly relayed as a *damage cross section*. The damage cross section, which is system specific, is expressed as:

$$I_m = I_{mo} \ \exp(-\sigma.I_p) \tag{3.5}$$

where $I_m$ is the recorded signal of the respective molecular species, $I_{mo}$ the original surface density of the respective molecular species, $\sigma$ the disappearance cross section, and $I_p$ the primary ion dose.

Sputtering of intact molecular species beyond the static limit, particularly from substrates bearing molecules with low cohesive energies (intermolecular bonds), is possible under large cluster ion impact. Such clusters can include, but are not limited to, $C_{60}^+$, $C_{84}^+$, $C_{24}H_{12}^+$, and $Ar_n^+$ with $n$ being an integer starting from unity and extending to several thousand. Indeed, this relatively new field has far reaching implications, the most notable being the possibility of imaging intact organic molecular species in all three dimensions (Vickerman 2009; Winograd and Garrison 2010; Mahoney 2013). This possibility arises from the fact that large cluster ion-induced sputtering occurs via a different mechanism, namely that of cooperative motion.

As discussed in Section 3.2.1.2, this mechanism describes the coherent motion of many atoms within a relatively small area as opposed to a linear sequence of collisions as assumed in knock-on sputtering. This, in turn, can allow for a significant enhancement in sputter yields relative to those implied by knock-on sputtering. Under ideal cases, this can also result in the removal of intact and fragmented molecular species from the entire damaged region (damage resulting from the sputtering event) such that the underlying surface is effectively damage free. In addition, these large cluster ions tend to evaporate from the sputtered region during the sputtering event. The fact that this is a relatively new field of research; therefore, the readers are advised to access the latest journal publications pertinent to their field of interest.

Lastly, with the exception of the less common form of sputtering referred to as potential sputtering (see Section 3.2.1.2), little dependence of sputtering yields is seen with primary ion charge.

### 3.2.2.1.1  *Dependence on Primary Ion Momentum*    A strong atomic sputter rate/yield dependence on primary energy and mass is noted for all of the most common forms of sputtering (all but the potential form of sputtering). This can be understood as arising from the kinematics in action. Indeed, such dependencies are noted in both atomic ion and molecular/cluster ion impact with simulations providing an effective insight into the dynamics in action.

*Sputtering of atomic species* from inorganic substrates arising from atomic ion and small molecular ion impact provides for some of the most definitive trends. Examples of this are shown in Figure 3.15 in which sputter yields arising from different projectiles as a function of impact energy from a polycrystalline Nickel substrate are shown (from Biesack and Eckstein 1984). A polycrystalline substrate is used as this effectively averages out crystal orientation-induced anisotropies (sputter rates vary with crystal orientation relative to the surface, and sputter yields vary with ejection angles relative to closely packed crystallographic orientations (Wehner 1955; Gnaser 1999)).

Two aspects noted from Figure 3.15 are of particular interest.

The first aspect is the fact that the sputtering threshold is greater than the surface binding energy, i.e. threshold energies range from 15 to 40 eV (Malherbe 1994), whereas surface binding energies lie between $\sim$3 and 5 eV (see Table 3.1). This stems from the fact that energy is lost (dissipated) by the impacting ion through momentum transfer to numerous atoms/ions contained within the substrate and that only following a sequence of multiple collisions does an atom/ion located close to the surface experience sufficient momentum transfer in the direction away from the substrate such that it can overcome its surface binding energy.

The second aspect is the fact that the maxima lie at projectile-specific energies. This can be understood from the depth into the substrate at which the projectile comes to rest and hence the depth at which the energy dissipation primarily occurs. If this is too deep, sputtering from the respective surface will not occur. The *Projected Range* (the average depth at which the projectile comes to rest), more commonly defined as $R_p$, increases with incident energy for a projectile of constant $Z$ within a specific substrate.

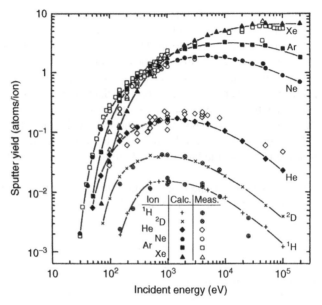

**Figure 3.15**  Atomic Nickel sputter yields of from a polycrystalline Nickel surface for the projectiles listed over the respective impact energy range. The symbols represent empirical data, whereas the lines represent simulated data. Reprinted with permission from Eckstein (1989) Copyright 1989 John Wiley and Sons.

These trends with the primary ion's projection range can be approximated for normal incidence and within the energy regime of interest in SIMS via:

$$R_p = C_1((m_2/m_1)m_2).\left[\left(\frac{(Z_a^{2/3} + Z_i^{2/3})}{(Z_a.Z_i)}\right)E_p\right]^{2/3} \tag{3.6a}$$

where $C_1$ is an impact-specific parameter, $m_1$, $m_2$, $Z_a$, and $Z_i$, take on the same meaning as in Relations 3.1–3.3, and $E_p$ the impact energy in units of keV (Schiott and Dan 1966, Schiott 1970). Note: The product of $C_1(m_2/m_1)$ for $Ar^+$ impact on Silicon is found to equate to ~9 when $R_p$ is relayed in units of nm.

For $O_2^+$, $Ar^+$, and $Cs^+$ ion impact on Silicon, the projected range can be more easily approximated using the empirically derived relations (Wilson et al. 1989):

$$R_p(O_2^+) = 2.15 E_p \cos \Theta \tag{3.6b}$$

$$R_p(Ar^+) = 1.622 E_p^{0.34} \cos \Theta \tag{3.6c}$$

$$R_p(Cs^+) = 1.838 E_p^{0.68} \cos \Theta \tag{3.6d}$$

where $\Theta$ is the incidence angle in units of degrees relative to the sample normal. Note: This differs from $\phi$ used in Relations 3.2 and 3.3, which should not to be confused with the work–function represented by the nonitalicized $\phi$. In the case of $O_2^+$ primary ions, the energy must be divided by a factor of two as there exist two atoms per ion.

*Sputtering of intact molecular species* initiated via atomic or small molecular projectiles is generally limited to static SIMS conditions. This results from the fact that the damage induced by the kinematics of knock-on sputtering effectively destroys molecular information over the region impacted.

Sputtering of intact molecular species via large cluster ion impact past the static limit is, however, noted under appropriate conditions (large *reduced cluster ion energy* on soft substrates as depicted in Figure 3.2). This arises from a combination of the fact that less damage is imparted during the initial impact, and the increased sputter yields noted can result in the removal of a significant fraction of the damaged region. If a significantly large enough fraction of the damaged is removed, an undamaged under-layer can then be exposed for subsequent analysis. This opens up the possibility of depth profiling or even three-dimensional imaging of specific substrate-based molecular species.

For an example of the enhancement in sputter yields resulting from large cluster ion impact, see Table 3.2. This lists sputter yields of $H_2O$ molecules from a 500-nm film of amorphous ice deposited on a Silver substrate resulting from $Au^+$, $Au_2^+$, $Au_3^+$, and $C_{60}^+$ ion impact at 40° with respect to the sample normal as defined via Molecular Dynamics simulations (Szakal et al. 2006).

Table 3.2 also reveals the nonadditive nature of the sputter yields arising from large cluster ion impact, i.e. the sputter yields noted from a single $C_{60}^+$ ion is well in excess of the sum of 60 individual $C^+$ ions of 1/60th of the impact energy of $C_{60}^+$. Note: This division of the impact energy by the number of atoms making up the cluster, termed *energy per atom*, allows for effective comparisons of sputter yields from different cluster ions of the same family.

Also of note is the fact that the sputtering threshold for large cluster ion impact occurs at much greater overall impact energies than noted for atomic ion impact, i.e. these occur in the keV energy range with respect to 15–40 eV noted for atomic ion impact. This can be understood on the basis that the energy is now split between the many atoms making up the cluster, hence the parameter *energy per atom*.

Indeed, an interesting experiment noted in the literature relays the dependence of secondary ion intensities of various intact molecular species on the energy per atom.

**TABLE 3.2   Sputter Yields of $H_2O$ Molecules as Determined by Molecular Dynamics Simulations from a 500-nm Film of Amorphous Ice on Silver Resulting from the Listed Projectiles all Incident at 40° (Szakal et al. 2006).**

| Projectile | Energy | Energy per Atom | Sputter Yield |
|---|---|---|---|
| $Au^+$ | 25 keV | 25 keV | 94 |
| $Au_2^+$ | 25 keV | 12.5 keV | 570 |
| $Au_3^+$ | 25 keV | 8.33 keV | 1200 |
| $C_{60}^+$ | 25 keV | 0.417 keV | 2510 |

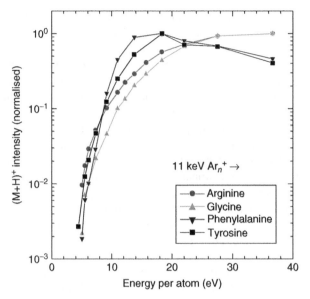

**Figure 3.16**    Normalized intensities of the listed protonated molecules as a function of impact energy per atom for $Ar_n^+$ cluster impact (overall impact energy is of 11 keV). The impact energy per atom is controlled by adjusting the value of $n$. Reprinted with permission from Gnaser et al. (2012) Copyright 2012 John Wiley & Sons.

The result, shown in Figure 3.16, suggests a threshold below 3 eV for $Ar_n^+$ cluster impact on various amino acid substrates, i.e. a value consistent with the cohesive energies of the respective molecules. The decrease in various intensities with increasing energy per atom can be understood as resulting from the deposition of excessive energy, which then results in the fragmentation of these molecules. The energy per atom in this experiment was varied by adjusting the value of $n$ from 300 to 2200 of the impacting 11 keV $Ar_n^+$ cluster ions (Gnaser et al. 2012). Note: These profiles represent the sum of the sputtering and ionization processes. Ionization processes are discussed in Section 3.3.

*3.2.2.1.2   Dependence on Primary Ion Incident Angle*    A strong atomic sputter rate/yield dependence on primary incident angle is noted for all of the most common forms of sputtering (all but the potential form of sputtering). This can be understood as arising from the location relative to the surface at which the kinematics responsible is taking place. Such dependencies are most common for atomic and small molecular ion impact with simulations again providing an effective insight into the dynamics in action.

*Sputtering rates/yields of atomic emissions* resulting from atomic and small molecular primary ion impact exhibit trends that can be understood on the basis that a greater component of the incoming ion's energy is distributed within a region closer to the surface of the solid when smaller incidence angles with respect to the surface normal are used. This effect, however, maximizes at around 60–80°. The

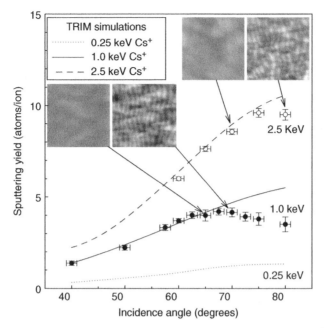

**Figure 3.17** Silicon sputter yields from a Silicon substrate as a function of $Cs^+$ ion energy and angle of incidence (with respect to the surface normal). The symbols represent experimentally derived values, whereas the line relay the calculated values based on TRIM simulations (see Section 3.2.1.3). In the insets are AFM images from the surfaces following sputtering under the respective conditions indicated by the arrows. Reproduced with permission from van der Heide et al. (2003) Copyright 2003 Elsevier.

introduction of increased scattering of incoming ions (illustrated in Figure 3.7) and the formation of surface topography (discussed in Section 3.2.3.3) will influence sputter rates thereafter.

An example of this is shown in Figure 3.17. This shows the variation in sputter yields from Silicon sputtered by $Cs^+$ ions over the 0.25–2.5 keV energy range at impact angles from 60 to 80°. The symbols represent experimentally measured sputter yields, whereas the lines were defined from TRIM calculations. In the insets are shown Atomic Force Microscopy (AFM) images collected from the crater base regions following sputtering. These show the introduction of ripples when excessively large incidence angles in combination with low-impact energies are used (these are discussed further in Section 3.2.3.3). In all cases, the surfaces were atomically smooth before sputtering. As TRIM simulations assume a smooth surface, the deviations at larger impact angles can thus be partially assigned to the introduction of surface topography.

Although sputtering of intact molecular species via atomic or small molecular projectile impact is limited to static SIMS conditions, the same general trends tend to apply.

*Sputtering of intact molecular species* beyond the static limit, particularly from substrates bearing molecules with low cohesive energies (intermolecular bonds) with cluster ions ($C_{60}^+$, $C_{84}^+$, $C_{24}H_{12}^+$, $Ar_n^+$, and so on where $n$ is an integer starting from several hundred and extending to several thousand) provides a myriad of trends that are still under study. For this reason, the reader is advised to access the latest literature.

### 3.2.2.2 Sputter Yield Dependence on Substrate

As with Section 3.2.2.1, this section will concentrate on the general trends noted under the more routinely used conditions applied in SIMS, i.e. those noted under primary ion impact within the 0.1–50 keV energy range with incidence angles ranging from 0 to 80° relative to the surface normal.

As also indicated, all dependencies tend to be substrate specific. But what should be noted is the fact that ion beam impact under sufficiently energetic conditions will induce modifications of their own to the substrate surface. These tend to scale with dose, at least until steady-state sputtering conditions prevail. The subsequent two sections will thus cover substrate-specific variations (those typically noted following steady-state sputtering conditions) and those resulting from the ion bean-induced modifications. The former are covered in Section 3.2.2.2.1, whereas the latter are covered in Section 3.2.2.2.2.

For the sake of clarity, single-component substrates of amorphous or polycrystalline structure are only discussed unless otherwise specified. Multicomponent systems introduce additional effects as will be discussed in Section 3.2.3.4. Lastly, the bulk of the discussion is centered on the use of atomic and/or small molecular ion impact, since there are too numerous a number of molecular ion–substrate combinations, many which still lack a full understanding of the trends exhibited. For specific cases, the reader is advised to access the latest literature.

### 3.2.2.2.1 Dependence on Substrate Type

*Atomic sputter yields* arising from atomic and small molecular ion impact at fixed energy and incidence angle can vary by as much as a factor of 15. An example of this is illustrated in Figure 3.18, in which the sputter yields from various polycrystalline elemental solids resulting from 1 keV $Ar^+$ impact at normal incidence are plotted versus atomic number.

In its simplest form, such sputter yields can be shown to vary primarily as a function of:

1. the collisional cross section
2. the masses of the colliding nuclei
3. the surface binding energy

The dependence on the collisional cross section is impact energy dependent (see Relation 3.1). At a specific energy, the cross section also varies according to the atomic numbers of the collision partners. This dependence takes the form of a larger collisional cross section for heavier elements, which is the reason why heavier ions are used in sputtering (the small cross section of $He^+$ allows this ion to travel a

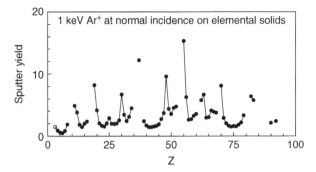

**Figure 3.18**  Atomic sputter yields from elemental substrates under 1 keV Ar$^+$ impact at normal incidence (Magee 1981). Reprinted with permission from van der Heide (2012) Copyright 2012 John Wiley and Sons.

significant distance below the surface before interacting, thereby minimizing the probability of sputtering).

The variation on atomic mass stems from the fact that the momentum transfer between collision partners increases as the masses of the two collision partners approach each other. This dependence, revealed in Relation 3.2, will modify the form and location of the collision cascade. Thus, if the energy transferred to subsequent recoils remains close to the surface, an increased sputter yield will occur.

The variation in surface binding energy (discussed in Section 3.2) stems from the fact that this represents the minimum energy required to remove atoms from a solid. Such energies are usually approximated as some function of the sublimation or cohesive energy. Although other parameters come into play, the dependence of sputter yield on surface binding energies explains:

1. the lower sputter yields of oxides with respect to their base metals (oxides have larger surface binding energies than the base metal),
2. the preferential sputtering of light elements such as O and N from multicomponent solids such as oxides and nitrides (this arises as O and N form anions, which have larger radii than the neighboring cations, thereby reducing surface binding energies and increasing collisional cross sections),
3. to a lesser extent, the variation in sputter yields with crystal orientation. Note: Channeling effects, and so on, can also come into play.

*Sputter yields of molecular ions* from atomic and small molecular primary ion impact exhibit a multitude of trends that, in most cases, can be related back to the substrate type. For specific examples, the reader is advised to examine the latest literature.

Sputtering of intact molecular species also exhibit a multitude of trends, many of which are substrate specific. This is particularly evident when large cluster ions are used ($C_{60}^+$, $C_{84}^+$, $C_{24}H_{12}^+$, $Ar_n^+$, and so on, with $n$ being an integer starting from several hundred to several thousand for $Ar_n^+$) on substrates bearing molecules with low cohesive energies (intermolecular bonds), i.e. organic substrates.

Note: The sputtering of amorphous materials, materials that become amorphous under energetic ion impact (i.e. Silicon), or polycrystalline materials is assumed throughout this text unless otherwise specified. This is assumed as the sputtered population from single crystal materials exhibit strong anisotropic angular distributions that can be directly related back to the surface crystallographic orientation relative to the incident beam direction (for an example, see Scymczak and Wittmaack 1993). Indeed, the increased emissions noted at specific angles have been noted for some time in the atomic ion emissions resulting from atomic ion impact, with the moniker *Wehner spots* applied (after Wehner 1955). Such trends have also been reproduced through simulations appropriate to the system in question.

*3.2.2.2.2 Dependence on Sputtered Depth*    Sputter rates, as well as sputter yields, do not vary over time within a homogeneous volume once steady-state sputtering conditions have been realized. Steady-state sputtering describes the equilibrium setup between the incoming ion flux and the sputter rate. This is apparent for all but the large cluster primary ions. Note: The incoming ion flux and the sputter rate controls the amount implanted and therefore the modification experienced by the substrate under study (this is also system specific). As a result, there exists a delay time before the sputter rates stabilize, i.e. before the development of steady-state sputtering. This is particularly evident during the initial stages of sputtering or when sputtering across interfaces with chemically reactive ions such as $O_2^+$ or $Cs^+$.

This is realized because, as revealed in Figure 3.7, a significant fraction of the incoming ions are not scattered, at least not under standard analytical conditions (0–60°). Rather these ions travel within the solid to depths in excess of that from which sputtering occurs. As a result, these become implanted below the surface to an average depth defined as the projected range normal to the surface whereupon they can modify the substrate and thus the sputter rate. These can be approximated using Relation 3.6a or 3.6b. The concentration of the implanted ions, and thus their effects, continues to increase until the sputtering front reaches the projected range of initially implanted ions. Beyond this, the concentration tapers off and shortly thereafter reaches a *steady-state* value. This occurs at a depth of approximately 2–2.5 times the projected range. Steady state refers to the condition in which the incoming primary ion current matches the removal rate of previously implanted primary ions.

An example of the buildup in the concentration of implanted ions is shown in Figure 3.19 as a function of the apparent depth sputtered. This will vary according to the conditions used (ion energy, type, and incidence angle and whether any potential field exists within or above the sputtered surface). The data in Figure 3.19 was measured for $Cs^+$ impact on Silicon under the conditions used in Figure 3.7 (1 keV $Cs^+$ at 60°). The term *apparent depth* is used as modifying the solid's composition can alter the sputtering yield and thus the sputter rate. As a result, depth scales defined using the linear sputtering time to sputter depth relation (see Section 5.3.1) will be in error. The depth over which such errors may be present is, however, shallow. In the case of SIMS, the affected region is referred to as the *transient region*.

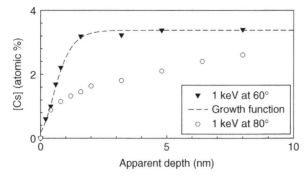

**Figure 3.19** XPS derived Cesium concentration in Silicon as a function of depth for 1 keV Cs$^+$ ion impact at 60° and 80° relative to the surface normal. The dashed line is to aid the eye. Reproduced with permission from van der Heide (2004) Copyright 2004 Elsevier.

Concomitant to this is the reduction in sputter rates/yields noted with increasing sputtering time (van der Heide 2004; Berghmans et al. 2008). Some examples are shown in Figure 3.20(a–d), with the sputter rates measured from a Boron delta-layered structure under the listed conditions. Such variations cease once steady-state sputtering conditions ensue, i.e. once sputtering beyond the transient region. This inverse relationship with Cesium content ([Cs]) has been recognized for some time (Deline et al. 1978). This was originally approximated as:

$$\frac{SR \propto 1}{([Cs] + 1)} \tag{3.7}$$

Similar sputter rate variations are also apparent under $O_2{}^+$ primary ion impact on Silicon. Lowering the $O_2{}^+$ energy further complicates matters owing to the surface oxidation induced (this results in surface swelling) and the potential for ripple topography growth (Wittmaack 1996; Wittmaack 2000; Homma 2003).

Owing to the effects of ion impact-induced recoil implantation, cascade mixing, diffusion, and segregation (Gibbsian and radiation induced), more complex variations can exist over interfacial regions, particularly within multicomponent substrates (solids in which more than one type of atom exists in a particular layer). All such processes will modify the solid during the sputtering process, as discussed in more detail in Section 3.2.3. Multicomponent substrates are discussed in Section 3.2.2.2.3.

These same effects will not be as prevalent during molecular depth profiling with large cluster ion beams. Figure 3.2(d) provides a flavor of the modifications expected. Also of note is the fact that the remnants of many such projectiles are removed from the surface (a result of the sputtering process or through evaporation).

*3.2.2.2.3 Multicomponent Substrates*  Although single-component substrates are most effective in describing the physical principles of the sputtering processes,

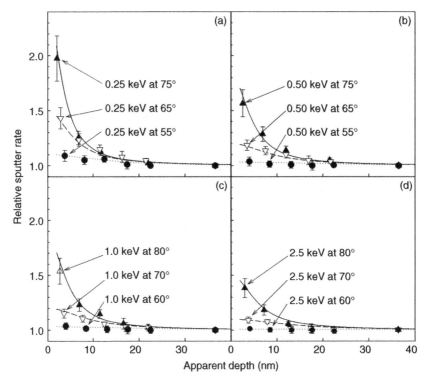

**Figure 3.20**   Measured differential sputter yields (normalized steady-state values) versus apparent depth for (a) 0.25 keV, (b) 0.5 keV, (c) 1.0 keV, and (d) 2.5 keV $Cs^+$ impact on Silicon at the listed angles. Reproduced with permission from van der Heide et al. (2003) Copyright 2003 Elsevier.

multicomponent systems are more commonly encountered in day-to-day analysis. Multicomponent substrates are those in which there exist:

1. regions of different solid-state phases,
2. regions of different chemistry, i.e. inclusions, alloys, and so on,
3. a mix of different elements within a specific region, i.e. $SiO_2$, NaCl, and so on.

The more commonly encountered issues noted in sputtering of multicomponent substrates concern the sputter-induced modifications of the region of interest, which can occur spatially and/or as a function of sputtering time. These tend to manifest themselves in the form of compositional variations occurring during the course of sputtering. The primary cause relates back to various fundamental processes, which are substrate and analysis condition specific. These include, but are not limited to, variations in:

1. recoil implantation
2. cascade mixing

3. diffusion (radiation enhanced)
4. segregation (Gibbsian and radiation induced)
5. sputter rates

The first four arise from ion beam-induced substrate damage. These are discussed in Section 3.2.3.

Sputter rate variations in multicomponent substrates, often referred to as *differential sputtering*, have long been recognized as one of the potential issues that can result in erroneous quantification (quantification is covered in Sections 5.4.2 and 5.2.3). This is realized as differential sputtering alters the sputter yield, and thereby the surface composition of deeper layers. As will be discussed in Section 3.2, the surface chemistry is one of the primary factors governing secondary ion yields. In short, severe complications are introduced as a result of the fact that sputter yields and ionization yields are usually intrinsically intertwined.

The two approaches to solve such issues include:

1. Assigning substrate and analysis condition-specific corrective quantification procedures. These are best applied once steady-state sputtering conditions prevail. Note: This will only be effective if accurate reference samples are available (quantification is discussion in Sections 5.4.2 and 5.2.3).
2. Understanding the dynamics of the differential sputtering process that is in effect for the substrate of interest under the analytical conditions applied. Note: This is a potentially complex undertaking in condensed matter physics (for a review, see Gnaser 1999) requiring the delineation of the sputtering and ionization phenomena.

As will be illustrated in Chapters 5 and 6, SIMS is able to effectively circumvent many of these complications through the optimization of analytical methods and post-processing procedures suited to the substrate and information of interest. Note: Understanding the potential sources of error is, however, of prime importance in setting up such optimized analytical conditions, which in most cases are substrate and information content specific.

### 3.2.3  Sputter-induced damage

During the course of sputtering, a significant amount of kinetic energy can be dissipated via elastic (nuclear collisions) and inelastic (electronic excitation) processes into the substrate (only a small fraction of this energy is removed in the emission of photons, electrons, sputtered atoms/ions, and molecules). The remainder induces a myriad of processes that culminate in the modification of the composition and electronic structure of the solid. These can occur for any sufficiently energetic impacting ion.

One example of energetic ion-induced modification is the implantation of chemically active ions (see Figures 3.8 and 3.19). In specific cases, this can also result

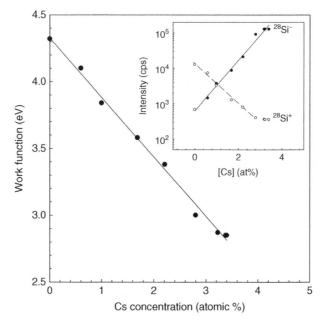

**Figure 3.21**   Silicon substrate work–function variation as a function of Cesium content introduced through implantation from 1 keV Cs$^+$ primary ion impact at 60°. In the inset is the response of the Si$^+$ and Si$^-$ secondary ion intensities. The work–function value was adjusted through the deposition of Cesium onto and into the Silicon substrate (Cesium, like all alkali metals, reduces the work–function value of all metals and semiconductors). Reproduced with permission from van der Heide (2004) Copyright 2004 Elsevier.

in the modification of the substrate's work–function (see Figures 3.21 and 3.22). Even inert gas ions such as Ar$^+$ and Xe$^+$ can induce modifications to the lattice structure that may or may not be limited to:

1. recoil implantation
2. cascade mixing
3. diffusion, and
4. segregation (Gibbsian and radiation induced).

These processes are of importance in multicomponent solids as these will result in the redistribution of atoms, thereby altering any compositional gradient that may initially be present. These are discussed in greater detail in Section 5.3.2.4.1.

Sputtering can also modify the lattice structure in the form of amorphization and re-crystallization, and introduction of surface roughening. This can occur on both single-component and multicomponent solids. Again, single-component solids are discussed only for the sake of simplicity.

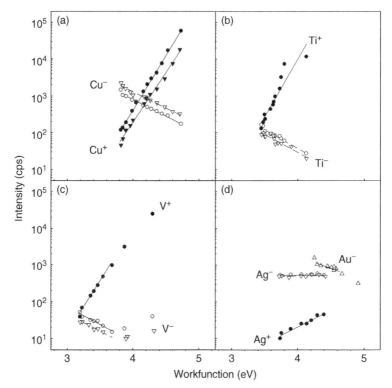

**Figure 3.22**    Secondary ion intensity variations as a function of the work–function from sputter cleaned ($Ar^+$) polycrystalline metal substrates of (a) Copper, (b) Titanium, (c) Vanadium, (d) Silver and Gold under 1 keV $Cs^+$ impact at 60° (the data was collected over the respective transient regions). Reprinted with permission from van der Heide (2005) Copyright 2005 Elsevier B.V.

### 3.2.3.1    Recoil Implantation, Cascade Mixing, Diffusion, and

**Segregation**    The deposition of energy greater than $\sim 15$ eV will result in the displacement of an atom bound within a lattice to some neighboring position within the lattice. If this occurs within a crystalline lattice, stable *Frenkel defects* (stable interstitial-vacancy pairs) can be formed. As sputtering represents many billions of collision cascades, all of which contain ions and recoil atoms with more than the required energy to displace lattice atoms, significant damage of the lattice below the sputtering front will ensue. This damage is generally discussed in terms of recoil implantation, cascade mixing, radiation-enhanced diffusion, and radiation-induced segregation.

*Recoil implantation* describes the *anisotropic* redistribution of atoms within the solid resulting from energetic ion impact. In other words, this is an ion–atom knock-on event that transports atoms preferentially in the direction the ions are initially traveling in. This results in extensive defect formation. As indicated by Relation 3.3, recoil implantation depends primarily on the masses of the collision partners and the impact energy. As collisional cross sections decrease with increasing impact energy, the probability of direct knock-on events scales with impact energy. Thus, lighter elements in multicomponent solids exhibit greater susceptibility to recoil implantation at higher impact energies. Lowering the impact energy increases collisional cross sections. This enhances the scattering of heavier elements and reduces the probability of knock-on events such that at impact energies below 100 keV, recoil implantation becomes minimal.

*Cascade mixing* describes the isotropic redistribution (redistribution in every direction) of atoms within solids resulting from energetic ion impact. This process dominates over recoil implantation at impact energies less than 100 keV. The isotropic nature arises from the greater number of lower energy collisions present within a collision cascade. This also reduces the size of the affected volume and the number of defects formed. Indeed, redistribution can occur with no defect formation, i.e. atoms simply swap sites over what appears as a molten region in the core of the collision cascade, which cools within $\sim 10^{-12}$ s. Cascade mixing is primarily responsible for sputter-induced loss on depth resolution (see Section 5.3.2.4.1). Indeed, the Gaussian distribution noted for layers initially exhibiting atomically sharp interfaces have been approximated using diffusion-based models. The diffusion coefficients derived are, however, temperature independent. Lastly, cascade mixing can be shown to act as a feeding mechanism for the preferential sputtering of atoms of different masses.

*Diffusion* is a thermally induced mixing process that can occur in any solid. This occurs via a hopping mechanism in which an atom moves to displace a defect and so forth. The temperature dependence arises as the mobility of defects increases with increasing temperature. Diffusion coefficients (a parameter that relates the extent of diffusion) are thus a function of the number of defects present, their mobility, and their lifetimes. As sputtering introduces stable Frenkel defects (stable interstitial-vacancy pairs) in crystalline solids, this increases diffusion coefficients. This special form of diffusion, referred to as *Radiation-enhanced diffusion*, can result in the redistribution of elements over regions well outside that of the collision cascade. In addition, radiation-enhanced diffusion may or may not exhibit a temperature dependence. This depends on the recombination rate of defects formed during sputtering and the annihilation rate of these defects at extended sinks (long-range defects in the form of grain boundaries). At low temperatures, the aforementioned process dominates, yielding temperature-dependent diffusion coefficients. At higher temperatures, the temperature dependence of the diffusion coefficient vanishes. At even higher temperatures, normal diffusion comes into play.

*Segregation* is a chemically driven separation process that occurs in multicomponent solids. Gibbsian segregation is one form of segregation that describes the redistribution of elements to or from a surface in an effort to reduce the surface free energy. In other words, this describes the evolution of a surface toward its most stable state (Recall: A surface represents an abrupt termination of the long-range lattice structure). As with diffusion, Gibbsian segregation is a thermally driven segregation process that is enhanced in the presence of defects. Defects are introduced into the lattice during sputtering.

As sputtering removes surface layers, Gibbsian segregation is continually triggered. This is otherwise referred to as *Radiation-induced segregation*. Such segregation can also result in the redistribution of elements over regions well outside the collision cascade region. Unlike radiation-enhanced diffusion, radiation-induced segregation can drive an initially homogeneous multicomponent solid to a heterogeneous state (radiation-enhanced diffusion induces the reverse).

### 3.2.3.2 Substrate Amorphization and Re-crystallization

As covered in Section 3.2.1, energetic ion impact can introduce extensive damage in the form of Frenkle defects, and so on. If these damages are allowed to agglomerate into extended defects (dislocation loops), an initially crystalline solid may become amorphous. Indeed, this is observed in brittle solids such as Si and extends over the entire depth region at which ions become implanted.

An example of amorphous layer formation is shown in the Transmission Electron Microscopy (TEM) cross-sectional micrograph shown in Figure 3.23(a–c). Figure 3.23(a) shows an as-received unannealed 5 keV $Sb^+$-implanted wafer (implanted at $7°$ off normal) terminated with a 1-nm-thick native oxide. An amorphous layer over $\sim 10$ nm thick on top of the crystalline substrate is noted before sputtering. The amorphous layer arises from the $Sb^+$ implantation process. Figure 3.23(b) and 3.23(c) shows TEM cross sections of the same wafer following sputtering with a 1-keV $Cs^+$ ion beam at $80°$ to a depth of $\sim 20$ nm, i.e. well past the initial amorphous layer introduced through implantation. These reveal a thinner amorphous layer (the thickness will scale with the projected range normal to the surface of the impacting ion) that oxidizes on exposure to air ($SiO_2$

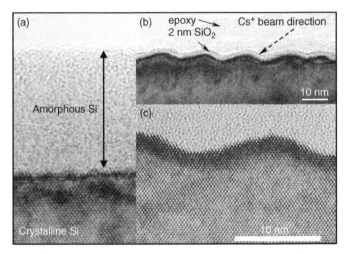

**Figure 3.23** TEM cross sections of an as-received 5-keV $Sb^+$-implanted Silicon wafer (a) before anneal or sputtering, and (b) and (c) following sputtering with a 1-keV $Cs^+$ primary ion beam at $80°$. Note: Epoxy was used in preparation of the TEM sample (hand polished). Reprinted with permission from van der Heide et al. (2000) Copyright 2000 Elsevier Science B.V.

layer), along with the evolution of ripple topography (the surface was initially smooth).

### 3.2.3.3 *Surface Roughening and Surface Smoothing*    The surface roughening observed in the form of ripples occurs only under specific conditions. Furthermore, the amplitude of the ripples stabilizes within a depth that is dependent on the projected range of the incoming ions normal to the surface. As this range is reduced (scales with impact energy and angle), this depth merges with the transient region (is of the order the projected range of the incoming ions).

Correlations suggest that the reduction in sputter rates observed stems from surface roughening. This would be expected as the face of the ripple facing the incoming ion beam will exhibit a reduced angle with respect to the ion beam. Any variation in the sputter rate will also modify the steady-state concentration of ions implanted during sputtering, which will further affect sputter rates.

The cause of ripple topography growth on Silicon is accepted to occur via the competing roughening and smoothing effects on/in amorphous solids (Cuerno and Barabasi 1975; van der Heide et al. 2003; Homma et al. 2003). Roughening is thought to stem from sputter-induced surface stress, which results in surface curvature on a microscopic scale. As revealed in Figure 3.17, any variation in the incidence angle of the incoming ion will result in a variation in sputter yields. This, in turn, will produce hillocks and valleys that will propagate toward the incoming ion beam. Any inhomogeneities present on the surface, i.e. Cesium agglomeration, oxide formation, and so on, may also play a role. Smoothing is believed to occur via diffusion apparent during sputtering, whether thermally or ballistically induced (atomic diffusion from the molten region in cascade mixing and/or radiation-enhanced diffusion).

In contrast to Silicon and other brittle solids, the distance over which Frenkel pairs can recombine in metals is so large that amorphization and thus ripple topography growth does not occur. Sputtering can, however, induce the formation of textured surfaces on an initially untextured polycrystalline surface. This arises from the variation in sputter rates with crystal orientation. Prolonged sputtering can result in the formation of cones, pyramids, and so on, within a particular crystal face. Examples of the texturing and pyramid formation observed on Zirconium under 17.5 keV $Cs^+$ ion beam incident at 20° are shown in the Atomic Force Microscopy (AFM) images presented in Figure 3.24(a) and 3.24(b), respectively.

The additional structure in the form of cones, pyramid, and so on is believed to stem from the presence of foreign heavier mass atoms at the outer surface of the respective solid. These heavier mass atoms exhibit a lower sputter yield than the surrounding surface. Hence, these are expected to form the tip of the cone, pyramid, and so on. In the case of the Zirconium surface shown, this would likely take the form of previously implanted Cesium.

Roughening, segregation, diffusion, cascade mixing, and recoil implantation of the sputtered surface will distort the shape of a profile collected, as well as the depth resolution. These issues are covered further in Section 5.3.2.4.1.

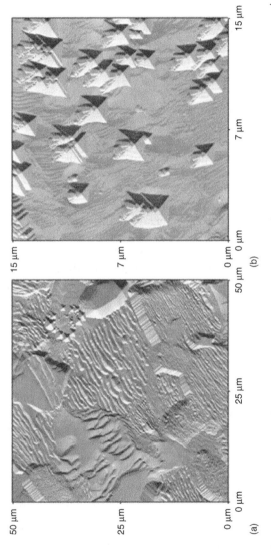

**Figure 3.24** AFM images of the surface roughness (texturing and pyramid formation) generated via prolonged 17.5 keV Cs$^+$ ion impact at 20° on a Zirconium-based substrate (a) over a polycrystalline region and (b) within a single crystal region of the same sample. Authors unpublished images.

For large cluster ion beam impact, there exist additional possibilities, all of which are summarized in Figure 3.2. These depend, largely, on the impact energy, or more precisely the energy per atom as well as the substrate type. Analysis conditions can thus be tuned to provide the favored outcome.

Two phenomena of interest that can be induced by organic or inert gas cluster ion beams are surface smoothing (the horizontal or lateral displacement of substrate material from regions in which surface atoms/molecules are seen to protrude from the average surface plane) and surface cleaning (the removal of foreign material (typically organics) from the substrate). Both occur from specific substrate types (generally, hard surfaces such as Silicon, Silicon Carbide, and inorganic oxides) when the cluster ion beam energy is reduced below some sample specific limit. Organic or inert gas cluster ion beams are recommended as these readily evaporate off from such surfaces following their *soft landing*. The energies required are within the soft landing regime, i.e. well below that at sputtering, implantation, or any other substrate modification. Molecular Dynamics simulations (see Section 3.2.1.3) have proved effective in elucidating the mechanisms responsible. Examples of Molecular Dynamics simulations of such *soft landing* processes are illustrated in Figures 3.13 and 3.14.

As noted, cluster ion beam-induced surface smoothing requires that the energy per incoming atom (that within the incoming ion beam) to be around 10 eV. This is realized as this energy is sufficient to move substrate atoms around the surface but not from the surface. In other words, this energy is above the surface binding energy (see Table 3.1) but below the sputtering threshold (15–40 eV (Malherbe 1994)). In the case protrusions, direct impact appears to result in the displacement of the protruding substrate atoms to surrounding areas. Depressions are then filled with the originally protruding atoms. Although some substrate repair also occurs, this is not necessarily complete. The degree of damage is dependent on the dynamics (cluster ion energy and size) and the dose of the impacting cluster ion beam as well as the substrate type.

Cluster ion beam-induced surface cleaning, on the other hand, requires that the energy per incoming atom (that within the incoming ion beam) to be around 5 eV, i.e. close to the binding energy of the substrate atoms but more than the energy binding the unwanted material to the substrate, hence, the effectiveness of this process for removing organic residuals from hard substrate surfaces.

## 3.3    IONIZATION/NEUTRALIZATION

Secondary ions are the population recorded in SIMS. This represents the ionized portion (atomic and molecular) emanating from a solid's surface following an energetic ion impact. Recall: SIMS is not capable of measuring the sputtered neutral population.

As mentioned in Section 3.1.1, the departing ionized population is generally accepted to arise following a two-step process akin to that depicted in Figure 3.1. This assumes kinematic processes resulting in sputtering that can be delineated from electronic processes resulting in the departing ionized population, i.e. that electron transfer between the surface and the sputtered particle occurs following the sputtering process.

This two-step scenario is generally assumed as:

1. Secondary ions display little to no dependence on the charge state of the primary ion species. The one exception is noted when potential sputtering is in effect.
2. Electron transfer rates are significantly faster than sputtering rates, i.e. sputtered particles appear as slow-moving entities relative to the electron transitions involved.
3. The energy transfer from kinematic (elastic) to electronic (inelastic) processes is inefficient, a fact partially attributable to the differing channels and time scales involved.
4. This allows for the simplification of discussions concerning secondary ion populations.

Kinetic sputtering is initially considered because this is conceptually the simplest form of sputtering and the best understood of the sputtering processes. This is also the most common form of sputtering in SIMS as used in elemental analysis. Discussion of the other sputtering processes will follow.

Although kinematic sputtering can be delineated from the events leading up to secondary ion formation and survival, sputtering can also have a significant impact in that it can severely distort the electronic structure, and hence the local band structure of the sputtered surface. This distortion results as this sputtering process proceeds through momentum transfer, which in itself induces significant variations in the distances between the nuclei of the collision partners (also known as bond lengths), and thus the energies of the interacting electronic states. As depicted in Figures 2.2 and 2.3, energy states within the valence region are a direct function of interatomic distances, with the equilibrium bond distance defined as that in which the valence electrons (those involved in bonding) attain the lowest energy.

Before delving into the various possible mechanisms, a few salient features concerning recorded secondary ion emissions are first outlined. These are as follows:

1. The vast majority of secondary ions are singly charged. Of the multiply charged ions, most are of a positive polarity (negative multiply charged ions have very short lifetimes).
2. Secondary ion angular distributions tend to peak around the sample normal with a generally noted Cosine function. Exceptions are seen when crystallographic surfaces are present.
3. Secondary ion energy distributions for both atomic and molecular species peak at around a few electron volts. This is generally higher for atomic versus molecular secondary ions.

4. Atomic secondary ions display high energy tails that extend to a sizable fraction of the primary ion energy, i.e. to 500 eV or more. This is not noted for molecular secondary ions.

5. Secondary ion populations tend to, but not always, exhibit a strong dependence on:

   a. the electronic properties of the parent atomic or molecular species,

   b. the electronic properties of the substrate's surface from which they emanate,

   c. the polarity of the secondary ions formed,

   d. the primary ion species used, specifically how these modify those of the substrate's surface when implanted into the substrate.

6. In most cases, no correlation between secondary ion yields and primary ion polarity are noted.

The sensitivity to the surface chemistry reveals that the vast majority of singly charged secondary ions form in close proximity to the surface from which they were sputtered, as opposed to within the gas phase. In order to contend with the complexity introduced by both the sputtering process and the fact that every chemically distinct solid acts/reacts differently upon ion impact, ion–solid interactions (those in which the sputtering process is removed) are first introduced in Section 3.3.1. Ionization/neutralization relevant to secondary ion emission is then presented in Section 3.3.2. For the sake of clarity, this discussion is limited to yield trends noted from relatively simple systems, i.e. those from metallic surfaces, including Silicon, resulting from $Cs^+$ and $O_2^-$ primary ion impact. Some of the more commonly accepted models, along with those of historical significance, that are used in describing atomic and molecular secondary ion emission trends from the various different substrate types are presented in Section 3.3.4.

### 3.3.1  Ion–Solid Interactions

Interactions among atoms, ions, and solid surfaces are not limited only to those in SIMS. They are also noted and even used in other analytical techniques with examples including:

1. Low-Energy Ion Scattering (LEIS)
2. Ion Neutralization Spectroscopy (INS)
3. Metastable De-excitation Spectroscopy (MDS), also known as Penning Ionization Electron Spectroscopy (PIES) and Metastable Quenching Spectroscopy (MQS)

These are of interest as they simplify the ionization/neutralization processes active on an ion, atom, and/or molecule in close proximity to a solid's surface, by removing the sputtering event. In the case of LEIS, a low-energy ion beam is directed at a solid's surface such that electron interaction (that resulting in charge transfer) between the ion and the surface occurs as the ion scatters from the aforementioned surface. The processes active are examined by measuring the outgoing ion population $(I_o)$ relative to the incoming population $(I_i)$. In INS and MDS, electron emissions resulting from low-energy ion interaction or low-energy metastable atom interaction with a solid's surface are measured, respectively. An example of the type of emissions measured in INS is shown in Figure 3.6(b).

In these cases, the type and probability of the electron interaction taking place may depend on:

1. the surface chemistry and crystalline structure,
2. the electronic properties of the scattered atom/ion,
3. the interaction time as defined by the scattered atom/ion angle of incidence and velocity,
4. the excitations present within either or both the surface and the scattered atom/ion,
5. the processes in effect, i.e. the type of transitions.

Note: The interaction time is controlled through both the incidence/exit angle of the atom or ion beam with respect to the surface and the velocity, which itself is energy and mass dependent $\left(E = \frac{1}{2}mv^2\right)$.

The charge transfer processes (electron transitions between the ion/atom and the surface) active as an ion/atom approaches/recedes a solid's surface are modeled at the atomic level via quantum mechanics. The most prevalent processes, illustrated in the potential energy diagrams in Figure 3.25(a–c), are referred to as:

1. Resonant charge transfer
2. Quasi-resonant charge transfer
3. Auger charge transfer

*Resonance charge transfer* describes the transfer of electrons between electronic states of the same energy, one of which must be vacant so as to accept an electron. This transfer will occur via quantum mechanical tunneling of electrons between atoms/ions making up the substrate surface and the atom/ion in close vicinity ($\sim$1 nm) to the surface. As the wave function of electronic states closest to the Fermi edge $(E_F)$ extend out the greatest distance from the respective atom or ion's nucleus, only these levels are considered within this process (these are the first and last to come into contact with each other).

As this charge transfer can occur in either direction (to or from the departing atom or ion), both positive and negative ion formations along with neutralization can occur. This is illustrated for positive ion formation/survival in Figure 3.25(a). The outcome depends on:

1. the relative electron populations within the substrate and the departing ion/atom

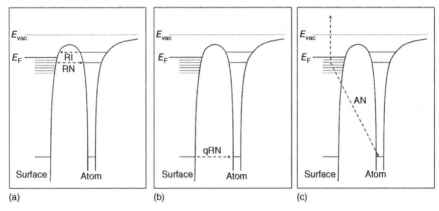

**Figure 3.25**   Potential energy diagrams representative of (a) resonant charge transfer (RI = Resonant Ionization of a neutral atom and RN = Resonant Neutralization of a positive ion), (b) Quasi-resonant charge transfer (qRN = quasi-Resonant Neutralization of a positive ion) and Auger charge transfer (AN-Auger Neutralization of a positive ion). The dashed arrows represent electron transfer from populated to vacant electron levels, whereas the horizontal lines represent the allowed electron levels, otherwise referred to as stationary states.

2. the energy separating the filled and empty states within both
3. the distance over which the interaction takes place (the image field discussed in Section 2.2.2.3 affects the position of the interacting energy levels within the departing atom or ion).

This can occur for any solid surface, with the only prerequisite being the presence of allowed energy states, one of which must be empty. Indeed, the lack of this charge transfer process from insulator surfaces is argued to arise from the lack of allowed energy states within the band gap. Both the electron tunneling and the bond breaking models discussed in Section 3.3.3 assume resonance charge transfer.

*Quasi-resonance charge transfer* is similar to resonance charge transfer, except for the fact that core levels in both the substrate surface and the ion or atom in close proximity to the substrate's surface are now involved. This is illustrated for positive ion neutralization in Figure 3.25(b). As a result, one of the core levels must be vacant to accept the electron to be transferred. As core holes describe an energetically unfavorable case (see Section 2.2.2.4.4), electron transfer will only be seen following highly energetic atomic collisions. As a result, this process is far less efficient relative to the resonance charge transfer process and occurs only for specific substrate–ion/atom combinations. As the wave functions of core holes do not extend out as far from the respective atom or ion's nucleus as do valence levels, such processes occur in much closer proximity to the surface relative to resonance charge transfer.

*Auger electron transfer* describes charge transfer between electronic levels of very different energies. In general, that within the substrate's surface is situated close to $E_F$, whereas that within the ion or atom in close proximity is some transient vacant core level. As one of the two levels is a core level, such processes occur in closer proximity to the surface than that of resonance charge transfer.

An example of positive ion neutralization via this process is illustrated in Figure 3.25(c). Negative ion formation via analogous arguments can also be envisaged. As core holes describe an energetically unfavorable case, electron transfer can occur following highly energetic atomic collisions. Although such processes can occur from any surface, this will be less efficient than the resonance charge transfer process.

Closely related to the Auger process is auto-ionization. Auto-ionization describes the emission of electrons from an atom or ion into the gas phase once it has departed a sufficient distance from the surface. As illustrated in Figure 3.6(c), this can be initiated through energetic collisions of the form occurring in knock-on sputtering. Indeed, this process appears to be responsible for explaining the multiply charged secondary ion trends from specific light elements as discussed in Section 3.3.3.3.

Other gas phase processes are also possible. These, however, are not discussed in detail within this text because of the low yields generally apparent, i.e. these will play a minor role, if any, in SIMS.

*Band structure variations* and their influence on ions in close vicinity to the solid surface serve to further complicate matters. This is realized as the effects of the band structure of a solid do not terminate abruptly at the substrate's surface. Rather, when exposed to an ion's charge, an electrostatic field extending a small distance into the gas phase from the substrate's surface is formed. This is otherwise referred to as the *image field*. As introduced in Section 2.2.2.3, this influences energies of the ionization levels/electron affinity levels of ions in close proximity to a solid surface.

Such ionization potential variations can be crudely approximated by Relation 2.3 for singly charged ions close to a metallic surface over distances greater than that at which a quasi-molecular orbital (temporary orbital) is formed, i.e. less than $\sim 1$ nm. A similar but opposite trend is noted for affinity levels as can be approximated by modifying Relation 2.3, i.e. replacing the ionization potential with the electron affinity and changing the negative sign to a positive sign. This effectively increases the affinity-level energy of the ion as the distance between the surface and the ion decreases. These interrelations are illustrated in Figure 2.5. For further discussions, see reviews by Tully and Tolk 1977 and Rabalais 1994 to name a few.

### 3.3.2  Secondary Ion Yields

The secondary ion yield represents the number of ions produced per sputtered atom or molecule, whichever is of interest. This should not be confused with the sputter yield, which represents the number of atoms, ions, or molecules sputtered per incoming primary ion (these are discussed in Section 3.2.2). Secondary ion yields are derived empirically by:

1. Relating the secondary ion signal (the number of ions striking the detector once corrected for the instrument transmission function) to the volume sputtered. This is an absolute approach.

2. Relating the secondary ion signal to the elemental or molecular concentration (the instrument transmission function is not needed). This is a relative approach.

3. Measuring and relating both the secondary ion and the sputtered neutral populations. This requires specialized instrumentation as discussed elsewhere (Wahl and Wucher 1994).

Discussions on the methods used in quantifying the depth and concentration scales are presented in Sections 5.4.2 and 5.4.3. For the sake of clarity, results from low-energy ion emissions are discussed throughout this text unless otherwise specified.

Secondary ion yields for atomic emissions span many orders of magnitude. An example of this is shown in Figure 3.26(a) for positive secondary ions of elements at concentrations below 1 atomic % present in Silicon under $O_2^+$ primary ion impact and in Figure 3.26(b) for negative secondary ions under $Cs^+$ primary ion impact.

Figure 3.26(a) and 3.26(b) was derived following steady-state sputtering through normalization of Relative Sensitivity Factors (RSF) of $Cs^+$ and $F^-$ secondary ions to unity. RSFs are discussed in Section 5.3.2.1. The trends noted in Figure 3.26(a) and 3.26(b) illustrate the fact that some elements readily form ions, whereas others do not. Their preference is also polarity specific, i.e. alkali metals (electropositive elements) readily form positive ions, whereas halogens (electronegative elements) form negative ions. In addition, there exist elements such as Nitrogen and the inert

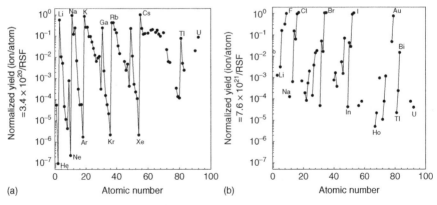

**Figure 3.26** Normalized monatomic yields for (a) positive secondary ions from Silicon under $O_2^+$ primary ion impact and (b) negative secondary ions from Silicon under $Cs^+$ primary ion impact. The yields are derived from RSF values listed in Wilson et al. 1989 on the assumption that $Cs^+$ and $F^-$ have unitary yields (the values $3.2 \times 10^{20}$ and $7.6 \times 10^{21}$ represent the arbitrary scaling factors used to obtain the unitary yields for $Cs^+$ and $F^-$).

gas elements that prefer not to exist as either. In short, secondary ion yield trends can in general be related to the elements preference to gain or lose electrons. The same can be said for molecular emissions. These correlations are discussed further in Section 3.3.2.1.

Note: Secondary ion emissions from elements present at concentrations below 1 atomic % are used to derive in Figure 3.26(a) and 3.26(b) as, if above this concentration, the presence of the respective element can and often does alter the respective secondary ion yield. Indeed, this effect, termed the *matrix effect*, is the primary reason why quantification remains a difficult task in SIMS. The matrix effect is discussed along with the effects of the substrates work–function, the secondary ion emission velocity, and the effects of electronic excitation in Section 3.3.2.2.

Molecular secondary ions formed within the surface region tend to follow similar trends.

### 3.3.2.1 Ionization Potential and Electron Affinity

As introduced in Section 2.2.1.3, ionization potentials ($I$) and electron affinities (EA) are numerical parameters that describe the preference of an atom or molecule to attract or repel an electron from or to itself. Although these values pertain to the removal (ionization potential is used) or addition (electron affinity is used) of an electron to an isolated atom or molecule in the gas phase, the same trends are noted in moving between the gas and the solid phases. The condition of the surface can introduce additional effects.

As atoms, ions, and molecules can gain or lose more than one electron, there exist state-specific ionization potentials and electron affinities, i.e. the first ionization potential represents the minimum energy needed to remove an electron from a neutral ground-state atom, whereas the second ionization potential represents the minimum energy needed to remove the second electron from a neutral ground-state atom, and so forth. As singly charged ions are the most prevalent in SIMS, the parameters, ionization potential, and electron affinity will be used throughout this text to refer to the first ionization energy and first electron affinity energy, respectively. Multiply charged ions noted in SIMS tend to be noted only in the positive spectrum. This arises as negative multiply charged ions are much shorter lived such that they are rarely observed in SIMS.

Re-plotting the relative yield data for the singly charged secondary ions, which is shown in Figure 3.26(a) and 3.26(b), against the ionization potentials or electron affinities of the respective elements being ionized reveals trends consistent with the expectations, i.e. elements with higher ionization potentials are less likely to form positively charged secondary ions, whereas elements with higher electron affinities are more likely to form negatively charged secondary ions. These trends are shown in Figure 3.27(a) and 3.27(b) for the positive secondary ions and the negative secondary ions from Silicon, respectively.

The trends noted in Figure 3.27(a) and 3.27(b) reveal a general exponential dependence over some limited range, i.e. over ionization potentials from ~5 to

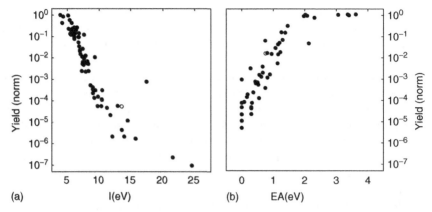

**Figure 3.27** Normalized yields of (a) positively charged secondary ions versus the relative element's ionization potentials ($I$) and (b) negatively charged secondary ions versus the element's electron affinities (EA). The yield numbers are from Figure 3.26(a) and 3.26(b), respectively.

10 eV and electron affinities from 0 to $\sim$2 eV. This implies the relations:

$$Y(M^+) \propto \exp(-I) \tag{3.7a}$$

$$Y(M^-) \propto \exp(EA) \tag{3.7b}$$

where $Y(M^+)$ defines the positive secondary ion yield and $Y(M^-)$ defines the negative secondary ion yield, respectively.

Deviations noted for secondary ion yields for elements below ionization potentials of $\sim$5 eV and above electron affinities of $\sim$2 eV are believed to be due to saturation effects. Deviations for secondary ions of elements with ionization potentials above $\sim$10 eV, on the other hand, appear to arise from the inclusion of additional ionization mechanisms facilitated through excitation modes, and/or the suppression of secondary ion neutralization processes. Excitation effects are discussed in Section 3.3.2.3.

The fact that the correlations noted in Figure 3.27(a) and 3.27(b) are only rough approximations is ascribed to the fact that ionization potential and electron affinity values apply to individual atoms or molecules existing in the gas phase with no account of the effects of the image field considered (see Section 2.2.2.3). This and other effects are encapsulated in what is referred to as the *matrix effect*. As mentioned, the matrix effect describes the dependence of secondary ion yields on the solid's surface, or more specifically, surface chemistry along with any distortions to the surface chemistry induced through the sputtering process.

Indeed, altering the matrix from which the secondary ions emanate tends to alter the dependence (slope of the trends noted in Figure 3.27(a) and 3.27(b)) in a manner that is considered consistent with the bonding occurring between the atoms situated at the outermost surface of the respective solid. Efforts to relate RSFs to ionization potentials or electron affinities of the substrate have met with some limited success, with the systematic trends noted reported elsewhere (Wilson et al. 1989). From such trends, the possibility of matrix-transferable RSFs has been suggested (Grasserbauer et al. 1989), although rarely used. These are covered further in Appendix A.9.1.

**3.3.2.2  Matrix Effects**  The matrix effect (the effect of the matrix on the secondary ion yield of a specific element or molecule from which it was emitted) represents one of the primary causes for the difficultly in quantifying secondary ion signals. This is realized as this effect is difficult if not impossible to accurately predict even when analyzing emissions from within a well-defined single matrix.

Reasons for this stem from the fact that the charge transfer process responsible for secondary ion formation/survival, which ever process it may be (see Section 3.3.1 for possibilities), is highly sensitive to the chemistry of the outermost surface of the substrate at the instant or shortly after the sputtering event. Indeed, the information needed to accurately predict the ionization/neutralization processes active would preclude the necessity of analyzing the respective solid.

The surface chemistry present at the time of analysis will be a function of the substrate itself (Recall: Any element present within the substrate at greater than $\sim 1$ atomic % will induce its own effect) and any modifications induced before or during the course of analysis. Examples of modifications that can affect secondary ion yields from a Silicon matrix include:

1. Adsorption of chemically active species before or during analysis. Note: $O_2$ can exist even under ultra-high vacuum analysis conditions as discussed in Section 4.2.1.1.
2. Implantation of chemically active primary ions such as $O^-$, $O_2^+$, $Cs^+$, and $SF_5^+$ during analysis. Examples of that resulting from the use of $Cs^+$ are covered in Section 3.2.2.2.2.
3. Primary ion beam-induced damage along with associated modifications to the lattice structure. Examples include diffusion, segregation, and surface topography, as covered in Section 3.2.3.

As a result, the most effective method for quantification is to analyze matrix-matched reference samples containing a known amount of the element/molecule of interest at the same time as the sample of interest. Analysis at the same time is needed to account for any variations in the instrumental parameters used, of which, as covered in Chapter 4, there are many. This method of quantification is referred to as the RSF method, a topic that is covered in detail in Section 5.4.3.1. This method is most effective within specific matrices as opposed to their interfaces.

Matrix effects noted on moving from one matrix to another will be further affected by primary ion impact-induced atomic mixing, segregation, diffusion,

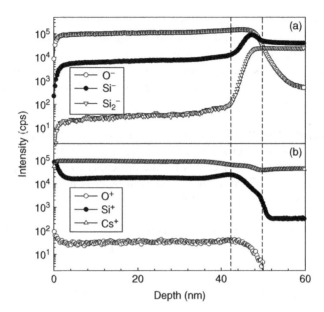

**Figure 3.28**   Depth profiles resulting from $Cs^+$ primary ion impact through a 42-nm $SiO_2$ film on Silicon of (a) the secondary ions of $O^-$, $Si^-$, and $Si_2^-$ and (b) the secondary ions of $O^+$, $Si^+$, and $Cs^+$. Vertical dashed lines designate the oxide to Silicon interface as noted during SIMS analysis. Author's unpublished data.

and so on. Sputter-induced damage is covered in Section 3.2.3. As a result, quantification over these regions proves more difficult if not impossible.

An example of this form of matrix effect is shown over the ~40–50 nm depth range in Figure 3.28(a) and 3.28(b). This shows various positive and negative secondary ion intensities as a function of depth through a 42-nm $SiO_2$ film on Si using a $Cs^+$ primary ion beam. Of particular note are the variations in the $Si^-$ and $Si^+$ secondary ions as these display trends that are inconsistent with the known concentration gradients. Only the $O^-$, $O^+$, and $Si_2^-$ signals trend in a manner consistent with the structure, although the latter does so to a much greater extent (actual Si content variation is 33%–100%).

There also exist secondary ion yield variations during the initial stages of sputtering as noted over the first ~5 nm in all of the signals portrayed in Figure 3.28(a) and 3.28(b). Such trends are particularly evident when using chemically reactive primary ions such as $O^-$, $O_2^+$, $Cs^+$, and $SF_5^+$. This effect, more commonly referred to as the transient effect, is most evident in the secondary ion intensities from elements present in the substrate at some constant concentration.

In the case of $Cs^+$, this effect is relatively easy to examine as the enhancement/suppression induced scales with the concentration of $Cs^+$ implanted into the substrate's surface. This is illustrated in Figure 3.29(a) for 1 keV $Cs^+$ impact on Silicon and Figure 3.29(b) for $Cs^+$ impact on $SiO_2$ with exponential trends noted between the Cesium concentration and the yield enhancement/suppression noted.

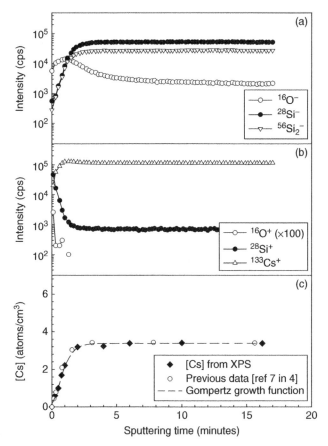

**Figure 3.29** Plots of the listed negative secondary ion intensities (top plot), the listed positive secondary ion intensities (middle plot), and the Cesium concentration as derived via XPS (bottom plot) during the initial stages of sputtering with a constant current 1-keV $Cs^+$ primary ion beam on (a) a Silicon substrate and (b) an $SiO_2$ substrate. The dashed line is a fitted Gompertz growth function. Reprinted with permission from van der Heide (2004) Copyright 2004 Elsevier Science B.V.

As no Cesium is present in the substrate before analysis, the enhancement or suppression effect continues to increase with increasing sputtering time until steady-state sputtering is reached. Steady state refers to the condition where the primary ion implantation rate equates to the sputtering rate of implant primary ions.

The general rule of thumb is that the sputtered depth over which *transient effects* prevail (referred to as the *transient width*) scales as $2-2.5$ times the primary ion projected range (projected range values can be approximated via Relations 3.6(a–d)). This and exceptions to this trend noted at high impact angles are illustrated in Figure 3.30. This shows that the *transient width* decreases on increasing the $Cs^+$ incident angle from 40° to 60° (this decreases the projected range).

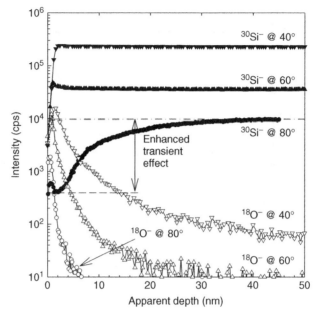

**Figure 3.30**   Overlaid plots of the Si$^-$ and O$^-$ secondary ion intensities during the initial stages of sputter depth profiling of Silicon with 1-keV Cs$^+$ ions (constant current) at the listed angles. Note: The enhanced transient effect extends to a depth of ∼40 nm. Author's unpublished work.

However, increasing the incidence angle further (to 80°) introduces a new type of *transient effect* sometimes termed the *enhanced transient effect*. This effect is noted only on substrates that experience ripple topography growth during sputtering (this occurs only on substrates that are amorphized during sputtering as is discussed in Section 3.2.3.3). This has been shown to slow the rate at which the substrate reaches steady-state sputtering, i.e. ripple topography slows the sputter rate as is illustrated in Figure 3.17, which in turn, allows for a greater buildup of Cesium within the substrate's sub surface.

Note: Owing to the large array of possible matrix effects/transient effects, this discussion concentrates only on the matrix/transient effects experienced by atomic emissions from Silicon and SiO$_2$. These surfaces were chosen because:

1. these present well-defined surfaces (single crystal Silicon presents an atomically smooth surface, which readily amorphizes on sputtering, whereas SiO$_2$ is already amorphous),

2. these surfaces react in a well-understood manner under reactive ion impact (these surfaces have been examined more than any other surface),

3. the resultant trends appear well understood, i.e. the systematic trends noted can be explained within the context of specific substrate parameters, which have been used in formulating the models presented in Section 3.3.3.

TABLE 3.3 Positive Secondary Ion Yields ($Y(M^+)$) from the
Listed Clean and Oxygen Covered Metal Surfaces Under 3-keV
$Ar^+$ Primary Ion Impact. Data reproduced with Permission from
(Benninghoven 1975).

| Metal | $Y(M^+)$/Clean Metal | $Y(M^+)$/Oxygen Covered |
|-------|---------------------|-------------------------|
| Mg | 0.01 | 0.9 |
| Si | 0.0084 | 0.58 |
| Ni | 0.0006 | 0.045 |
| Nb | 0.0006 | 0.05 |
| Mo | 0.00065 | 0.4 |
| W | 0.00009 | 0.035 |

Transient and interface matrix effects noted when using $Cs^+$ primary ions have been attributed, largely, to the strong reduction of the substrate work–function (van der Heide 2004; Gnaser 1999; Yu 1991). Correlations between the work–function and secondary ion yields are discussed in the following section.

The transient effects experienced when sputtering with $O^-$ or $O_2^+$ primary ion beams are more complex. This is realized as oxidation of the surface can also be induced, which can then enhance diffusion, segregation, and so on. The oxide formed will depend on the analysis conditions used (primary ion energy, primary ion incidence angle, and $O_2$ partial pressure) and the substrate examined, all of which can induce further matrix effects. What is particularly interesting is that both the positive and the negative atomic secondary ion yields of Silicon from a Silicon matrix are enhanced when Oxygen is introduced. This is not seen when Cesium is introduced, i.e. $Cs^+$ enhances the $Si^-$ yields but suppresses $Si^+$ secondary ion yields (see Figure 3.28(a)). Further examples of the enhancement of various positive secondary ion yields from various elemental substrates through the adsorption of Oxygen are listed in Table 3.3.

As for other primary ions, matrix-specific trends are noted that scale with the chemical reactivity of the respective primary ion. None-chemically active primary ions can also induce transient effects, although much weaker, that can generally be related to the damage imparted as a result of energetic ion impact (see Section 3.2.3).

*3.3.2.2.1 Work–Function* The work–function ($\phi$) represents the minimum energy needed to extract an electron from a solid, i.e. it can be envisaged as the solid's ionization potential ($I$). As the value of the work–function equates to the difference between the Fermi edge ($E_F$) relative to the vacuum edge ($E_{vac}$), it can be used to define the position of the Fermi edge. The work–function is a global parameter, i.e. this is noted only for solids containing a sufficient number of bound atoms such that the individual electronic levels merge to form a well-defined band structure as discussed in Section 2.2.2.1. In other words, this represents the valence electron density of a solid.

The work–function is of interest as strong correlations between this and the low-energy emissions of various positive and negative atomic secondary ion yields have been noted from both conductive and semiconducting surfaces, including that of Silicon when sputtered with $Cs^+$ primary ions (Anderson 1970; Yu 1978, 1982, 1983; Yu and Lang 1983; Gnaser 1996a, 1996b; van der Heide 2005). Cesium is of interest, as like all alkali metals, this induces a strong reduction in the $\phi$ when deposited on or within conductive and semiconducting surfaces. This reduction in the substrate work–function stems from the high electron density imparted by Cesium. This, in turn, is a result of the fact that Cesium is the largest and most electropositive element in the periodic table. Indeed, the influence of Cesium appears to cover a sphere of $\sim 1.5$ nm in diameter (Yu 1991).

From a mechanistic perspective, the dependence on work–function has been used as evidence that charge transfer processes resulting in the recorded secondary ion signals stem primarily from resonance processes active during or just following the sputtering event. This can be understood when examining the effect any change in the position of the Fermi edge will have on the direction of charge transfer and hence the respective secondary ion yields as shown in Figure 3.25(a). Indeed, this affects not only the direction of electron transfer but also the probability of charge transfer. As an example, the further the Fermi edge rises above the vacant ionization level of an ion in close proximity to the surface (occurs with decreasing work–function), the greater the reservoir of electrons that can be transferred. This will enhance neutralization of the aforementioned positive ion.

An example of the reduction in the work–function imparted by the presence of Cesium on and within a Silicon surface (that induced by the fraction implanted in Silicon on sputtering with a $Cs^+$ primary ion beam) is shown in Figure 3.21. The inset of Figure 3.21 shows the resulting variations in the $Si^+$ and $Si^-$ secondary ion intensities from the aforementioned Silicon substrate as a function of concentration of Cesium implanted into the Silicon substrate. As can be seen, the positive secondary ions are quenched in an exponential-like manner with increasing Cesium content, whereas the positive secondary ions are enhanced, also in an exponential-like manner. These trends can be described by the relations:

$$Y(M^+) \propto \exp(\phi) \tag{3.8a}$$

$$Y(M^-) \propto \exp(-\phi) \tag{3.8b}$$

Here, $Y(M^+)$ and $Y(M^-)$ again define the respective secondary ion yield dependence.

Further examples of the effect of $Cs^+$-induced variations in the work–function on the low-energy atomic secondary ion emissions from other substrates are shown in Figure 3.22(a)and 3.22(d). There are, however, also cases where the work–function dependence appears to vanish. An example of this is seen for the low-energy (up to 10 eV) $Ag^-$ secondary ions emanating from a Silver surface under $Cs^+$ impact in Figure 3.22(d). This unusual case is believed to be due to the prevalence of surface excitation processes limited to specific ion–solid combinations as discussed further in Section 3.3.2.3.3.

As discussed in Section 3.3.2.1, there also exists a strong dependence on ionization potentials or electron affinities of the interacting ion (which of the two is dependent on the polarity). As recognized (Yu 1978, 1991), this can be incorporated into these relations as:

$$Y(M^+) \propto \exp(\phi - I) \tag{3.9a}$$

$$Y(M^-) \propto \exp(EA - \phi) \tag{3.9b}$$

These relations, however, extend only over a limited range, with the range depending on the respective secondary ion ionization potential or electron affinity relative to that of the substrates Fermi edge at the time of emission as discussed below.

The dependence of secondary ion intensities over a finite work–function range can be understood when examining Figure 3.25(a) and considering the effect that the position of the Fermi edge has if it were to lie far from either the ionization or affinity levels of the departing atom or ion. In these situations, no dependence would be seen.

Indeed, the probability of Cesium to form $Cs^+$ secondary ions is, in the majority of cases, independent of the position of the Fermi edge, which itself is due to the very low ionization potential of Cesium, i.e. this places the Cesium ionization level well above the Fermi edge of most surfaces. This insensitivity to the position of the Fermi edge can thus explain the trends in $Cs^+$ secondary ions from most $Cs^+$ sputtered metal and semiconductor surfaces and why these are seen to increase over the transient region, with the increase simply reflecting the increased Cesium content implanted into the respective substrate.

Examples of this increase are illustrated for $Cs^+$ secondary ions collected as a function of $Cs^+$ sputtering time over the transient region from both Copper and Gold surfaces in Figure 3.31(a). The suggested reasoning illustrated in Figure 3.31(b) is that there are no filled states within the substrate's surface above the Fermi edge from which electrons can be transferred, even after extensive Cesium loading. The increase in the $Cs^+$ secondary ion intensities would thus reflect the increase in the amount of Cesium being implanted into the respective substrate.

If, however, the position of the Fermi edge were to cross either the affinity or ionization level of the interacting ion, a work–function sensitivity would be introduced. In the case of $Cs^+$ emission over the transient region, this situation is noted from a select few substrates, i.e. those with sufficiently low work–function values. An example of this is noted in the $Cs^+$ secondary ion intensities collected as a function of $Cs^+$ sputtering time over the transient region from a Titanium surface in Figure 3.31(a), with the suggested reasoning again illustrated in Figure 3.31(b). The reasoning being that, as the Cesium content increases, the position of the Fermi edge rises to the point that it crosses the ionization level of Cesium, thereby allowing neutralization of $Cs^+$ to ensue. This has been demonstrated on semiconductors and metals alike (Yu 1978, 1982, 1983, 1991, Gnaser 1996a, b, 1999, van der Heide 2005).

The work–function dependence has also been found to vary according to the emission velocity (emission energy) of the secondary ion in question, with a

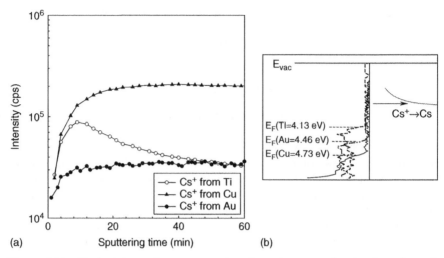

**Figure 3.31**    To the left are shown sputter time-dependent secondary ion intensity variations noted during the initial stages of sputtering in the low-energy Cs$^+$ secondary ion intensity from Titanium, Copper, and Gold surfaces. To the right are shown the expected charge transfer process with the represented valence states (defined before sputtering via XPS) and the positions of the Fermi edge (defined before sputtering via Kelvin probe measurements) relative to the Cesium ionization level (approximated on the basis of $I$(Cs) and Relation 2.7). Note: If the position of the Fermi edge does not rise to or above the ionization level of Cesium, the transition shown cannot occur. The reversal of trends for the Cs$^+$ emissions from Titanium suggests that this does occur on this surface at a sputtering time of ~10 minutes. Reproduced with permission from van der Heide (2005) Copyright 2005 Springer-Verlag.

weakening of the dependence noted with increasing emission velocity (Yu 1991). For this and other reasons, it is also worthwhile examining the secondary ion energy distributions from which velocity distributions can be extracted. This is discussed further in Section 3.3.2.2.2.

The variations in the secondary ion yields noted when sputtering with O$^-$ or O$_2$$^+$ cannot easily be explained within the context of the surface work–function or any variations induced on sputtering, particularly over the transient region. This is realized because, as is seen in the sputtering of Silicon, enhancements in both the Si$^+$ and the Si$^-$ secondary ion yields are noted. Furthermore, there are situations where adsorbed oxygen induces an increase (Nickel, Silicon, and Tungsten) or decrease (Magnesium, Niobium(110), and Molybdenum(100)) in the substrates work–function, a fact related to whether adsorbed Oxygen atoms reside above (Nickel, Silicon, and Tungsten) or beneath (Magnesium, Niobium(110), and Molybdenum(100)) the top most atomic layer (from Yu 1991). Further discussion on these variations can be found in Section 3.3.3.2.

Likewise, there are other matrix effects from other primary ion–substrate combinations that cannot be explained within the simplified work–function variation picture (for examples, see Yu 1991; Wittmaack 1996). Many of these also tend to fit with the prevalence of specific surface excitation processes that tend to be limited to specific ion–solid combinations. The effects of various surface excitations on secondary ion emissions are discussed in Section 3.3.2.3.

### 3.3.2.2.2 Secondary Ion Emission Velocity

As introduced in Section 3.3, the sputtered atomic and molecular populations move at a relatively slow speed compared to the electronic interactions that take place between the aforementioned emissions and the solid's surface. It is during this departure that charge transfer to/from the atomic/molecular emissions and the surface occur. Insight into the interaction region (the region over which charge transfer can take place) can be derived by measuring ion yields as a function of their emission velocity.

Indeed, the neutralization of atomic ions in close proximity to a solid's surface as noted in LIES and INS does exhibit a velocity dependence (see for example Tully and Tolk 1977, Rabalais 1994 for further discussions). In accordance with theoretical studies, the neutralization rate appears to drop off in an exponential manner with increasing velocity. This is realized through measurements of ion yield variations as a function of the ion beam's energy and/or angle relative to the surface. Note: The angle influences the emission velocity normal to the surface ($v_\perp$) and hence the interaction time.

In a similar manner, measurements of the secondary ion velocity dependence have also been carried out (see for example Krause and Greun 1980; Vasile 1984; Wucher and Oechsner 1988). This, however, is no trivial task, a fact realized in that neither the secondary neutral population nor its distribution as a function of emission energy is measured in SIMS. Note: These are required as the secondary ion yield relates the number of secondary ions produced per sputtered neutral of the same energy emitted. The secondary neutral distribution can be derived either by modeling the secondary neutrals as a function of emission energy/velocity (this is accomplished using the Sigmund–Thompson relation as shown in Relation 3.4 and applied Figure 3.9) or through the use post-ionization methods (these employ specifically tuned lasers to ionize the sputtered neutral population as demonstrated by Wah and Wucher 1994).

The majority of the velocity distribution studies have used the Sigmund–Thompson relation as this is the easier of the two methodologies, and the relative values derived tend to agree with the studies in which the secondary neutral population is measured (see Figure 3.9).

Derivation of the secondary ion velocity distribution through the use of modeled secondary neutral distributions (Relation 3.4) is carried out in the following manner:

1. Collecting the energy distribution with a relatively narrow energy window over a relatively wide energy range (to 200 eV or more).
2. Accounting for the emission angle and the transmission function of the instrument (the transmission of secondary ions varies with energy).
3. Dividing the transmission-corrected signal by the sputtered neutral signal (this provides values directly proportional to the secondary ion yield).

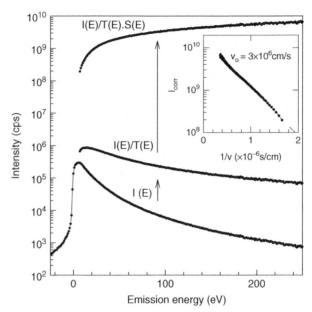

**Figure 3.32** Example of the derivation of velocity distributions from the collected energy distribution ($I(E)$) after correcting for the transmission function of the instrument ($T(E)$) and the sputter yield ($S(E)$) assuming a linear collision cascade took place (Sigmund–Thompson relation is assumed). In the inset is shown the velocity distribution plotted as a function of inverse velocity ($1/v_\perp$). The line applied in the velocity distribution shown in the inset assumes an exponential yield dependence on $1/v_\perp$ hence the reason for plotting such distributions in this manner. Reproduced with permission from van der Heide and Karpusov (2000) Copyright 2000 Elsevier.

The outcome is referred to as the *corrected intensity* ($I_{corr}$). If actual yields are known, this corrected intensity can then be scaled to provide absolute ion yields as a function of emission velocity. Note: However, similar to the sputtered neutral distribution, absolute secondary ion yields are difficult to derive. An example of the procedure described earlier is illustrated in Figure 3.32 for $Cu^+$ secondary ions resulting from 7.5 keV $O^-$ impact. This combination is shown as the relatively low energy and mass of $O^-$ should result in a full isotropic linear cascade as assumed by the Sigmund–Thompson relation. The derived velocity distribution is shown in the inset of Figure 3.32.

Like the secondary neutral energy distributions (see Figure 3.9), the recorded atomic secondary ion energy distributions exhibit a peak below 10 eV along with a high energy tail that extends to several hundred electron volts. Molecular distributions exhibit an energy distribution peak that appears at an even lower energy (typically around 2 eV or less) with a much reduced high energy tail. Note: Secondary ions formed an appreciable distance from the surface display apparent negative energies when measured in instruments employing high extraction fields (potential gradients). In such instruments, the extraction field is set at the zero energy mark in the resulting secondary ion energy distributions. Although these energy-deficient ions are generally formed via auto-ionization, other possibilities exist.

Of note when examining Figure 3.32 is the fact that the corrected intensities (these scale with yield) increases as the emission velocity increases. Furthermore, the noted exponential trend is consistent with the results from ion–solid interactions. As covered in Section 3.1.1, these generally assume that the resonance or Auger charge transfer processes are the dominant modes in charge transfer (see Figure 3.25(a)).Including this velocity dependence into the work–function and ionization potential or electron affinity dependence already discussed (see Sections 3.3.2.1 and 3.3.2.2.1, respectively) implies that secondary ion emission from conductive and semiconducting matrices can be described via the relations:

$$Y(M^+) \propto \exp(-(I - \phi)/v_\perp) \qquad (3.10a)$$

$$Y(M^-) \propto \exp(-(\phi - EA)/v_\perp) \qquad (3.10b)$$

Consistent with Relations 3.8(a–b) and 3.9(a–b), these relations apply only over limited ranges of work–function, ionization potential, or electron affinity.

The relations reveal that the slope of the inverse velocity plots can be related to all parameters present in the numerator of the respective exponential function. The parameters within the numerator are also sometimes encapsulated within a parameter termed the *characteristic velocity* ($v_o$). In various theoretical studies, the denominator in these relations is sometimes replaced by system-specific parameters encapsulating the velocity dependence. These parameters are expressed as $\varepsilon_p$ for the positive secondary ions or $\varepsilon_n$ for the negative secondary ions. Further discussion on these aspects can be found elsewhere (Yu 1991).

There are instances where strong deviations from the exponential velocity dependence over all emission energies when conditions necessary to induce a full isotropic linear cascade are in effect (Barth et al. 1986; Wucher and Oechener 1988; van der Heide 1994). In such cases, the deviations reveal a weakening of the velocity dependence at emission energies less than $\sim 50$ eV (higher inverse velocities). Some examples from the same substrate under different sputtering conditions are illustrated in Figure 3.33. These were derived on the assumption that the secondary neutral energy distribution follows the Sigmund–Thompson relation. Similar deviations have also been reported for experiments in which the secondary neutral distributions were measured (Wahl and Wucher 1994; Marazov et al. 2006).

Theoretical studies in which velocity distributions of the same form (a weakening of the velocity dependence at lower emission energies) have suggested that such deviations stem from the inclusion of various effects on the ionization/neutralization process active on the sputtered population that arises from surface excitation (considered within Relations 3.10(a) and (b)). Possibilities include introduction of additional charge transfer processes at lower emission energies whether via the thermalization of the surface during sputtering (this effectively populates electronic states above the Fermi edge as displayed in

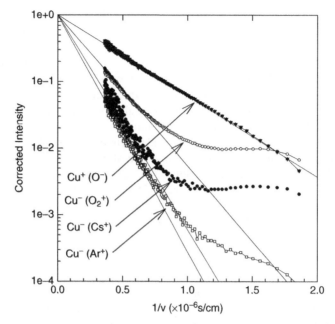

**Figure 3.33** Overlay of the apparent velocity distributions derived on the assumption that the sputtered neutral population follows the Sigmund–Thompson distributions (Relation 3.4) for $Cu^+$ and $Cu^-$ secondary ions emanating from a polycrystalline surface under 17.5 KeV $Ar^+$ impact, 14.0 KeV $Cs^+$ impact, 17.5 KeV $O_2^+$ impact, and 7.5 KeV $O^-$ impact. The same calculations were applied to all data sets and with all plots arbitrarily normalized to unity at zero $1/v_\perp$. The lines represent the trends relayed by Relation 3.10(a) or (b) fitted to the lower $1/v_\perp$ (higher emission energy) populations. Reproduced with permission from van der Heide and Karpusov (2000) Copyright 2000 Elsevier.

Figure 2.6) or the introduction of an additional sputtering pathway not described by the Sigmund–Thompson relation (a nonlinear sputtering process leading to the generation of heat spikes as described in Section 3.2.1.2).

Interestingly, the likelihood of heat spikes is expected to increase as denser matrices are sputtered by heavier primary ions or molecular ions ($Cs^+$ and $O_2^+$, respectively in Figure 3.33). Likewise, surface thermalization will have a greater impact on the lower energy secondary ion emissions. Both possibilities can also reduce or even possibly remove the work–function dependence typically expected. In addition, the possibility of other excitation-mediated processes is apparent. One such example is in the formation of plasmons. These are suspected to be responsible for the insensitivity of $Au^-$ secondary ion emissions from Au surfaces to the value of work–function when produced via $Cs^+$ impact (see Figure 3.22(d)). This is discussed further in Section 3.3.2.3.3.

Note: The inclusion of processes not described by the Sigmund–Thompson relation will introduce inaccuracies into the corrected intensity values derived in such velocity distributions at higher $1/v_\perp$ (lower emission energies) when using

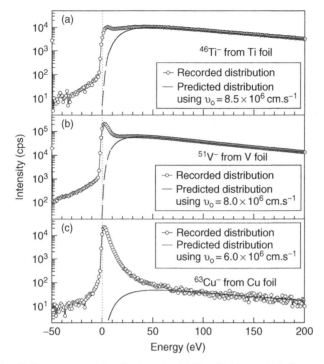

**Figure 3.34**   Collected energy distributions for (a) Ti⁻, (b) V⁻, and (c) Cu⁻ secondary ions resulting from 14.5-keV Cs⁺ impact on the respective elemental matrices. The lines represent the extrapolation of the velocity dependence exhibited by the higher energy secondary ions where the characteristic velocity was derived from the slope of the respective velocity distributions (see lines in Figure 3.33). The reverse calculation was then applied (reverse to that shown in Figure 3.32) on the bases that a linear cascade prevails (that approximated by the Sigmund–Thompson relation). The deviations between these lines and the collected energy distributions reflect the deviations from the velocity dependence implied by Relation 3.10(b) and the collected data at lower emission energies. Author's unpublished work.

the Sigmund–Thompson relation. The general trends implied do, however, remain informative, i.e. reveal a weakening of the velocity dependence. This is vindicated by the fact that similar deviations are noted in experiments in which the secondary neutral distributions are measured.

The deviations induced by what appears to be the effect of surface excitation can be depicted in the original energy distributions by simply back tracking the relations used in deriving the velocity distributions once the high energy dependence is known, i.e. once the value of the characteristic velocity has been derived. This procedure is illustrated in Figure 3.34(a–c), for the secondary ions of Ti⁻, V⁻, and Cu⁻ resulting from 14.5 keV Cs⁺ impact on the respective elemental matrices.

The lines shown in Figure 3.34(a–c) are representative of the velocity dependence implied by Relation 3.10(b). The deviation in the recorded data (signified by the symbols) from the dependence implied by Relation 3.10(b) at lower emission energies thus reflects the likelihood of surface excitation type processes on moving from left to right in the periodic table for the first row transition metal ions.

Similar trends (a strong velocity dependence at higher emission energies with deviations apparent as the energy decreases) are also noted for the lighter positive multiply charged atomic ions (Franzreb and van der Heide 1998). Note: Negative multiply charged ions are not recorded effectively in SIMS because of their short lifetimes, whereas heavier positive multiply charged ions face a far greater neutralization probability because of their slower rate of departure (a fact stemming back to their larger mass). Similar to the atomic singly charged positive secondary ions, mechanisms for multiply charged ion formation appear to be centered on both resonance and Auger charge transfer with auto-ionization processes playing a greater part. These are discussed further in Section 3.3.3.3.

Lastly, there remains a significant lack of information concerning the velocity dependence for molecular secondary ion emission, a fact that can be related back to the greater variability in secondary neutral distributions (see Section 3.2.1 and subsections contained within). This, and the remaining questions concerning the velocity distribution of atomic secondary ions, thus signifies the need for further studies in this area.

### 3.3.2.2.3 Secondary Ion Mass Fractionation

Mass fractionation, also referred to as isotope fractionation, describes a physical process that acts to separate isotopes of the same element over time. Isotope fractionation occurs when velocity, diffusivity, or bond strength-dependent processes are in effect. As a result, mass fractionation is of particular interest in the areas of chemistry, cosmology, and geology (in particular, chronology).

As SIMS can be used to measure the intensities of the isotopes of the same element, as well as carry out analysis to extreme sensitivity, it has been applied to the examination of mass fractionation in the atomic emissions whether intrinsic to the material of interest (that existing before analysis) or extrinsic (that arising as a result of the analysis being carried out).

Analysis-induced mass fractionation occurs in the sputtering and ionization/neutralization events leading up to the secondary ion signal of interest. Note: Sputtering arising from knock-on processes is defined by the momentum transfer taking place (see Section 3.2.1.1), with momentum being the product of the mass and velocity of the collision partners. Ionization/neutralization probabilities are also a function of the emission velocity of the sputtered emissions (see Section 3.3.2.2.2) with lighter isotopes of the same emission energy departing the surface with a greater velocity.

Analysis-induced mass fractionation is encapsulated in what is referred to as the *Instrument Mass Fractionation* (IMF). Because this is matrix dependent, this can only be assessed through comparison of known isotope compositions of some standard reference material with the isotopic composition derived via SIMS analysis of the above-mentioned reference. The fact that a matrix-dependent IMF is observed leads to the realization that examination into the IMF can provide the much-needed information on the fundamentals of the secondary ion formation/survival process.

Indeed, secondary ion yield enhancement of the lighter isotope can partially be understood as resulting from enhanced sputter yields. This stems from the fact that the diffusivities as well as bond strengths, inclusive of surface binding energies, are mass dependent (these variations stem from the fact that vibration frequencies are mass dependent). This enhanced sputter yield also results in a sputter time-dependent enrichment of heavy isotopes over the surface region following the onset of sputtering. This stabilizes after steady-state sputtering conditions are reached.

Ionization/neutralization variations are also to be expected and indeed noted with values typically greater than those observed from sputtering alone. The dependence of the IMF on mass and emission energy can be understood as resulting from the interrelation of $E = \frac{1}{2}mv^2$ and how this affects the time the emitted ion/atom will spend in the interaction zone before its departure from the surface. As mentioned in Section 3.3, this interaction zone is believed to extend over distances of less than $\sim 1$ nm. The IMF is defined, on a per mil basis, as:

$$\text{IMF} = \left( \frac{R_m^s}{R_t^s - 1} \right) \times 1000 \tag{3.11}$$

where $R$ is the abundance ratio of the two isotopes, with the subscripts $m$ and $t$ referring to that measured or known ($t$ for true), respectively, and the superscript referring to the standard reference material used. With the IMF defined, the isotopic composition of the sample of interest can be derived through rearrangement of Relation 3.11 if, and only if, the instrument is operated under the same conditions as that used to derive the IMF and the matrix analyzed is the same as the reference material used in defining the IMF.

IMF trends can, in general, be summarized as follows:

1. Sputter yields and ion yields favor lighter isotopes relative to the respective heavier isotopes of the same element (as the lighter isotopes have the higher velocity, the ion yield trend is also evident in the velocity distributions discussed in Section 3.3.2.2.2)
2. The magnitude of the secondary ion yield variation arising from the IMF scales with:
   a. the mass difference of the isotopes measured and
   b. the emission energy (emission velocity) of the secondary ions measured.

An example of the IMF resulting from the variation in secondary ion yields with emission velocity is shown for the Silicon isotopes from Silicon under $O^-$ impact in Figure 3.35. Note: The secondary ion mass fractionation observed ($F$) is depicted in units of permil. This exhibits trends over higher emission velocities (inverse velocity values less than $1.5 \times 10^{-6}$ s/cm) consistent with the velocity dependence defined in Relation 3.10(a) (Gnaser and Hutcheon 1987). The deviations at lower emission energies appear to be consistent with those noted in Figure 3.30 with speculated reasons for these deviations covered in Section 3.3.2.2.2.

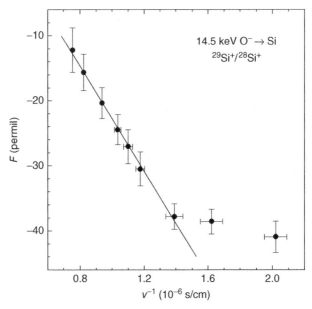

**Figure 3.35**    Measured IMF (represented in this figure as $F$ [permil]) for the Silicon isotopes as a function of $1/v_{\perp}$ as defined from the secondary ion emission energy. The data is reproduced with permission from Gnaser and Hutcheon (1987) Copyright 1987 Springer.

Deviations from the trends outlined in Figure 3.35 will be expected if alternate pathways via which the ionization and/or neutralization of the sputtered atomic or molecular emissions were to be introduced. Two such examples can include:

1. The influence of excitation processes on the ionization and/or neutralization processes/s active on the atomic emissions produced through sputtering. These have been used to explain the deviations from the velocity dependence defined in Relations 3.10(a) and (b), with examples displayed in Figure 3.30 for the $Cu^{-}$ secondary ion emissions arising from $Ar^{+}$, $O_2^{+}$, or $Cs^{+}$ impact. Excitation processes are discussed in detail in Section 3.3.2.3.

2. The enhanced prevalence of a quasi-resonance charge transfer process such as that illustrated in Figure 3.25. This possibility is mentioned as oscillatory trends in the IMF as a function of emission energy has been reported in various ion scattering (Tully and Tolk 1977; O'Connor et al. 1988) and SIMS experiments (Abramenko et al. 1988; Urazgil'din and Borisov 1990). A constant periodicity is then noted when the IMF is re-plotted against the emission velocity. How this influences the IMF is discussed below.

Quasi-resonance charge transfer will accompany resonance charge transfer if a core hole exists within the departing atomic emission of the substrate. For this to occur, a significant amount of energy must be imparted during the atomic collisions in knock-on sputtering. When present, an additional channel for electron transfer between the surface and the departing atom/ion can be envisaged assuming that the two levels (atomic and substrate levels) are of the same energy. However, because the image potential adjusts the energy of the electronic level of the departing ion (see Section 2.2.2.3), and the fact that core-level wave-functions extend a much lesser distance from the surface than valence levels, there exists a much shorter time window over which such transitions can occur. Oscillatory trends can be expected if this time approaches electron transfer rates. This is realized as any change in the time, introduced through emission velocity, will turn on or off the transition or even allow for an even or odd number of transitions.

From an analytical perspective, measurement of isotope ratios has become an effective technique for studying mass fractionation. Indeed, ratios to 1–2 permil precision levels can be obtained once measurement-induced variations are accounted for. This level of sensitivity, along with the small volumes analyzed, has thus made SIMS the technique of choice in numerous Earth Science applications.

### 3.3.2.3 Electronic Excitation

There is a significant amount of energy imparted during knock-on sputtering events leading up to secondary ion formation/survival (see Section 3.2.1.1). Sputtering can also occur via one of the several inelastic processes (see Section 3.2.1.2). As a result, the possibility of electronic excitation has long been recognized and, indeed, has been used in explaining specific secondary in emissions including, but not limited to:

1. positive multiply charged secondary ions from elements with $Z$ less than 15,
2. stimulated desorption of $F^+$, $Cl^+$ from Aluminum and Silicon surfaces,
3. the low-energy atomic secondary ions displaying deviations from velocity dependencies implied by Relations 3.10a and 3.10b (see Figures 3.33 and 3.34), and
4. plasmon-assisted emission of $Ag^-$ secondary ions from Silver surfaces.

As noted in Section 2.2.2.4, electronic excitation can result in the production of:

1. Phonons
2. Excitons
3. Plasmons
4. Core holes

Which of these processes can potentially play a role in secondary ion formation/survival depends on:

1. the energy deposited upon during/following the sputtering event,
2. the most likely channel/s via which electron excitation can/will occur, and
3. the time and spatial scales involved.

As an example, phonons (lattice vibrations) are generally not considered because the frequency of the lattice vibrations induced is much too small (slow) to influence the atom/ion being sputtered. In other words, the lattice appears stationary during atom/ion emission. Note: This may not be the case in large cluster ion emission as these have much lower emission velocities. In addition, the cluster secondary ion emission has been postulated as resulting from inelastic sputtering processes (see Section 3.2.1.2).

Electronic excitations resulting from phonons (vibrationally and/or rotationally excited states) are also not considered because such electronic excitations relax within time scales much shorter than that of atom/ion emission. In other words, the emitted atom/ion now appears stationary with respect to the electronic transitions.

At this point, it is worth recalling the fact that sputtering entails the movement of atoms/intrinsic ions (in ionic lattices) from their equilibrium position within the lattice. This will alter the electronic band structure over the localized region from which the atom/ion emanates. This is understood on the basis that the band structure represents the culmination of many valence states within the lattice. As discussed in Section 2.2.2.1, the energies of valence states are strongly dependent on the distances between the atoms and/or ions making up the lattice.

The fact that the vast majority of atomic secondary ion yields from metallic and semiconducting surfaces scale with the work–function of the respective surface implies that the final process from these surfaces is, however, not associated with the bond breaking process but, rather, is associated with a process or processes active over longer distances. One such process is that of electron tunneling (that prescribed in resonance charge transfer). This does not appear to be the case in secondary ion emission from insulating surfaces, i.e. their yields tend to scale with the properties associated more closely with the bond/s being broken, a fact that can be associated with the greater localization of the valence electrons in ionic substrates and the larger band gaps present.

This brings us back to the possible electronic excitation mechanisms that can play a role in secondary ion formation. Section 3.3.2.3.1 describes those in which secondary ions appear to form as a result of the de-excitation of some sufficiently long-lived core hole. This includes a discussion on the formation of multiply charged ions from elements lighter than Phosphorus and the formation of $F^+$ and $Cl^+$ ions from Aluminum and Silicon surfaces. Section 3.3.2.3.2 describes the various valence-level excitations, whether resulting from core hole production or other less energetic means. These can lend themselves to the formation of various low-energy atomic secondary ions, i.e. those deviating from the trends implied by Relations 3.10a and 3.10b. Section 3.3.2.3.3 describes a specific example of long-lived valence-level excitation (plasmons) and how these are believed to result in the enhanced secondary ion yields of $Ag^-$ from Silver surfaces from that expected from Relation 3.10b, in particular, the negative velocity dependence exhibited by the $Ag^-$ secondary ions resulting from $Cs^+$ impact on Silver.

*3.3.2.3.1  Core-Level Excitation*   As mentioned in Section 2.2.2.4.4, core holes represent the vacancy left behind when an electron is removed from its equilibrium core-level stationary state within the associated atom or ion. This can occur via the promotion of the electron to some vacant level of lesser binding energy (states closer to the Fermi edge) or the emission of the electron from the atom or ion into the gas phase, i.e. to some levels above the vacuum energy. For this to occur, an amount of energy must be imparted onto the respective electron that is greater than the energy binding the electron to its nucleus, i.e. greater than its binding energy.

Evidence for the presence of core holes in sputtered atoms/ions can be identified in the spectra from several techniques, with the most direct being that of Ion-induced Auger Electron Spectroscopy (IAES). This is understood as IAES spectra arises from the de-excitation of core holes formed as a result of atomic collisions induced through energetic ion impact on the solid's surface of interest. De-excitation in the sputtered population is, however, only noted in spectra collected at close to glancing angles (reasons are outlined in Section 3.2.1.2). This is noted as:

1. Auger signals from the de-excitation of atoms/ions within the substrate exhibit an anisotropic angular distribution, i.e. peak in the direction normal to the surface (although produced with an isotropic distribution, those emitted from the surface in the normal direction are detected in far greater numbers).
2. Auger signals from sputtered atoms/ions display an isotropic angular distribution.

Examples of Auger emissions from Neon and Aluminum are shown Figure 3.6(a). As can be seen, these result in discrete peaks over a broader background peak. The discrete peaks arise from core hole de-excitation occurring within the sputtered atoms/ions, whereas the broad background peak stems from the same de-excitation process but occurring within atoms/ions still present within the substrate. Additional information that can be acquired from such spectra, include the identification of:

1. The elements displaying de-excitation in their sputtered populations (this is noted for all but the lightest elements through to Phosphorus)
2. The incoming ion threshold energy required to induce core hole formation ($\sim$900 eV for $Ar^+$ impact on Aluminum (Valerie 1993)). Note: This is ion and substrate specific
3. The charge state of the sputtered population suffering de-excitation (Auger emissions have been noted from both neutral and positively charged ion emissions)
4. The specific core hole present and the specific Auger process active in de-excitation

Indeed, such spectra have been used as one source or evidence supporting the Kinetic Emission model for multiply charged positive atomic ions from elements lighter than Phosphorus (Joyes 1973). This mechanism, discussed in Section 3.3.2.4, concerns the formation of core holes via atomic collisions with multiply charged ions formed as a result of the de-excitation of these core holes when present in the sputtered atomic/ionic populations. Owing to the short lifetimes of the above-mentioned core holes and the fact that complete neutralization of ions formed within the substrate occurs rapidly, such a process applies only to atomic/ionic emissions with a sufficient velocity to escape the surface region, i.e. this allows the core hole to survive atom/ion departure and thus any subsequent neutralization processes that would otherwise occur with the surface. This explains the limitation of this process to elements lighter than Phosphorus.

Atomic multiply charged ions from elements heavier than Phosphorus are formed at a much greater distance from the substrate surface via what appears to be some type of collisional process active in the gas phase (van der Heide et al. 1993). This is revealed through the observation that these emissions exhibit energy-deficient populations in their respective energy distributions when collected under high extraction fields (the high extraction field allows for the separation of this population from those formed at the substrate's surface). Note: Owing to the extensive amount of energy required to produce a core hole, such mechanisms would result in the fragmentation of molecular emissions.

The influence of core holes on secondary ion formation/survival is, however, not limited to that described by the Kinetic Emission model. Indeed, other processes can include, but are not limited to, the formation of anti-bonding states, which can then lead to the stimulated desorption of various atoms, ions, and/or molecules from various surfaces, and excitation of valence states, which can then de-excite back into their original levels or other levels dependent on the substrate in question.

Stimulated desorption, whether photon, electron, or ion induced, is an inelastic sputtering process as it is the energy associated with the formation of a core hole that results in the emission of the element in question. Indeed, the formation of $F^+$ and $Cl^+$ ions on electron irradiation of Aluminum and Silicon surfaces is accepted to arise through core hole formation followed by ejection through the Coulombic repulsion induced. In the case of ion-irradiated surfaces, it has been suggested that stimulated emission arises from Auger electrons formed in relatively distant neighbors (Williams 1981).

The excitation/de-excitation of valence electrons, whether core hole induced or otherwise, are discussed in Section 3.3.2.3.2.

*3.3.2.3.2  Valence-Level Excitation*  Core-hole formation can also lead to excitation of valence electrons. These electrons can then relax back into their original state or some other state. Which level they relax into depends on the electronic structure of the system and the initial bonding present. Note: Such transitions are the basis of Raman spectroscopy and explain the core-level satellite features seen in X-ray Photoelectron Spectroscopy (XPS). This is particularly evident in the first row transition metal oxides, with the type of transitions appearing to fall into one of the two categories. Which category depends on whether the oxide displays what are referred to as *Mott-Hubbard transitions* ($V_2O_5$ is an example) or *Charge transfer transitions* (CuO is an example). Schematic illustrations of the transitions believed to be induced on core-hole formation in $V_2O_3$ and CuO are shown in Figure 3.36(a) and 3.36(b).

The underlying reason as to why oxides such as CuO are believed to suffer charge transfer can be traced back to their optical properties or more precisely their

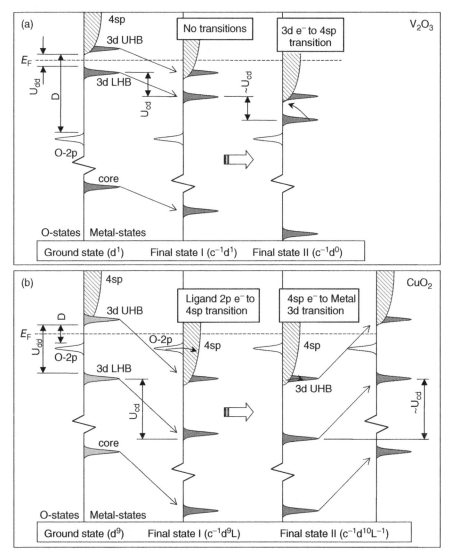

**Figure 3.36** Schematic energy diagram of the electron transitions believed to result in the final states following photoelectron emission from (a) Vanadium initially in its ground state in $V_2O_3$ and (b) Copper initially in its ground state in CuO. Ligand levels are portrayed as white (2p), whereas the metal levels are portrayed as gray (3d) and hatched (4s). The electronic transitions, depicted by the curved arrows, proceed to fill levels below the Fermi edge/vacate levels above the Fermi edge (Aufbau principle). All levels are arbitrarily referenced to a common Fermi edge, and all definitions are described in the text. Reproduced with permission from van der Heide (2008) Copyright 2008 Elsevier.

band gaps ($E_g$). This is asserted on the basis that the transitions believed responsible can be represented as $d^9L \rightarrow d^{10}L^{-1}$ (Imada et al. 1998) with the nomenclature covered in Section 2.2.1.2.1. The energy between the $d^9L$ and $d^{10}L^{-1}$ states is generally represented by the theoretical parameter $\Delta$ (~4 eV for CuO from Zannan et al. 1985). The difference between the band gap energy and $\Delta$ arises as $\Delta$ does not account for hybridization or transitions into the 4sp band. Optical methods are used in defining the band gap energy as core holes and their effects are not introduced, and these oxides have band gap energies within the ultraviolet range (<6.7 eV). Note: The 4sp band is often left out of such descriptions, as well as many *ab-initio* calculations used to model these *many body* effects (many electrons), as the simplified two-level approximation ($d^9L \rightarrow d^{10}L^{-1}$) appears sufficient. In other words, an electron in the delocalized 4sp band has essentially the same effect as if it were to remain in the O-2p band. In reality, the transfer appears to proceed via the 4sp band (Imada et al. 1998).

Not discussed thus far is the effect of the *on site Coulomb repulsion* of electrons within a specific band (Mott 1949; Hubbard 1964). This effect splits the 3d band into what is referred to as a Lower Hubbard band (LHB) and an Upper Hubbard Band (UHB), with electrons in the LHB becoming more localized, whereas those in the UHB becoming more delocalized. The latter can also rise above the Fermi edge. The energy difference between these bands is generally represented by the theoretical parameter $U_{dd}$ (~7 eV in the case of CuO). As the value of $U_{dd}$ decreases slightly with the number of electrons in the 3d band while the value of $\Delta$ increases, transitions between these states are believed to become the dominant factor in defining the band gap energy in the early transition metal oxides (Zannan et al. 1985; Imada et al. 1998). When optically induced, this transition can be represented as $d_i^n + d_j^n \rightarrow d_i^{n-1} + d_j^{n+1}$ where the subscripts $i$ and $j$ refer to adjacent metal ions and $n$ the initial 3d population. Core holes will modify this transition to: $c^{-1}d_i^n + c^{-1}d_j^n \rightarrow c^{-1}d_i^{n-1} + c^{-1}d_j^{n+1}$, with the $c^{-1}d_i^n$ and $c^{-1}d_i^{n-1}$ configurations. Note: It can be argued that these transitions proceed via the 4sp band as this would delocalize these electrons before being trapped on an adjacent metal ion site.

As Mott–Hubbard transitions appear to define the value of band gap energies in the early transition metal oxides and charge transfer defines the band gap energy in the late transition metal oxides, the former are referred to as Mott–Hubbard compounds, whereas the latter are referred to as charge transfer compounds (Zannan et al. 1985). Compounds with similar $\Delta$ and $U_{dd}$ values exhibit optical properties that lie between these extremes. These are referred to as *intermediate compounds*, with $Fe_2O_3$ and MnO being two examples (Zannan et al. 1985; Imada et al. 1998; Hufner 2003). Of note is the fact that 3d electrons are lost from photoelectron emitting ions suffering Mott-Hubbard transitions, whereas 3d electrons are gained by photo-electron emitting ions suffering charge transfer. Also, $U_{cd}$ reduces the energy needed for charge transfer (this does not appear to effect Mott-Hubbard transitions as strongly).

The extremely short lifetimes of such excitations are, however, expected to preclude the possibility that the above processes could affect the formation/survival of secondary atoms/ions formed within the same linear collision cascade sequence (see Section 3.2.1.1). In other words, any such excitations formed during the course of the linear collision cascade responsible for the emission of a specific atom/ion is expected to dissipate well before this specific atom/ion were to depart even a small distance from the effected surface.

If, on the other hand, the sputtered atom/ion were to result from a different collision sequence from that responsible for the excitation, or the excitation lifetime is prolonged for one reason or another (one possible outcome being the formation of an exciton), then there exists a small but finite possibility that the two could interact. For this to occur, the spatial and time spans of the excitation/s must overlap with the emission of the sputtered atom/ion. As this overlap would increase as the emission velocity of the sputtered atom/ion decreases, an increased interaction probability could be expected with decreasing emission energy.

At this point, it is worth considering that overlapping collisions appear to be apparent during polyatomic ion impact on dense matrices (see Section 3.2.1.2), and the fact that valence electron excitation has been recognized as pivotal in the emissions of molecular ions in MALDI (see Section 3.3.4.2). Akin to this are the valence band excitations resulting in thermalization of a solid's surface, as this is one of the underlying premises behind the *heat spike* type models (such models have been used to explain the higher than expected sputter rates, and increased excitations noted in the sputtered atomic emissions from oxidized surfaces relative to their metallic counterparts (Betz 1987)). Also of note is the fact that ion beam impact-induced excitation/s of valence electrons that by themselves result in the emission of atoms/ions. This process, otherwise referred to as stimulated desorption, is typically described as resulting from a thermalization process. Owing to the dynamics involved, the energy distribution of any atoms and ions resulting from such a process would peak at lower values than that from purely kinetic sputtering process. Such excitations also occur at energies as low as the respective band gap energies, which would suggest that little or no energy threshold would be apparent.

Understanding the influence of valence-level excitations should thus provide a more detailed insight into the higher than expected secondary ion yields noted in the atomic emissions from ionic/insulating surfaces relative to metallic surfaces. Further work is, however, needed before a more comprehensive picture can be arrived at.

*3.3.2.3.3  Plasmon Formation in Ag*   The influence of plasmons appears to be unique to specific secondary ion-substrate combinations, with their effect on secondary ion yields noted in only a very few cases. One of the more fully studied cases in which a plasmon-mediated process appears to be active lies in the energetic ion beam-induced emissions of $Ag^-$ from Silver substrates. This case is of interest as:

1. The energy distribution of the $Ag^-$ secondary ions has been reported to peak at a much lower energy than normally observed (Berthold and Wucher 1996; Berthold and Wucher 1997).

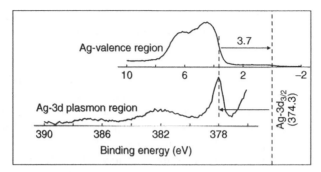

**Figure 3.37** X-ray-induced electron spectra of Silver collected over the valence region (top) and the energy loss region from the Ag-3d peak (bottom). The arrow in the valence region defines electron transitions from the Ag-4d level to vacant states at the Fermi edge, as defined by the 5s-5p band. This transition reduces the energy of a small fraction of the core photo-electrons (the Ag-$3d_{3/2}$ energy loss region is shown) by equivalent amounts. The first plasmon is depicted by the arrow in the opposite direction in the core-level spectra. Reproduced with permission from van der Heide (2005) Copyright 2005 Elsevier.

2. These emissions appear to exhibit a negative velocity dependence when produced through keV Cs$^+$ sputtering conditions (van der Heide 1999).
3. A strong work–function dependence is not noted in the emissions produced through keV Cs$^+$ sputtering conditions (van der Heide 2005).
4. Silver displays a low surface plasmon energy (see section 2.2.2.4.3).

The reduction in the plasmon energies relative to that implied by the surface plasmon variant of Relation 2.8 (see Section 2.2.2.4.3) arises from the fact that the interband transition energy for Silver (that between the 4d and 4p-5s levels) approaches that of the free electron plasmon energy. As a result, long-lived plasmons are predicted to form in Silver at energies equivalent to the interband transition energy (~3.7 eV). This is observed in XPS studies that show equivalent separation between the 4d peak edge to the Fermi edge (3.7 eV) and the position of the first plasmon peak relative to the Ag-$3d_{3/2}$ peak (3.7 eV) as illustrated in Figure 3.37 (van der Heide 2006).

Such a process can then be used to explain the above trends using the arguments depicted in Figure 3.38. In short, the process would be as follows; long-lived plasmons would allow the Silver 5s level within the substrate to be populated more efficiently than would otherwise be expected. Populating levels closer to the Fermi edge would enhance the probability of resonance charge transfer of electrons to departing Silver atoms with the enhancement increasing the longer the Silver atom spends in the region close to the Silver surface (this would explain the negative velocity dependence observed).

Such a mechanism could also be applied to the other group IB elements (Copper and Gold) and would explain the greater deviations noted in the velocity

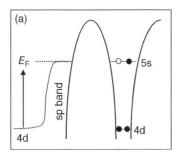 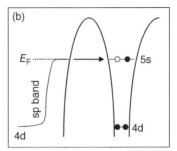

**Figure 3.38**   Pictorial illustration of the mechanism believed responsible for negative Silver ion formation through (a) plasmon excitation followed by (b) charge transfer to a vacant 5s level in the departing Silver atom. Reproduced with permission from van der Heide (2005) Copyright 2005 Elsevier.

distributions of the lower energy negative secondary ions of these elements, particularly when derived using $Cs^+$ primary ions (see Figures 3.30 and 3.34).

### 3.3.3   Models for Atomic Secondary Ions

Although an extensive amount of research has been carried out on the fundamentals of atomic secondary ion emission, there is still not yet a complete accepted understanding of the active processes. In addition, there still remains conjecture concerning the accuracy/validity of some of the published data.

What is generally accepted is that resonance charge transfer (see Figure 3.18(a)) is the primary process responsible for atomic secondary ion formation/survival. This charge transfer occurs between a sputtered particle and the solid's surface as the sputtered particle departs the surface, i.e. during or shortly after the bond is broken between the two. The velocity dependencies observed for the higher emission energy secondary ions (see Figure 3.30) tend to exhibit trends that agree with this assertion.

Some secondary ions with lower emission energies/velocities exhibit deviations from the strong velocity dependencies. This has resulted in the suggestions that additional excitation-based mechanisms whether in the form of heat spikes (resulting from inelastic energy transfer produced in nonlinear collision cascades as discussed in Section 3.2.1.2) and/or Auger de-excitation processes occurring within the sputtered particles while at/or close to the surface (see Figure 3.18(c)) may also be contributing.

To further complicate matters, different models assuming the same charge transfer process (resonance charge transfer) have been developed for different substrate types, i.e. a bond-breaking approach was developed for insulating substrates (ionically bound systems), whereas an electron tunneling approach was developed for conducting and semiconducting matrices (metallic or covalently bound systems).

This difference primarily arises from the fact that in ionic matrices, the valence levels are heavily localized around the respective ions making up the ionic lattice, whereas for metals and semiconducting surfaces, the valence electrons are heavily delocalized. As a result, secondary ion formation/survival from ionic surfaces must occur closer to the surface, i.e. over distances comparable to bond lengths, whereas secondary ion formation from metallic and semiconducting surfaces must encompass this bond-breaking effect as well as the longer range effects imparted by the delocalized sea of electrons that blanket the entire surface and that extend out the greatest distance from the surface. As the departing secondary ions experience the effect of this delocalized blanket of electrons last, the trends exhibited by this population should predominantly relay the properties of the sea of electrons, which is otherwise relayed in the work–function of the aforementioned surface.

In an attempt to simplify the foregoing discussions, only a select few models are covered. This starts, for historical reasons, with a brief overview of the *Local Thermal Equilibrium model*. This is covered in Section 3.3.2.1. The *Bond Breaking model* is then discussed in Section 3.3.2.2, followed by the *Electron Tunneling model* in Section 3.3.2.3. For completeness sake, the *Kinetic Emission model* is presented in Section 3.3.2.4 as this appears to be responsible for the production of multiply charged atomic ions from the elements lighter than Phosphorus. Although many other models have also been put forward, only these are covered as the latter three, in particular, represent those currently accepted for the respective systems described.

Some of the models not covered include, as listed in chronological order:

1. The Band Structure model (van der Weg and Bierman 1969)
2. The Work–function model (Anderson 1970)
3. The Molecular model (Thomas 1977)
4. The Surface Polarization model (Williams and Evans 1978)
5. The Perturbation model (Norskov and Lundqvist 1979)
6. The Surface Excitation model (Williams 1979)

As a final note, it should be emphasized that such models are not intended to describe the details of the ionization process in effect but rather to provide a generalized framework for understanding the secondary ion yield variations noted as some parameter is changed, i.e. the work–function, ionization potential, electron affinity, or even the emission velocity in the electron tunneling model. To understand the details of secondary ion emission, i.e. which electrons are going where, and when, would require a full quantum mechanical treatment that would have to be specific to the system under study as well as a complete understanding of the substrate's composition (this would otherwise negate the need for SIMS analysis). Owing to the complexity involved (outside the scope of this text), few such calculations have been carried out.

**3.3.3.1** *LTE Formalism* During the 1970s, it was argued that atomic secondary ions are emitted from a dense plasma that extended over the surface region impacted during knock-on sputtering events. As a local thermal equilibrium was assumed to exist over a sufficiently long period to allow secondary ion formation/survival, the mechanism suggested was termed the *Local Thermal Equilibrium* model or the LTE model for short (Anderson and Hinthorne 1973). This model was conjectured as the trends predicted were found to approximate to actual secondary ion yields from various metals, alloys, and oxides.

The LTE formalism applies Boltzmann statistics (thermodynamic arguments) for defining the ionization equilibrium within a hot plasma. This was argued as secondary ion emission can be viewed as a statistical process. In short, positive and negative secondary ion yields are derived using *Langmuir–Saha surface ionization theory* on the assumption that the dissociative reactions: $M^o \rightarrow M^+ + e^-$ and $M^- \rightarrow M^o + e^-$ describe the formation/survival of the respective secondary ions. The expressions used for defining the secondary ion formation probabilities ($P^+$ and $P^-$ for positive and negative secondary ions, respectively) were:

$$P^+ = \frac{N_+}{N_o} = \left(\frac{1}{N_e}\right) \cdot \left(\left(\frac{2\pi}{h^2}\right) \cdot \left(\frac{(M_+.M_e)}{M_o}\right) KT\right)^{3/2} \left(\frac{(B_+.B_e)}{B_o}\right) \exp\left(-\left(\frac{I}{kT}\right)\right)$$

(3.12a)

$$P^- = \frac{N_-}{N_o} = \left(\frac{1}{N_e}\right) \cdot \left(\left(\frac{2\pi}{h^2}\right) \cdot \left(\frac{(M_-.M_e)}{M_o}\right) KT\right)^{3/2} \cdot \left(\frac{(g_o.g_e)}{g_-}\right) \exp\left(-\left(\frac{EA}{kT}\right)\right)$$

(3.12b)

where $N_o$, $N_+$, and $N_e$ are the densities of atoms, ions, and electrons; $M$ and $g$ represent the statistical weight of the ground state of the atoms, ions, or electrons as defined via the subscripts o, +, −, and e; $h$ is Planks constant; $k$ Boltzmanns constant; $T$ temperature; $I$ the ionization potential; and $EA$ the electron affinity of the respective secondary ion element.

In short, positive ion yields are expressed as an exponential function of ionization potential of the respective element, whereas negative ion yields are expressed as an exponential function of the electron affinity of the respective element. Interestingly enough, the bond-breaking model incorporates the same exponential dependencies on ionization potential and electron affinity.

The oxygen-induced enhancement of positive secondary ion yields is explained as resulting from an increase in the work–function, which then reduces the probability that an electron can overcome the potential barrier required to neutralize positive ions, i.e. the decreasing $N_e$ results in increasing $P^+$. As for the effect of Cesium, the LTE model explains the increased negative secondary ion yields noted as resulting from the decreasing work–function, which, in turn, increases the values of $N_e$ and thus $P^-$.

Although it is now commonly conjectured that a local thermal equilibrium does not effectively describe the ionization/neutralization process/es active in atomic secondary ion formation/survival, the possibility of relating modified variations of the LTE concept to the trends resulting from thermal spike-type processes (see Section 3.2.1.2) has also since been proposed. Likewise, LTE-based arguments have attracted interest in modeling-specific molecular secondary ion yields (see Section 3.3.4).

### 3.3.3.2 Bond Breaking Model

The *Bond Breaking* model, first proposed by Slodzain and Henneyuin (Soldzian and Henneguin 1966), was introduced to explain the secondary ion emissions from ionic solids. Williams (Williams 1979) then extended this model to compounds exhibiting partial ionic character. A more quantitative formalism was subsequently proposed by Yu and Mann (Yu and Mann 1986; Mann and Yu 1987).

This model presently represents the accepted concept for describing secondary ion formation/survival from ionic/insulating substrates. Similar to the electron tunneling model, a resonance charge transfer process is assumed to occur as the bond between cations and anions making up the lattice break. Note: Valence electrons in ionic lattices are highly localized around the respective anions (negatively charge lattice ions) with the cations (positively charged lattice ions) being electron deficient.

The electronic interaction occurring as a cation departs the surface was stipulated as being analogous to ion-pair dissociation in the gas phase, with the latter having been described theoretically and experimentally by Laudau, Zener, and Stuckelberg (Landau 1932; Zener 1932; Stuckleberg 1932). This describes the interaction as resulting from the crossing of the diabatic potential energy curves for the covalent pair stabilized by being in an excited state and the ionic pair. This describes the equilibrium bonding state with the lowest energy defining the bond distance. Note: Diabatic means that there is no electronic interaction between the two covalent and ionic pairs.

An example of the Diabatic potential energy curves for the NaCl molecule is shown in Figure 3.39(a). This suggests that as the Sodium atom/ion separates from the Chlorine atom/ion, the potential curves representing the neutral and charged

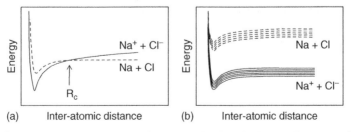

**Figure 3.39**   Potential energy curves of (a) the NaCl dimer in the gas phase as explained by the Laudau, Zener, and Stuckelberg model and (b) a Sodium atom departing an NaCl surface.

pairs cross at some distance defined as $R_c$. At this crossing point, electronic interaction in the form of resonance charge transfer is presumed to occur. This interaction, which results in the neutralization of some portion of the separating ions, has a probability defined by the value of $R_c$ and the time the ion spends in this interaction zone. The interaction time in turn is dependent on the velocity of the separating ions with relative to each other. Laudau, Zener, and Stuckelberg also pointed out that the transition probability is dependent on the wave functions and the shape of the diabatic curves at $R_c$. At infinite distance, the ionic curve lies above the covalent curve by 1.5 eV, as defined by the ionization potential of Sodium (5 eV) and the electron affinity of Chlorine (3.5 eV).

Secondary ion emission, however, concerns the removal of charged particles from a solid's surface. To contend with this, the bond-breaking model assumes the removal of a positively charged Sodium ion from the NaCl solid. This requires an energy that is equal to about half the lattice energy ($\sim$4 eV). The removal of a neutral Sodium atom, on the other hand, can occur only if an electron is left behind on the surface at the vacancy site with a lifetime close to the sputtering time, i.e. $10^{-13}$ s. This requires an energy that is close to the band gap energy (8.5 eV). The resulting diabatic energy curves for this system are shown in Figure 3.39(b). Of note here is the fact that these curves do not cross. This would imply complete suppression of the neutralization of outgoing Sodium ions, hence the very high $Na^+$ secondary ion yields commonly observed. Indeed, complete suppression of the neutralization process is implied for all ionic substrates displaying band gaps greater than half the lattice energy.

Although the situation for charged particle emission from compounds displaying partial ionic character is more complicated, Williams (Williams 1979) contends that the Laudau, Zener, and Stuckelberg concept can still be applied, albeit on a much more localized manner. Likewise, the oxygen-induced positive secondary ion yield enhancement noted from many metals has been explained as arising from the increase in the crossing distance $R_c$, which reduces the neutralization probability of outgoing positive secondary ions.

The Laudau, Zener, and Stuckelberg relation for positive secondary ions is expressed as:

$$P^+ \approx G \exp\left(-\left(\frac{(2\pi H^2{}_{12})}{(v_\perp |a|)}\right)\right)_{R=R_c} \tag{3.13}$$

where $P^+$ is the secondary ion yield, $G$ the ratio of degeneracies of $M_+$ and $M_0$ (the ionized and neural cation states), $H_{12}$ the transition-matrix element, $v_\perp$ the emission velocity normal to the surface at $R_c$, and $a$ the difference in the first derivative of the potential energy curves.

The Laudau, Zener, and Stuckelberg relation predicts a direct relation between $P^+$ and $v$ at $R_c$. Indeed, at high emission energies (low $1/v_\perp$), this approaches an exponential-like trend with $v_\perp \sim (2E/m)^{1/2}$. At low emission energies, $P^+$ converges to a constant finite value with $v_\perp \sim [2(1-EA)/m]^{1/2}$ with the potential of a power law dependence being introduced at intermediate emission energies.

Although the trends implied by the bond-breaking model appear consistent with observed secondary ion yield trends, calculations of ionization probabilities to an acceptable degree of accuracy will, however, only be arrived at once a more detailed interpretation of the relevant parameters are derived.

Lastly, many of the concepts used in the bond-breaking model show similarities to both the LTE formalism and the electron tunneling model. In the case of the LTE formalism, this is noted as the bond-breaking concept reproduces the same Boltzmann-like exponential dependence on the ionization potential. As for the electron tunneling model, the dependence of the neutralization probability of the outgoing ion on the work–function is similar to the effect the energy of the trapped electron in the cation vacancy site has on the neutralization of the departing cation.

#### 3.3.3.2.1 *Electron Tunneling Model*  The *Electron Tunneling* model was first proposed by Yu (Yu 1978) and further developed (Yu 1981; Yu and Lang 1986) to explain secondary ion emission from metallic and semiconducting solids. This model presently represents the best-accepted concept for describing secondary ion yield variations from conducting and semiconducting substrates under energetic ion impact.

The electron tunneling model describes a one-electron resonance charge transfer process between spin independent states situated around the Fermi edge at the solid's surface (empty and filled states) and states of the same energy within the departing ion/atom (again these may be empty or filled). As resonance charge transfer can proceed in either direction, ionization and neutralization can result, with the outcome dependent primarily on the position of the Fermi edge and the ionization or affinity levels of the departing atom/ion.

Global parameters (the Fermi edge, which is defined by the solid's work–function) are used as there exists a *sea* of valence electrons present in conducting and semiconducting solids that interact as a whole. Thus, if one electron is moved, another replaces it and so forth (this also screens any excitations present and causes their fast decay). The ionization/neutralization can then be described by the interaction of the ionization or affinity levels of the departing atom/ion with a sea of those electrons whose distributions extend out the greatest distance from the solids, i.e. those that reside closest to the Fermi edge. Note: These wave-functions extend a greater distance than the valence electrons in ionic compounds.

As is illustrated in Figure 3.40, complications arise when trying to describe the influence of the surface potential on the position of the ionization or affinity levels of the departing atom/ion. Recall: The ionization and affinity levels are influenced by the electronic structure of the surface as discussed in Section 2.2.2.3, with the influence effectively approximated via Relation 2.3 until molecular orbitals are formed. The formation of molecular orbitals (bonding) starts to occur as the distance between the nuclei of the atoms defining the solid's surface and the departing atom/ion decreases below $\sim 1$ nm (Note: Typical bond lengths are in the 0.2–0.3 nm range).

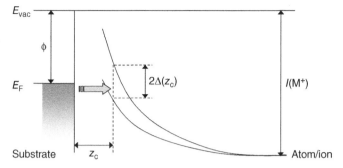

**Figure 3.40**    Pictorial illustration (energy diagram) describing the neutralization of a positive ion on departing a conducting or semiconducting surface as surmised by the electron tunneling model. Note: The direction of electron transfer, as represented by the large arrow, is defined primarily by the positions of the Fermi edge ($E_F$) and the ionization level at a distance at which the interaction occurs ($z_c$). The probability, on the other hand, is defined by the width of the ionization level ($2\Delta$) at $z_c$ as well as the emission velocity normal to the surface ($v_\perp$).

As electron transfer will occur only when the energy of the ionization or affinity levels of the departing atom/ion coincide with the Fermi edge, other parameters come into play such as:

1. The distance ($z$) between the departing atom/ion and the surface when coincidence occurs ($z_c$). This affects the likelihood of electron transfer
2. The velocity ($v_\perp$) at which the departing atom/ion moves away from the surface (this affects the time over which the departing atom/ion resides in the interaction region)
3. The deviation of the ionization or affinity levels in the departing atom/ion from Relation 2.3 when bonding occurs, i.e. when quasi-molecular orbitals start to form

Indeed, the closer the departing atom/ion to the surface is, the more likely electron transfer will occur, assuming the energy levels are in coincidence. This is typically expressed as:

$$\Delta(z) = \Delta_o \exp(-\gamma z) \tag{3.14}$$

where $\Delta(z)$ and $\Delta_o$ define the broadening of the ionization or affinity level in the departing atom/ion as a function of distance ($\Delta(z)$ is at some distance and $\Delta_o$ is the initial value) and $\gamma$ is a parameter termed the characteristic length ($\sim 0.2$ nm). This parameter will depend on the specifics of the bonding present. The exponential dependence arises from the fact that electron wave functions drop off exponentially with distance from the respective nuclei. This also implies that the interaction between the surface and the departing atom/ion varies as an exponential function of the emission velocity (this controls the length of time the departing atom/ion spends in the interaction zone, i.e. that in which coincidence occurs).

Assuming the energy of the ionization or affinity levels of the departing atom/ion coincide with the Fermi edge, the probability of electron transfer (this can occur in either direction depending on the relative population of electrons in the interacting states) can thus be expressed as:

$$P(\pm) \cong \exp\left[\frac{-2\Delta\ (z_c)}{(-\hbar\gamma v_\perp\ (z_c))}\right] \tag{3.15}$$

where $\hbar$ is Planks constant divided by $2\pi$ and $v_\perp$ the emission velocity normal to the surface.

There, however, remain difficulties in defining some of the parameters used in Relation 3.15. As a result, any changes in the secondary ion yield ($\Delta Y(M^+)$ and $\Delta Y(M^-)$ where $M^+$ and $M^-$ represents the respective secondary ion) have been rewritten as:

$$\Delta Y(M^+) \propto \exp\left[\frac{-(I-\phi)}{\varepsilon_p}\right] \tag{3.16a}$$

$$\Delta Y(M^-) \propto \exp\left[\frac{-(\phi-\mathrm{EA})}{\varepsilon_n}\right] \tag{3.16b}$$

This signifies a strong dependence on the Fermi edge, as represented by the substrates work–function, and the ionization and affinity levels of the departing atom/ions, as represented by the parameters, $I$ or $A$. All other parameters in Relation 3.15 are then combined into system-specific parameters termed, $\varepsilon_p$ and $\varepsilon_n$. These relations, however, only prevail if $I < \phi$ or EA $> \phi$. Of note is the fact that these relations (Relations 3.16a) and (b)) are in close agreement with Relations 3.10(a) and (b).

The strongest evidence for the validity of the concepts used in the electron tunneling model is noted in the change of secondary ion yield observed when either of the above restrictions ($I < \phi$ or EA $> \phi$) are crossed. For example, if the work–function of a substrate is reduced such that the Fermi edge rises above the ionization level of electropositive elements such as Cs (this has an ionization level that is above the position of the Fermi edge of most all substrates), then complete neutralization of the departing $Cs^+$ ions should then occur.

As is illustrated in Figure 3.31, the above trends are observed, i.e. when a Titanium substrate is sputtered for a short time with $Cs^+$ ions, a drop in the recorded intensities is noted. Recall: Sputtering of conductive or semiconductive substrates by $Cs^+$ primary ions results in the decrease of the substrate's work–function (owing to the finite fraction of $Cs^+$ ions implanted into the substrate). Also relayed in Figure 3.31 is the fact that only Titanium, and not Gold or Copper, exhibits this drop in $Cs^+$ secondary ion emissions following a short sputtering time. This can be explained by the Fact that the larger work–function of Gold and Copper do not allow for the Fermi edge to cross the Cesium ionization level.

In addition, the exponential-like dependence on emission velocity relayed in Relation 3.13 is noted at higher emission energies (low $1/v_\perp$) for many secondary ions. Deviations from this dependence at lower emission energies are, however, not described within this model. These deviations have been speculated to arise as a result of the ascendance of a competing excitation-based process. An example of an expected velocity distribution resulting from a combined electron tunneling/excitation approach is illustrated in the work by Urazgilden (Klushin et al. 1996).

Lastly, the dependence of the neutralization probability of the outgoing ion on the work–function can be viewed as similar to the effect the energy of the trapped electron in the Cation vacancy site has on the neutralization of the departing Cation as described by the bond-breaking model. Both models, however, describe secondary ion formation/survival occurring on very different surfaces (ionic versus metallic) with neither extendable in the intended forms outside the surface for which the model was developed for.

### 3.3.3.3 *Kinetic Emission Model*    The *Kinetic Emission* model proposed by Joyes (Joyes 1963, 1969a, 1969b) was developed to explain the yield trends exhibited by the positive multiply charged atomic ions from elements lighter than Phosphorus. This mechanism describes the formation of multiply charged ions as being formed as a result of the de-excitation of core holes present in the sputtered atomic/ionic populations, i.e. an auto-ionization process.

The core holes responsible are believed to be formed as a result of sufficiently energetic atomic collisions occurring within the collision cascade. As a result, this should display a primary ion energy threshold as is indeed observed ($\sim$900 eV for $Ar^+$ impact on Aluminum surfaces). The threshold is present as the energy transfer from atoms/ions involved in collisions with electrons is an inefficient process. Also of note is the fact that collision threshold energy scales with the masses of the colliding partners. Indeed, those between the atoms of the same mass, otherwise referred to as symmetric collisions, yield the lowest threshold. Collisions between atoms of different masses are referred to as asymmetric collisions. This variation can be traced back to the fact that the efficiency of momentum transfer increases as the atomic masses approach each other (see Section 3.2.1.1 and relations contained within).

The most widely accepted model for the production of core holes is the Electron Promotion model of Fano, Lichten, and coworkers (Fano 1965; Lichten 1967 and 1980; Barat and Lichten 1972). This model describes the formation of Molecular Orbitals (MOs) from the respective Atomic Orbitals (AOs) of the two colliding atoms as they approach each other to distances significantly smaller than those of equilibrium bond distances. As the energies of the MOs formed are defined by the interatomic distance (see Section 2.2.2.4.4 and Figure 2.8), variations in their energies are noted during the course of the collision process. If two diabatic curves (those from two distinctly different levels) were to cross, then there exists the possibility that an electron from one of these levels transfers to the other level. If this occurs as the two atoms are receding from each other (the latter part of the atomic collision process), then there exists the possibility that a core hole is formed and remains in one of the two colliding atoms.

As core holes are short lived, the formation of multiply charged secondary ions can occur only if the collision resulting in the core hole takes place close to the surface, with the core hole residing in the singly charged ion being sputtered. Multiply charged ion formation then occurs as the core hole relaxes (via an Auger process) once the ion travels some distance from the surface. Note: If the de-excitation process were to occur while the singly charged ion was within the solid or close to the surface, neutralization of the multiply charged ion would then ensue. Owing to the relation $E = \frac{1}{2}mv^2$, this limits this process to the emissions from elements lighter than Phosphorus, i.e. those with higher emission velocities. Indeed, as noted in Figure 3.6(a), Auger emissions arising from the above process have been observed from neutral atomic and singly charged atomic ions from Aluminum, and so on. The fact that such emissions have been observed from atomic emissions reveals that the above mechanism can also result in singly charged ions. Yields from this process are, however, much less than those observed from other surface-based mechanisms as described by the *electron tunneling* and *bond breaking models*.

Positive multiply charged atomic ions from elements heavier than Phosphorus appear to be formed via a collisional process active in the gas phase, i.e. at distances greater than 10 nm from the surface (van der Heide et al. 1993). This is realized through the observations that these emissions reveal energy-deficient populations in their respective energy distributions when collected under high-extraction fields.

Although mechanisms describing multiply charged molecular emissions are less clear, there does appear evidence that these are formed via auto-ionization (electron emission via Auger de-excitation). These would occur following atomic collisions with other atoms/ions a short distance from the surface, i.e. less than 10 nm (this distance is implied by the respective kinetic energy distributions). Observations of negative multiply charged ions are far more spare, a fact attributable to their shorter lifetimes. Indeed, most such observations concern the larger molecular ions (Gnaser 1999). Molecular and Cluster secondary ion emission models are discussed further in the following section.

### 3.3.4  Models for Molecular Secondary Ions

The possibility that a complete molecule can survive the sputtering event and be emitted in an ionized state has been empirically demonstrated since the early work of Benninghoven (Benninghoven 1973). Also recognized was the fact that emission of pre-formed ions can occur only if there is not a significant amount of energy imparted onto the respective pre-existing molecule during emission. The understanding of the molecular/cluster secondary ion emission is, however, far less clear than that for atomic secondary ion emission. This state of affairs can be attributed in part to:

1. The far greater theoretical complexity associated with describing the emission of bound atoms moving at a slower velocity relative to their atomic ion counterparts
2. The greater difficulty associated with recording sputtered neutral molecular/cluster distributions of the respective secondary ions of interest (that being said, improvements in post-ionization methods are noted Marazov et al. 2006)
3. The far fewer empirical studies carried out in this area. Note: The recent advent of molecular depth profiling and imaging in all three dimensions has, however, spawned new interest in this area

In any case, a range of models have appeared in an effort to provide, at the very least, a qualitative description of the process resulting in molecular secondary ion emission along with reasons for the much narrower kinetic energy distributions exhibited relative to their atomic secondary ion counterparts. These have been derived from the initial assertion that molecular secondary ion emissions arise from either or both direct nonreactive emission (Benninghoven 1973) and/or recombination shortly after ejection (Oechsner 1974). The models believed to be of greatest historical significance include:

1. The *Selvedge* concept introduced by the Rabalais team (Honda et al. 1978; Lancaster et al. 1979; Murray 1981)
2. The *Gas Flow* model formulated by the Michl team (Orth et al. 1982)
3. The *Nascent Emission* model of Plog and Gerhard (Plog and Gerhard 1983a, 1983b)
4. The *Pre-cursor* model of Benninghoven (Benninghoven, 1983)
5. The *Desorption/Ionization* model of Pachuta and Cooks (Pachuta and Cooks 1987)

These along with further discussions on recombination processes common to most of these models are covered in Section 3.3.4.1. Recombination is of interest as this is believed to be responsible for many of the signals often used in SIMS, whether in the analysis of elemental or molecular distributions. As an example of the former, recombination is believed to be the primary mechanism responsible in the formation of Cesium cluster ions used in the analysis of elemental distributions (that in which Cesium combines with some element of interest to provide signals displaying reduced matrix effects). As for the latter, recombination appears to be dominant in the formation of protonated molecular ion emissions, as well as those associated with alkali metal elements, and certain transition metal elements. Note: These typically display significantly greater intensities than their molecular parent ions.

Also of interest are models from closely related fields, such as Matrix Assisted Laser Desorption/Ionization (MALDI) and Electro Spray Ionization Mass Spectrometry (ESI-MS). MALDI in particular attracts attention as this technique records the large molecular ion emissions resulting from the localized deposition of energy occurring on laser irradiation of the solid's surface. Indeed, there appear to be many similarities in the recorded emissions from SIMS and MALDI. Examples of areas of interest include processes described within the *Cluster*-based mechanisms and the *pooling* mechanism. These along with their applicability to SIMS are discussed in Section 3.3.4.2.

As a final note, there exists the possibility of ion formation in the gas phase an appreciable distance from the surface (more than ~10 nm from the surface) as a result of gas phase collisions. However, as populations arising from such collisions appear to be small relative to those that would be formed via one of the above models (a fact relayed through experimentally derived kinetic energy distributions), these processes are not discussed in further detail.

### 3.3.4.1 Models for Molecular Ion Emission in SIMS
As covered in Section 3.3.4, there exist a range of models introduced in an effort to explain the detection of molecular ion emissions resulting from energetic ion impact, albeit if only qualitatively. These models, of which some share many similarities, are described henceforth in a chronological order.

The *Selvedge* concept introduced by the Rabalais team (Honda et al. 1978; Lancaster 1979; Murray 1981) relies on the premise that the surface region from which emission took place experiences an adiabatic expansion following primary ion impact. Indeed, the *selvedge* term was introduced to represent the assumed formation of the resulting diffuse boundary between the solid being sputtered and the gas phase. From this region, it is asserted that complete molecular ions could be emitted once attaining a sufficient energy. Also recognized was the possibility that a sizable molecular ion population could experience fragmentation within the *selvedge*, and that recombination reactions between the atomic and the molecular emissions (neutral and ionized species) could ensue. This concept was initially developed to explain the prevalence of positively charged molecular ion emissions from alkali salt surfaces.

The *Gas Flow* model introduced shortly thereafter by the team of Michl (Orth et al. 1982) was developed in an effort to explain the cluster ion emissions recorded from frozen matrices. This model uses many of the same assumptions used in the *selvedge* concept, but divided into two time regimes ($<10^{-12}$ s and $>10^{-12}$ s). The first describes the release of fragment ions resulting from a classical collision cascade. The second describes the *dislocation* of an excited region akin to the formation of a heat or thermal spike within the sputtered volume. The emissions of pre-formed ions is thus viewed as resulting from what appears to be an outward moving flow of gas, hence the name, from the substrate's surface.

The *Nascent Emission* model formulated by Plog and Gerhard (Plog and Gerhard 1983a, 1983b) specifies that molecular ions are formed as a result of the sputtering of neutral molecular species (nascent species), which then dissociate on their departure from the surface. As the ions are specified to form as a result of a nonadiabatic

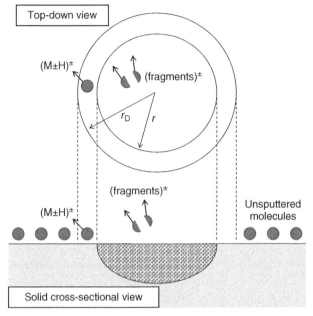

**Figure 3.41** Pictorial illustration of the *Pre-cursor model* of Benninghoven (adapted from Benninghoven 1983). This displays the prevalence of molecular emissions as arising from the periphery of the sputtered region (that defined by circles of radius of between $r$ and $r_D$ as is also signified by the vertical dashed lines). Emissions from within some radius $r$ fragment because of the greater energy imparted.

dissociation process outside the surface region, there exists a low probability of neutralization. The strong mass dependence also noted stems from the momentum transfer occurring during the collision cascade. This model was initially developed to explain the presence of positively charged molecular ion emissions from metal oxide and metal chloride surfaces.

The *Pre-cursor* model introduced by Benninghoven (Benninghoven 1983) concerns the emission of larger molecular ions. Sputtering is assumed to occur from a surface region exhibiting an energy gradient that dissipates rapidly on moving away from the point of primary ion impact. As depicted in Figure 3.41, molecules within some radius of the point of impact are assumed to attain a significant amount of energy such that only fragment secondary ions and atomic secondary ions are noted. Molecules exiting from this surface at and just outside of this boundary are, however, allowed to survive as a lesser degree of energy is imparted, i.e. these attain just the right amount of energy to force their departure from the surface, but not enough to result in their fragmentation. Note: This has become the most accepted model for explaining molecular ion emissions from most any surface in Static SIMS.

The *Desorption/Ionization* model proposed by Pachuta and Cooks (Pachuta and Cooks 1987) introduced the concept of isomerization. In short, this introduced the idea that translational and rotational modes of excitation are introduced through the energy transfer resulting from a collision cascade. This is asserted to result in pre-formed molecular and fragment ion emissions, with subsequent additional fragmentation and/or recombination then possible within the gas phase. Similarities with previous models are noted in that this can be thought of as a quasi-thermal process emission followed by selvedge reactions.

*Recombination* reactions, common to many of the above models, can occur between various sputtered atoms, ions, and/or molecules a short time following the sputtering event. The narrow low-energy peak in the kinetic energy distributions of the respective populations also reveal that such reactions must occur within a short distance from the surface, i.e. within ∼ 10 nm. Positive and negative secondary ion formation can occur depending on the atoms, ions, molecules and surface in question. Positive ion formation is, however, of greatest interest as this can increase the analytical capabilities of SIMS, depending on the atom, ion, molecule, and surface properties.

Positive ion formation through recombination is commonly referred to as *Cationization*. This describes the association of a cation, whether it is in the neutral or ionic state with another neutral atom or molecule/cluster to form a positively charged ion. The likelihood that association can occur is noted in Molecular Dynamics simulations, with an example depicted in Figure 3.42. These also show di-cationization as being possible. Note: The cation does not have to be an element preferring to reside in the positive ion state. Examples of elements noted in cationization reactions include, but are not limited to, Hydrogen, the alkali metal elements (Lithium through Cesium are the most common), and various transition metal elements with the more common being Copper, Silver, and Gold.

There then remains the question of how the respective positive secondary ions are formed. Two mechanisms prevail in the literature. The first being the association of sputtered neutral elements or molecules of interest with cations sputtered as a positive ion. This then results in a positively charged cluster displaying a sufficient lifetime to allow its detection and the association of sputtered neutral elements or molecules of interest with a cation sputtered as a neutral excited atom. The second describes de-excitation following association via electron emission (an autoionization process) which then results in a positively charged cluster displaying a sufficient lifetime.

The former appears to be applicable for cations displaying high positive secondary ion yields. As this has been noted for $Na^+$, $K^+$, and $Cs^+$ (see Figure 3.26(a)), it has become the mechanism accepted as primarily responsible for the generation of the $MCs^+$ and $MCs_2^+$ populations used in studying elemental distributions in SIMS, where M represents the element within the matrix of interest (Gnaser 1999). Evidence for this mechanism lies in the reduced matrix effects exhibited by these populations relative to the lone matrix (M) ions, and correlation between cationized ion yields and the atomic polarizability of the of element M.

The latter appears to be applicable for cations displaying relatively low positive secondary ion yields. As this is noted for the transition metal ions inclusive but not limited to $Cu^+$, $Ag^+$, and $Au^+$ (see Figure 3.26), this has been argued to be

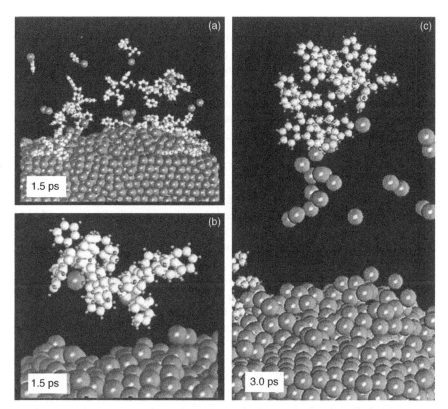

**Figure 3.42**    (a–c) Snapshots of Molecular Dynamics simulations at the times specified following 5-keV $Ar^+$ impact of the association of Silver atoms (larger dark gray spheres) with Polystyrene tetramers (smaller light gray spheres deposited on a Silver substrate) released through a concerted emission process. Reproduced with permission from Delcorte et al. (2002) Copyright 2002 American Chemical Society.

the primary mechanism responsible for positive ions formed via cationization of organic molecules deposited on Copper, Silver, or Gold substrates (Delcorte et al. 2002). Evidence for this mechanism lies in correlations noted between the molecular yields and cation excitation probabilities and the increased ion yields noted from the cationized molecules relative to the lone cations.

For the sake of completeness, anionization is the analogous process where ions preferring to reside in the negative ion state are complexed with a neutral atom or molecule/cluster to form a negatively charged molecular ion.

### 3.3.4.2  Models for Molecular Ion Emission in MALDI
As introduced in Section 3.3.4, MALDI is a closely related analytical technique to SIMS. This is realized as MALDI describes a localized adsorption of energy that induces phase-like explosions (a coordinated motion of many substrate atoms) into the gas phase. Although active on larger volume and time scales, there are similarities

between this and the simulations displayed in Figure 3.11, which describe a cluster ion ($C_{60}^+$)-induced fluid-like emission of intact molecules into the gas phase.

Indeed, recent computer simulations of secondary ion emission have revealed the presence of primary ion-induced cooperative motion between many substrate atoms/molecules, which can propel a complete molecule off the surface without the molecule attaining sufficient energy so as to cause its fragmentation, whether during the emission process or within the gas phase (see review by Garrison and Postawa 2008). Thus, if the molecule being sputtered can support a charge (larger molecules provide for increased charge stabilization), then there exists the possibility that emission in the ionized state can result.

However, following the discussions on atomic secondary ion emissions, it is generally assumed that the slower moving emissions face a much greater neutralization probability on their final departure. Indeed, one of the models debated in MALDI describes the emitted ion population as "lucky survivors." This model, referred to as the *Cluster model*, assumes the release/desorption of pre-formed cations, the large majority of which are neutralized on their departure. This model is actually a group of models developed within the MALDI community that describes the release of largely preformed ions via a desorption/ablation process (Knochenmuss 2010; Garrison and Postawa 2008). Pre-formed ions can be envisaged as those already existing within ionic matrices and possibly formed within the outgoing plume.

As an example, the ions within the salt-like crystals of $(NH_4)_2.SO_4$ are stabilized through the ionic bond (charge transfer) between neighboring cations ($NH_4^+$) and anions ($SO_4^{2-}$). Once separated (driven apart by the collective motions resulting from the sputtering event), their short lifetimes may be sufficient to allow for their detection. Indeed, signals from their singly charged counterparts are noted in positive and negative ion spectra collected under Static SIMS conditions.

A similar analogy can be drawn from covalent solids (for example, organics) in which intermolecular forces holding the respective molecules together are much weaker than intra-molecular forces. Indeed, as noted in Figure 3.16, the onset of emissions appears to scale with these energies. For ions to be formed during emission from such solids would, however, require the stabilization of a charge on the molecule, a fact that becomes increasingly likely the larger the molecule becomes.

Such reactions are speculated within the various MALDI cluster models, albeit occurring over much longer time scales (up to a micro second) than in SIMS (within tens of pico-seconds). The reactions occurring in the second step can take the form of electron, proton, or even ion transfer, with a thermodynamic equilibrium then envisaged to be dependent on the relative stabilities and mobilities of the respective charge carriers. In general, more cations are detected than anions (noted in both MALDI and SIMS), a fact that can be traced back to the relative stabilities of the respective molecular ions and the mechanisms for ion formation (cations can also form via proton attachment).

The possibility that electronic excitations, whether present in the substrate or departing emissions, play a role in molecular ion formation/survival in SIMS has also been recognized. Indeed, as first mentioned in Section 3.2.1.2, a significant amount of energy can be imparted during the sputtering event, whether from inelastic processes leading up to molecular ejection, or within the plume formed as a result of the phase explosion believed present in heat spike-based sputtering processes. Furthermore, as stated in Section 3.3.2.3, excitations have been used to explain specific cases in which higher than expected atomic ion yields have been observed. These include, but not limited to:

1. the formation of multiply charged atomic ions from elements lighter than Phosphorus,
2. stimulated desorption of $F^+$, $Cl^+$ from Aluminum and Silicon surfaces,
3. Plasmon-assisted emission of $Ag^-$ ions from Silver surfaces.

Owing to the lower energies involved, the stimulated desorption and plasmon-assisted routes appear to be the only possibilities that may play a role. In addition, these are speculated to be limited to highly specific cases, i.e. the emission of certain organics supported on metallic substrates. The mechanism responsible for multiply charged atomic ions from elements lighter than Phosphorus (the Kinetic Emission model) is discounted as this requires too great an energy to allow for the survival of complete molecular ions, i.e. energetic atomic collisions are required to introduce the core hole needed, and core hole de-excitation releases energies more than sufficient to cause fragmentation of any surviving molecules.

In the case of MALDI, the suggestion that ionized molecular emissions can form via excitations of valence band electrons has been postulated in what is termed the *Pooling mechanism*. As the name suggests, this mechanism describes the pooling of energy within a sufficiently localized area such as to induce the ionization of a molecule. It is also assumed that the molecules exist within the outgoing plume. The likelihood that two adjacent molecules will be in the required excited states within the time frame needed is, however, small. To contend with this issue, the possibility that the excitation is mobile was introduced. Mobile excitations of this form are otherwise termed *excitons*. Indeed, exciton motion and the pooling mechanism are related in that both describe the motion of energy via intermolecular interactions.

Experimental evidence for such a mechanism is relayed in time-delayed two photon irradiation experiments of 2,5-dihydroxybenzoic acid (DHB) with the energy of individual UV photons being less than that needed to cause the ionization of DHB, but the sum of the two energies being enough (Ludermann et al. 2002; Setz and Knochenmuss 2005). DHB was used as excitons can diffuse great distances with this matrix before being quenched (the time per exciton hop between molecules was known to be $\sim 50$ ps with the lifetime being $\sim 1$ ns). The observation of significant MALDI signals as a function of the delay time between the two photons also allowed for the intermolecular dynamics to be studied. Irradiation with either of the photons alone did not result in any appreciable MALDI signals. As a final note, numerical evaluation of the stepwise pooling mechanism identified in the above experiments has resulted in providing quantitative MALDI predictions (Knochenmuss 2002).

The pooling mechanism, however, only describes the first step. What follows are further interactions that can occur in the expanding plume. This second step includes the interaction of electrons, protons, and ions with each other which can result in the neutralization of ions already formed and/or the formation of new ions. As with SIMS, it is the final step that dictates the recorded trends. As the recorded emissions follow a Boltzmann-like distribution, thermodynamic type approaches have been applied. Indeed, the LTE concept appears well suited to this scenario as reactions taking place among electrons, protons, and ions within an expanding plume more closely match that of a dense plasma, as outlined in Section 3.3.3.1.

To extend this multistep mechanism to describe secondary ion emission of complete molecules along with their fragments in SIMS would, however, requires the assumption that a heat spike ensues during the sputtering of matrices in which sufficiently diffusive excitons can be formed. Heat spikes would have to be assumed as this would provide conditions similar to the plume noted in MALDI, whereas highly diffusive excitons must be assumed to allow energy transfer as described within the context of the pooling mechanism.

## 3.4   SUMMARY

There now exists a wealth of information on the mechanisms active in sputtering and secondary ion formation/survival. Although this is much more extensive for atomic secondary ion emission relative to molecular secondary ion emission, a framework describing the mechanisms is now available. In general, this can be considered a two-step process where sputtering precedes the electron transfer process responsible for ion formation and/or survival. Also of note is the fact that quite different mechanisms appear active in the formation of atomic secondary ions relative to the formation of large cluster ions.

In short, directing an ion beam at a solid's surface can result in a range of processes that include, but are not limited to:

1. surface sputtering, radiation damage, crater formation, buried clusters
2. cluster implantation, reflection (scattering), splatting, fragmentation
3. surface melting, alloy formation, plastic deformation, soft landing of clusters.

Which process occurs will depend on the energy imparted while directing the ion beam at the respective surface, the type of ion beam used, and of course the type of surface being irradiated. Of greatest interest in SIMS is surface sputtering.

Surface sputtering describes the act of removing atomic and/or molecular entities from the surface region of a solid. In SIMS, sputtering occurs as a result of directing a focused energetic (within the 0.1–50 keV range) ion beam at the surface of interest. This effectively etches the surface layers off, with the atoms, ions, and molecules making up this surface layer ejected into the gas phase. Sputtering can occur via kinetic means, potential means, or a combination of both.

The most accepted form of kinetic sputtering used in generating atomic emissions is referred to as knock-on sputtering. As the name suggests, this is a kinematic process describing pure momentum transfer occurring during individual collisions as initiated by the incoming ion and extending to many atoms/ions making up the lattice of the solid. This is referred to as a collision cascade, with the billiard ball analogy often used. If there are no overlapping collision sequences within a collision cascade, the process is well defined as an isotropic linear cascade. The degree of energy involved in this form of sputtering rarely allows complete molecules to escape before fragmentation. This, however, does allow for the formation of new molecules, some of which do not normally occur in nature. This is the primary form of sputtering used when atomic secondary ion emissions are of interest. Owing to the level of understanding of the dynamics in effect, these processes can be simulated to a high degree of accuracy.

The emission of unfragmented molecules requires a less energetic form of sputtering, of which there are several forms. The most common form is that occurring when there exists an overlap in the collisions making up the collision cascade. This can then result in the concerted motion of many atoms in the same direction, much like a wave, that can then result in the uplifting and removal of complete molecular units from the surface of interest. If channels for inelastic energy transfer open up, the process can also result in atomic and/or molecular emission through electronic excitation processes. As electronic excitation results in heat, these forms of sputtering are sometimes referred to as a heat spike of thermal spike type sputtering. This is referred to as kinetically assisted potential sputtering.

For completeness sake, there also exists a form of sputtering known as potential sputtering. This describes sputtering as occurring through purely inelastic processes, i.e. without the requirement of momentum transfer. As extensive energy transfer is involved, these typically result in atomic emissions alone. These are rare and tend only to occur on specific highly ionic matrices.

The formation/survival of ions formed as a result of the sputtering process or shortly thereafter is less well understood. What can be inferred from the extensive data that exists is that:

1. The vast majority of secondary ions are singly charged. Of the multiply charged ions, most are of a positive polarity (negative multiply charged ions have very short lifetimes)

2. Secondary ion angular distributions tend to peak around the sample normal with a Cosine function generally noted. Exceptions are noted on crystal surfaces

3. Secondary ion energy distributions for both atomic and molecular species peak at around a few eV. This is generally higher for atomic versus molecular secondary ions

4. Atomic secondary ions display high energy tails that extend to a sizable fraction of the primary ion energy, i.e. to 500 eV or more. This is not noted for molecular secondary ions

5. Secondary ion populations tend to, but not always, exhibit a strong dependence on:

   a. The electronic properties of the parent atomic or molecular species

   b. The electronic properties of the substrate's surface from which they emanate

   c. The polarity of the secondary ions formed

   d. The primary ion species used, specifically how these modify those of the substrate's surface when implanted into the substrate

6. In most cases, no correlation between secondary ion yields and primary ion polarity is noted.

Variations in secondary ion yields (ratio of ions per sputtered neutral), and hence the recorded secondary ion intensities, are contained within what are referred to as matrix effects, inclusive of transient effects. These effects describe the variations noted as the matrix is altered in one form or another, i.e. as some element is removed or introduced within the substrate matrix at levels greater than 1 atomic %. Transient effects describe the variations noted during the initial stages of sputter depth profiling. These primarily concern the variation resulting from the introduction of reactive primary ion species with two common examples being due to the introduction of Cesium (noted when a $Cs^+$ primary ion beam is used) or Oxygen (noted when $O_2^+$ or $O^-$ primary ion beams are used).

As for the introduction of Cesium, an exponential relation is noted between the Cesium-induced surface work–function changes and the atomic secondary ion yields emanating from these surfaces. The introduction of Oxygen, on the other hand, enhances both the positive and negative atomic secondary ion yields particularly on metallic surfaces, with the former being more heavily affected.

Recorded kinetic energy distributions of secondary ions (secondary ion intensity versus emission energy) display a peak at around $2-5$ eV emission energy whereupon they drop in intensity. Molecular ion intensities decrease more rapidly than those for atomic secondary ions. Indeed, the latter can extend out to several hundred eV. Decreased matrix effects are also noted for the higher emission energy secondary ions.

As for velocity distributions, the yields of atomic secondary ions are generally seen to decrease on reducing the emission velocity (an exponential trend is observed at higher emission velocities with some ions showing deviations at lower emission velocities). Also of note is the increased mass fractionation noted as the emission energy/velocity is decreased. As would be expected from $E = \frac{1}{2}mv^2$, this trend coincides with those noted in the respective velocity distributions. Note: The apparent discrepancy between recorded kinetic energy distributions and velocity distributions/mass fractionation arises from the fact that the former includes the sputter yield dependence with emission energy, whereas in both of the latter cases, this sputter yield dependence is removed.

As far as models for describing secondary ion yield variations are concerned, there now exist a range of models, with each applying to specific surface types (ionic versus metallic) and specific secondary ion emissions (singly charged atomic ions versus multiply charged atomic ions versus molecular ions). All have been developed to provide a mechanistic framework for the understanding of the secondary ion variations noted under the various different analytical conditions that can be applied (adsorption of Cesium, Oxygen, etc.) and from the various different matrices that are commonly examined (ionic through metallic).

The *Bond Breaking* model first introduced by Slodzian applies to singly charged atomic secondary ion emission from partially and fully ionic substrates, i.e. those in which the valence electrons are heavily localized around the respective ions making up the lattice. This model assumes the sputtering of pre-existing lattice ions. These can then face neutralization on their outgoing trajectory via resonance charge transfer at a distance $R_c$ from the surface. The extent of neutralization is defined by the crossing of the adiabatic curves of the ground-state ion pairs and the excited-state covalent pair.

The *Electron Tunneling* model introduced by Yu applies to singly charged atomic secondary ion emission from metallic and semiconducting substrates, i.e. those in which the valence electrons exist as a sea of electrons that extend out a greater distance from the respective atoms than those in ionic substrates and hence, blanket the respective surface. This model assumes sputtered atoms face the possibility of ionization/neutralization on their outgoing trajectory via resonance charge transfer at a distance $z_c$ from the surface. The extent of ionization/neutralization and the ion polarity is defined by the relative position of the ionization level (for positive ions) or affinity level (for negative ions) relative to the Fermi edge.

The *Kinetic Emission* model of Joyes applies to the formation of the commonly observed positive multiply charged atomic secondary ions from the elements lighter than Phosphorus. This model assumes core holes generated as a result of violent atomic collisions within the collision cascade survive the emission process such that de-excitation of the respective singly charged positive ion occurs within the gas phase at a short distance from the surface. Note: This only applies to the light elements owing to their greater emission velocities. Positive multiply charged ions from elements heavier than Phosphorus appear to be formed as a result of gas phase collisions. Negative multiply charged ions are less common.

The *LTE formalism* derived using Langmuir-Saha surface ionization theory assumes that ionization/neutralization takes place as a result of excitation of the sputtered surface region. This can be described thermodynamically using temperature parameters. Although this is now not commonly accepted to apply in the case of singly charged atomic secondary ion emissions following a linear cascade, there are areas in singly charged molecular secondary ion emissions where this approach appears applicable, i.e. those in which a plume of rapidly expanding gas occurs as speculated upon in heat spike-type sputtering.

Molecular ion models are less refined. Many of these derive from the concepts of Rabalais, which describes the direct emission of molecular secondary ions that can then experience fragmentation/recombination (association) within the so-called *Selvedge* region between the surface and the vacuum. To explain the detection of larger molecular secondary ions, particularly organics, Benninghoven introduced the *Pre-cursor* model. This describes the generation of an energy gradient on the substrate's surface around the point of primary ion impact. The energy close to this point is high enough to induce fragmentation of pre-existing molecular ions. Molecular ions emitted further away from this point are, however, allowed to survive owing to the energy gradient formed and thus be emitted/detected as complete molecular ions.

The cluster-based models developed within the MALDI community share the assumption that emission of unfragmented molecular ions occurs as a result of the collective motion of many atoms/molecules within the substrate. Neutralization of this population is then postulated to proceed as the molecular ions depart the surface. If this occurs within an expanding plume as a result of electron, proton, and ion interaction, it will be expected to display a Boltzmann-like distribution. The Pooling model is an example of an excitation-based mechanism derived to explain the higher than expected molecular ion yields in MALDI. This describes the pooling of energy within a sufficiently localized region such as to induce the emission of unfragmented or partially fragmented molecules. This population can then be subject to neutralization as a result of electron, proton, and/or ion interaction within the rapidly expanding plume. Such MALDI data has been quantitatively approximated using the LTE formalism.

Recombination reactions taking place just following the sputtering event can take the form of cationization and anionization. Cationization is the most useful of the two. It describes positive molecular ion formation through the association of a cation with the atomic or molecular secondary neutral emission within $\sim 10$ nm from the surface. Both would be uplifted from the surface during the same collision cascade event with similar momenta. The process resulting in positive ion formation may, however, occur via either the direct association of a positively charged cation (as is speculated for the alkali ions) or the association with an excited neutral cation, which then de-excites through the emission of one or more electrons. This is speculated (Delcorte et al. 2002) to occur for specific transition metals such as Copper, Silver, or Gold).

Although this may not be immediately obvious, the wealth of information collected/reported on secondary ion emissions from the various different types of surfaces has resulted in a consistent picture of the general processes active and overall trends expected. This is particularly evident for atomic secondary ion emissions.

The resulting models for sputtering are, however, more mature than those for the ionization/neutralization step. This and the fact that Newtonian mechanics effectively describes most sputtering processes, has resulted in the availability of highly effective simulation packages. Simulation packages describing ionization/neutralization processes are, however, more sparse. This can be partly attributed to the complexity involved in the quantum mechanics-based calculations required, and the fact that an all encompassing picture of the ionization/neutralization process responsible for atomic and molecular secondary ion emission from all substrate types still eludes the SIMS community.

Indeed, the fact that the extent of *a priori* information required in modeling the ionization/neutralization processes active in SIMS far exceeds the information provided by SIMS can be viewed as reason to relegate such understanding to academic curiosity. But, if it were not for such curiosity, many of the technologies that we take for granted today, inclusive of SIMS, would not have come to light.

# PRACTICES

# CHAPTER 4

# Instrumentation Used in SIMS

## 4.1 THE SCIENCE OF MEASUREMENT

As covered in detail in Chapter 3, Secondary Ion Mass Spectrometry (SIMS) derives compositional information from the solid under study by directing a focused energetic ion beam at the solid's surface. The ion beam directed at the solid's surface is termed the *primary ion beam*. On impacting the solid's surface, these ions induce the emission of atoms and molecules via sputtering, a small percentage of which exists in the ionized state. Ions from the sputtered region are then extracted/accelerated to form a beam that is otherwise referred to as the *secondary ion beam*. As first mentioned in Section 1.2.2, the name *Secondary Ion Mass Spectrometry* is derived from the fact that these secondary ions are analyzed by a mass spectrometer, with the mass spectrometer type defining the SIMS instrument category.

To make full use of the capabilities provided by SIMS, an understanding of the principles and practices commonly used in data collection is desirable. As with most micro-analytical techniques, two aspects make for a good result, these being the optimization of the analysis conditions to the signals of interest and pairing the instrumentation to the analysis need.

The instrumentation used in SIMS can be considered *the science of measurement*. The topic is covered within this chapter with a brief outline of the different forms of SIMS, i.e. Static and Dynamic SIMS, first given in Section 4.1.1. This is then followed by an overview of the methodologies used for attaining the required vacuum. Primary ion columns (that used for generating the probe beam) with the different sources available are outlined in Section 4.2.2 followed by a discussion on the various secondary ion columns (that used for collecting and filtering the signal/s of interest) in Section 4.2.3. Note: Much of the development/sophistication experienced by SIMS over the last few decades can be attributed to the understanding, development, and refinement of the hardware and software utilized in generating, filtering, and collecting the secondary ion signals of interest.

As with many techniques, optimization of the analytical conditions to the job at hand can be considered *the art of measurement*. Common aspects associated with the art of measurement pertinent to SIMS are covered in Chapter 5.

*Secondary Ion Mass Spectrometry: An Introduction to Principles and Practices*, First Edition.
Paul van der Heide.
© 2014 John Wiley & Sons, Inc. Published 2014 by John Wiley & Sons, Inc.

## 4.1.1 SIMS in Its Various Forms

Traditionally, SIMS is subdivided into two broad areas according to the volume probed per analytical cycle. An analytical cycle represents a single data collection step, which may be repeated multiple times. This data collection step may be in the form of an intensity data point, mass spectra, or a spatial image. As will be covered in Section 5.1.1.3, multiple such steps are required in the collection of a depth profile. The areas into which SIMS is subdivided are otherwise referred to as *Static SIMS* and *Dynamic SIMS*.

In Static SIMS, data is collected from less than 1% of the outermost monolayer of the solid's surface. In Dynamic SIMS, data is collected from many atomic layers situated at and below the outermost surface. The recent advent of molecular depth profiling using large cluster ions is also opening up what can be thought of as a third area. Technically, this is a form of Dynamic SIMS, but one in which molecular information can be derived as a function of depth (Note: Dynamic SIMS in its original form describes a methodology in which molecular information is typically not available). The definition of Static SIMS is presented in Section 4.1.1.1, whereas the definition of Dynamic SIMS in both its traditional form and that being opened up through large cluster ion impact is presented in Sections 4.1.1.2 and 4.1.1.3, respectively.

Although the instrumentation required and the information content provided in the above forms can be quite different, the basic instrument setup remains essentially the same. The major difference being that Static SIMS does not provide depth profiling capabilities, whereas Dynamic SIMS does.

A schematic of the major components that make up a SIMS instrument is shown in Figure 1.2. This pictorial also illustrates the commonly used data collection modes available when using the Dynamic and cluster ion methodologies. Static SIMS can be described using the same illustration on removing the depth profiling mode. To retain clarity, the additional primary ion beams, lenses, deflectors, apertures, and secondary ion energy filters, which are present in all SIMS instruments, are not shown. Likewise, for the numerous different instrument geometries that presently exist. These geometries and the associated instrumentation are covered in Section 4.2.

In all cases, mass spectra are used to identify signals of interest. Imaging, along with depth profiling, is used to define the location of the specific signals of interest. The collection of mass spectra, images, and depth profiles is covered further in Sections 5.1.1.1, 5.1.1.2, and 5.1.1.3.

**4.1.1.1 Static SIMS** Static SIMS describes the methodology for attaining information on the elemental and molecular distribution of the outermost surface of the solid of interest, i.e. the outermost monolayer (Benninghoven 1970). As sputtering is by nature a destructive process, the effect of the damage induced on and into the sample must be minimized to below detectable limits.

One way of controlling the effect of sputter-induced sample damage in the recorded signal is to ensure that the same localized area of the surface of interest is not impacted/sampled during the course of analysis. This can be implemented by minimizing the primary ion flux density of a kiloelectron-volt primary ion beam to less than $\sim 1$ nA/cm$^2$ over a period of 1000 s. This translates into a dose of $6.28 \times 10^{12}$ ions/cm$^2$ (1 nA equates to $6.28 \times 10^9$ singly charged ions per second). Indeed, if it is assumed that each impact damages an area of $\sim 20$ nm in diameter (dependent on the primary ion impact parameters and the substrate type) and impact regions do not overlap, the total damaged area will approach that of the total surface area. This would imply that each secondary ion can be emitted from an undisturbed region, with the secondary ions emitted, representing $\sim 1\%$ of the outer monolayer.

The fact that molecular secondary ion signals representative of the initial species present at the outer surface of the samples of interest are recorded in Static SIMS spectra attests to the fact that minimal damage is experienced. The damage induced during sputtering can be followed by recording the intensity of a specific signal, or signals of interest, from some monolayer on prolonged sputtering. The damage is then relayed in what is referred to as the *disappearance cross section* (see Relation 3.5). This damage will, however, be specific to the substrate of interest as well as the conditions used in characterizing the substrate's surface. Molecular sputtering yields are discussed within Section 3.2.2.1 and its subsections.

Mapping the spatial distributions can then be carried out. This is most commonly accomplished via the microprobe mode (see Section 5.3.2.2). As this relies on collecting the signal/s from specific locations, the spatial resolution is a direct function of the beam spot size, with detection limits also playing part. The beam spot size (beam diameter once striking the solid's surface) is typically much greater than 100 nm in diameter. The signal intensity can, however, result in the degradation of the spatial resolution to values in the micron range. This stems from the relatively low yields noted from many large molecular ions which, in turn, requires that data be collected over areas larger than the beam spot size in order to retain statistical significance.

The possibility of deriving molecular information is one of the strengths of Static SIMS. The primary parameters of interest in Static SIMS are:

1. *Sensitivity* and best possible *Detection limits*
2. *High Mass resolution*
3. *Spatial resolution*

Optimizing one of the above generally comes at the cost (minimizes) of one or more of the others. These aspects are discussed further in Chapter 5.

**4.1.1.2  *Dynamic SIMS***  Dynamic SIMS in its traditional form (that resulting from knock-on sputtering as described in Section 3.2.1.1) describes a methodology used in attaining information on the elemental distribution as a function of depth or over a prespecified volume from any solid of interest (Wittmaack 1980). Molecular information is typically not available as the damage imparted during the sputtering process is retained within the newly exposed underlying substrate. As covered in Section 3.2.3, this damage can also result in significant modifications of the underlying substrate in the form of ion beam-induced segregation, diffusion, amorphization, recrystalization, and so on.

As a greater volume is probed per unit time, this form of SIMS exhibits the best possible detection limits to the elements. As an example, sub parts per billion levels can be reached in optimal cases (a specific example is detailed in Section 1.2.2.1). This is made possible by recording the signals of interest as a function of sputtering time, assuming a beam of a sufficiently high energy and current density. Beam energies can range from ~0.1 to over 20 keV with current densities much greater than the static limit.

As sputtering removes atoms/molecules present at the outer surface of a solid and damage is of minor concern (opposite to Static SIMS), measurement of the secondary ion signal as a function of sputtering time provides the depth distribution of the signal measured. In other words, the removal of atoms from the outer surface that occurs during SIMS analysis exposes deeper layers, which then become part of subsequent analyzed populations. Profiles over depths ranging from several nm up to ~10 μm can then be collected. Under ideal conditions, the depth resolution can surpass 1 nm.

Images in all three dimensions can be constructed by stacking the spatial images collected at every depth. Imaging can be carried out via either the *microprobe* or the *microscope* modes (both are discussed in Section 5.3.2.2). When carried out using the microprobe mode, an ultimate spatial resolution approaching ~10 to 20 nm is possible (McPhail et al. 2010), although 50 nm and above is more common. This spatial resolution is, however, heavily dependent on primary ion spot size, and hence the primary ion current. When carried out in the *microscope* mode, the spatial resolution is fixed at ~1 μm irrespective of the primary ion spot size/current. Note: Improved detection limits also allow for improved spatial resolution.

The primary parameters of concern in Dynamic SIMS are:

1. *Sensitivity* along with the best possible *Detection limits* and *Dynamic range*
2. *High Mass resolution*
3. *Depth resolution*

Optimizing one of the above generally comes at the cost of one or more of the others. These aspects are discussed further in Chapter 5.

### 4.1.1.3  *Cluster Ion SIMS*

Dynamic SIMS in its traditional form typically does not provide molecular information. Recent studies have, however, revealed that using large cluster primary ions, molecular depth profiles and even images in all three dimensions can be collected.

The difference with large cluster primary ions lies in the fact that the damage imparted during the sputtering process is minimized, and the increased sputter yields allow for the entire damaged region to be removed during each sputtering event. As a result, the newly exposed underlying substrate remains essentially damage free (as described in Section 4.1.1.2, this is typically not the case in atomic or small molecular primary ion impact). The reduced damage suffered by the substrate is due to the fact that sputtering no longer occurs via the more traditional knock-on sputtering process (for further details see Section 3.2.1.2). Large cluster primary ions also tend to break up on impact with some even evaporating shortly thereafter. For some reviews on this subject, see Vickerman 2009 and Mahoney 2013.

Although this possibility has only been demonstrated on relatively soft materials, i.e. organic-based materials, this has opened up an area not previously accessible. Both depth resolution and spatial resolution are, however, poorer than in the traditional form of Dynamic SIMS. This can, in part, be attributed to the much poorer detection limits observed and the fact that there is room for development in this area. Indeed, there may be possibilities in the area of matrix-assisted secondary ion yield enhancements, as demonstrated in the closely associated analytical technique of Matrix-Assisted Laser Desorption/Ionization (MALDI) (this technique is introduced Appendix A.10.3.2).

## 4.2  HARDWARE

SIMS is a highly diverse analytical technique. This diversity pertains to both the capabilities supplied and the instrumentation available. In addition, there exists extensive freedom in the choice and conditions under which the instrumentation can be used. As an example, commercially available SIMS platforms all consist of several primary ion sources whose operational parameters can be freely adjusted over a wide range depending on the analytical need. These platforms also provide the possibility of switching from the collection of positive secondary ions to negative secondary ions. Most other analytical techniques generally use a single source operated under fixed conditions to produce the specific signal of interest. For example, the extremely effective surface analytical technique of XPS uses photons to produce and collect electrons. This reduced instrumental complexity removes the requirement for routine beam realignment, which is a standard prerequisite for effective SIMS analysis.

Secondary ions are collected to form mass spectra, depth profiles, or images, or a combination thereof, as illustrated in Figure 1.2 and discussed in detail in Section 5.1.1. The measurement parameters of interest, also discussed in Section 5.1.1, are, however, instrument dependent. These being:

1. *Mass resolution.* This describes the ability of the instrument to separate two adjacent signals of different mass (although the $m/q$ ratio is more correct, mass alone is used). This is defined as the mass ($m$) of the ion of interest divided by the difference in the mass ($\Delta m$).
2. *Spatial resolution.* This describes the ability of the instrument to separate two adjacent signals from different spatial locations. This is typically defined as the distance over which an intensity variation from 16% to 84% is noted for an abruptly changing signal.

3. *Depth resolution.* This describes the ability of the instrument to separate signals of the same ion coming from different depths. This is defined as the depth over which an intensity variation over one order of magnitude is noted for an abruptly changing signal.
4. *Energy resolution.* This describes the ability of the instrument to separate two adjacent signals of different energy. This is defined as the minimum energy at which a trough exits between two distinct signals.

An overview of the hardware found within a SIMS instrument, along with their typical areas of application, their attributes and limitations, and the manner in which the various parameters listed above are affected, is presented in Sections 4.2.2 and 4.2.3. Not to be forgotten is the importance of vacuum in SIMS and how this can affect the analytical results. This is covered in Section 4.2.1.

### 4.2.1 Vacuum

The word "vacuum" comes from the Latin word "vacuo," which means empty (vacuum and vacuo are to this day often interchanged). We, however, now know that there is no such thing as completely empty space (even the vacuum of outer space contains a few atoms/$m^3$). A revised definition for vacuum can thus be considered to be *a region in space containing less gas than its surrounding regions.*

A vacuum is defined by the pressure, or lack of, within the region of interest. The *SI* unit for pressure is pascal (Pa) and that most commonly used in vacuum science and technology is *torr*. The unit *mbar* is also heavily used. Conversion factors between these units and other units of pressure are listed in Table 4.1. Another term of interest is the *Langmuir* (L). This is a unit of exposure, i.e. 1 L equates to $10^{-6}$ Torr for 1 second. Note: the pressure–time integral required to form a monolayer is ~1 L assuming a surface with unit sticking coefficient.

The importance of vacuum in any surface analytical technique is twofold. This lies in the fact that the vacuum controls the distance a charged particle can travel in the gas phase, i.e. from the sample to the detector, which can span a meter or more, and the vacuum controls the rate of gas phase adsorption onto the sample of interest (sputtered areas generally exhibit greater adsorption rate than unsputtered regions). Although the former is a base requirement in any technique that uses charged particles (electrons or ions) to generate and/or derive the information of interest, it is less stringent than the latter. As outlined in Section 4.2.1.1, numerical values for these can be derived using the kinetic theory of gases. This reveals an upper limit for the former of ~$1 \times 10^{-4}$ Torr assuming a 1-m flight path and the latter of ~$1 \times 10^{-8}$ Torr assuming a surface with unit sticking coefficient ($1 \times 10^{-9}$ Torr or better is optimal).

**TABLE 4.1    Some Useful Conversion Factors for Pressure.**

| 1 torr (1 mmHg) | = 133 Pa | = 0.0193 psi | = 1.30 mbar | = 0.00132 atm |
|---|---|---|---|---|

The adsorption issue is of particular importance in surface analytical techniques as excessive adsorption during analysis will affect, in one form or another, the signals of interest. Indeed, this will introduce significant background levels to any secondary ions containing the same species. Note: It is commonly stated that SIMS does not suffer from a background signal, which is one of the reasons for the extreme detection limits possible with SIMS.

Examples in which this issue comes into play are in the trace analysis of Hydrogen, Carbon, Oxygen, and Nitrogen in the material of interest and/or any other species that are also present in the gaseous phase, and the analysis of any species that shares the sample nominal mass as any species adsorbed or formed as a result of adsorption. An example of the former is in the trace analysis of Oxygen in GaAs. Indeed, even under optimized Dynamic SIMS conditions ($^{16}O^-$ secondary ions examined under high sputter rate conditions with the vacuum extending into the $10^{-11}$ torr range), the detection limit is not much better than $1 \times 10^{16}$ atoms/cm$^3$ (this scales directly with vacuum). Various detection limits are listed in Appendix A.5. An example of the latter is in the often routine analysis of Phosphorus in Silicon. Again, under optimized Dynamic SIMS conditions (the same as above but with the $^{31}P^-$ secondary ions examined under high-mass resolution conditions), the detection limit can approach $1 \times 10^{14}$ atoms/cm$^3$. High mass resolution, as discussed in Section 5.1.1.1.1, is however needed as the $^{31}P^-$ secondary ions suffer an isobaric (mass) interference from $^{30}Si^1H^-$ formed as a result of $H_2$ adsorption on Silicon, even during analysis. These issues are further exacerbated in Static SIMS owing to the reduced sputter rates used and the greater propensity for molecular signals.

### 4.2.1.1  *Vacuum and the Kinetic Theory of Gases*

To understand the importance of vacuum, one first needs to realize that in any gas at room temperature and under atmospheric pressure (also defined as 1 atm, 101.3 kPa, 760 Torr, etc.), there exists $\sim 2 \times 10^{19}$ molecules flying around in random directions at velocities that are in excess of 100 m/s (this is mass and temperature dependent) with the average velocity best described using the *kinetic theory of gases*. As these molecules move, there is a finite probability that they will collide with some solid surface, i.e. container walls, sample surface, and so on. Indeed, there are over $10^{23}$ collisions/cm$^2$.s under atmospheric conditions, which is what results in the pressure measured.

The *kinetic theory of gases* implies that the resulting pressure ($P$) can be expressed as:

$$P = n.k_B.T \qquad (4.1)$$

where $n$ is the number of particles per unit volume (typically expressed in units of cm$^3$), $k_B$ Boltzmann constant in units of J.K$^{-1}$, and $T$ the temperature in units of kelvin. From Relation 4.1, the velocity ($v$) of the particles can be defined as:

$$v = \left( \frac{8k_B.T}{\pi.m} \right)^{1/2} \qquad (4.2)$$

where $m$ is the atomic weight in amu. The dependence of pressure on temperature results from the increased energy (velocity) contained within the randomly traveling molecules.

The *collision rate* $(Z_a)$ can be written upon substitution of Relation 4.2 as:

$$Z_a = \frac{P}{(2\pi.m.k_B.T)^{1/2}} \tag{4.3}$$

This is the *Hertz-Knudsen* relation expressed in *SI* units. This relation can be made more accessible by expressing pressure in units of *torr* and merging all the conversion factors into a single constant, i.e.:

$$Z_a = 3.51 \times 10^{22} \left( \frac{P}{(T.m)^{1/2}} \right) \tag{4.4}$$

This provides the collision rate in units of collisions/cm$^2$.s. Note: Although *torr* is not the *SI* unit for pressure (the *Pascal* is), the Torr is the most commonly used unit of pressure for many of the surface analytical techniques. Conversion factors between the different pressure units are listed in Table 4.1. These aspects are all covered in detail elsewhere (Redead et al. 1968, 1993).

By solving the Hertz–Knudsen relation and assuming interaction with a Silicon surface for which the sticking coefficient is close to unity and the surface density is $\sim 1 \times 10^{15}$ atoms/cm$^2$, one quickly realizes that a monolayer of Oxygen will form in $\sim 1$ ns at atmospheric pressure and room temperature (298 K). This drops to $\sim 1$ s at $10^{-6}$ Torr but really does not reach any kind of acceptable value for Static SIMS ($\sim 1000$ s) until into the $10^{-9}$ Torr pressure range.

The *mean free path* of the molecule in the gas phase is another useful parameter. This is the average distance traveled by a molecule before it strikes another molecule. As would be expected, this depends on pressure, temperature, and radii of particles. The collisional cross section $(\sigma)$ can be described by a cylinder of radius $(r)$ such that all particles that fall within an area of $\pi r^2$ collide with this particle. As both are traveling with respect to each other, a $\sqrt{2}$ factor is also inserted. In units of centimeter, the average distance traveled between collisions (the mean free path) is:

$$d_{M.F.P} = 7.50 \times 10^3 \left( \frac{k_B.T}{(P.\sqrt{2.\pi.r^2})} \right) \tag{4.5}$$

where pressure is again expressed in units of Torr, temperature in units of Kelvin, Boltzmann constant in units of J.K$^{-1}$, and radius in units of centimeter. Note: Although not all are in *SI* units, these are the most commonly used units in vacuum technology. Typical values for radii (derived from the Lennard-Jones potential), the collisional cross section, and the pressure at which the mean free path equals 1 m are listed in Table 4.2. As would be expected, larger molecules exhibit larger collisional cross sections.

Solving the above equations for $O_2$ gas molecules reveals a mean free path of $\sim 78$ nm at 1 atm and 298 K. From this, it becomes clear that if charged particles

**TABLE 4.2 Mass, Collisional Cross Section Radius, Collisional Cross Section, and Pressure at which the Mean Free Path Equals 1 m for the Specified Molecules.**

| Gas molecules | $m$ $(u)$ | $r$ $(\times 10^{-8}$ cm$)$ | $\sigma(\times 10^{-15}$ cm$^2)$ | $P(\times 10^{-5}$ Torr$)$ |
|---|---|---|---|---|
| $H_2-H_2$ | 2 | 2.556 | 2.05 | 8.03 |
| $N_2-N_2$ | 28 | 3.681 | 4.26 | 5.07 |
| $O_2-O_2$ | 32 | 3.433 | 3.70 | 5.84 |
| $CO_2-CO_2$ | 44 | 3.996 | 5.02 | 4.30 |

Numbers reproduced from (Redead et al. 1968, 1993).

are to be used as the source of information and/or for generating the information (such charged particles include primary and secondary ions), a mean free path in excess of 1 m would be needed. Such a mean free path is only attained at a vacuum of $\sim 1 \times 10^{-4}$ Torr or better. Note: The value of 1 m is used as this represents the minimum distance from a substrate surface to a detector in most presently available SIMS instruments. Some even reach close to 10 m (see Appendix A.8 for common SIMS instrument geometries).

The dependence of collision rate and average distance traveled for $O_2$ molecules and ions as a function of pressure are shown in Figure 4.1. The dotted line represents a minimum acceptable vacuum. The regions specified as HV and UHV refer to *High Vacuum* and *Ultra High Vacuum*, respectively. The mean free path of $O_2^+$ ions is $\sim \sqrt{2}$ times that of $O_2$ molecules. Similarly, the mean free path for electrons equates to $\sim (4 \times \sqrt{2})$ times that derived via Equation 4.5.

Also shown in Figure 4.1 are the definitions used to describe the various pressure ranges (vacuum ranges). As these are not universally accepted, those from the American Vacuum Society (AVS) are used. These are defined in Table 4.3.

Reasons why the first commercially available SIMS instruments did not appear until the 1960s can be traced back to the fact that the ability to create such vacuums did not come about until the 1950s.

**TABLE 4.3 Pressure Range Definitions as Used by the American Vacuum Society (AVS) in Units of Pa (the SI Unit for Pressure) and Torr. See Table 4.1 for Conversion Factors.**

| Nomenclature | Pressure (Pa) | Pressure (Torr) |
|---|---|---|
| Low Vacuum (LV) | $1 \times 10^5 \leftrightarrow 3.3 \times 10^3$ | $760 \leftrightarrow 25$ |
| Medium Vacuum (MV) | $3 \times 10^3 \leftrightarrow 1 \times 10^{-1}$ | $25 \leftrightarrow \sim 1 \times 10^{-3}$ |
| High Vacuum (HV) | $1 \times 10^{-1} \leftrightarrow 1 \times 10^{-4}$ | $\sim 1 \times 10^{-3} \leftrightarrow 1 \times 10^{-6}$ |
| Very High Vacuum (VHV) | $1 \times 10^{-4} \leftrightarrow 1 \times 10^{-7}$ | $\sim 1 \times 10^{-6} \leftrightarrow 1 \times 10^{-9}$ |
| Ultra High Vacuum (UHV) | $1 \times 10^{-7} \leftrightarrow 1 \times 10^{-10}$ | $\sim 1 \times 10^{-9} \leftrightarrow 1 \times 10^{-12}$ |
| Extreme High Vacuum (XHV) | $< 1 \times 10^{-10}$ | $< \sim 1 \times 10^{-12}$ |

Note: Other definitions of the respective pressure ranges also exist.

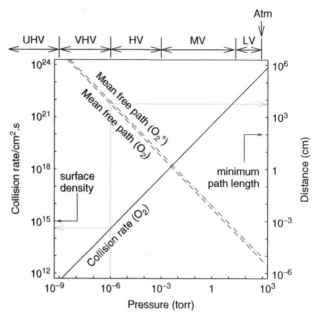

**Figure 4.1** Surface collision rates (left ordinate) and mean free path (right ordinate) as a function of pressure for $O_2$ molecules and ions. These were derived from Relations 4.4 and 4.5. Reprinted with permission from van der Heide (2012) Copyright 2012 John Wiley and Sons.

#### 4.2.1.2 Pumping Systems

Although the vacuum pump was invented in 1650, it took close to 300 years to produce vacuum conditions better than $\sim 10^{-7}$ Torr.

A vacuum is produced by reducing the amount of molecules within a confined space relative to its surroundings. Indeed, it is the presence of these molecules, or more precisely their collisions with their surroundings that produce pressure. Reaching UHV conditions, however, requires a highly specialized pumping system and a chamber displaying exceedingly low outgassing characteristics. Outgassing describes the removal of surface adsorbates generated when exposing any surface to atmospheric gasses.

Vacuum chambers are typically constructed out of high-grade stainless steel. This material is used as it is cost-effective and displays:

1. low outgassing rates (once surface adsorbates have been removed)
2. low corrosion rates (this is further minimized under UHV)
3. low vapor pressures (stainless steel has a high melting point)
4. excellent structural integrity (under both ambient and thermal conditions), and
5. good machining characteristics.

Use of brass, borosilicate glass, epoxy resins, adhesive tapes, rubber O-rings, and other high vapor pressure materials is minimized.

To further enhance the structural integrity, chambers capable of supporting UHV conditions tend to be manufactured in cylindrical or even spherical shapes. However, it cannot be said that square/rectangular chambers are not in use. Access is then provided through ports attached using specialized welding procedures. Other structural units are then connected using specifically designed flange/gasket combinations.

The most common flange/gasket type used in UHV systems is what is referred to as the *Conflat* system. This system, trademarked by Varian and Associates, comprises of Oxygen-Free High Conductivity (OFHC) Copper gaskets, which are pressed into flanges bearing a knife edge. When compressed (bolted together), the flanges press into a Copper gasket generating a UHV tight seal. Such gaskets must, however, be replaced after each use. Other soft gasket materials that have been used in producing UHV conditions include Aluminum, Gold, and Indium with the latter two being rare in SIMS instruments. Carbon-based O-rings are commonly used in HV applications as these rings tend to be more cost-effective, i.e. these are reusable.

Introduction of samples from atmosphere is carried out through an attached self-contained introduction chamber that can be sealed off via a series of valves from atmosphere and/or the analysis chamber. This allows the sample to be pumped down from atmosphere to a pressure of $\sim 10^{-7}$ Torr within 10–20 min depending on the outgassing rates of the sample. Once pumped down, the valve between the introduction chamber and the analysis chamber is opened to allow the transfer of the sample.

The plumbing used in SIMS instruments comprises of multiple pumps, valves, tubing and, of course, vacuum gauges for registering the vacuum produced. A typical example of a vacuum system is displayed in Figure 4.2. This reveals the use of several different types of vacuum pumps that together with the valves are operated in specific sequences.

Vacuum pumps can be broadly subdivided into three groups, these being:

1. Positive displacement pumps
2. Momentum transfer pumps
3. Entrapment pumps.

*Positive displacement pumps* are best for evacuating from atmospheric pressure to $\sim 10^{-3}$ Torr and for supporting (backing) momentum transfer pumps. Those most commonly found on SIMS instruments are *Rotary vane* or *scroll pumps*.

*Momentum transfer pumps* can produce a vacuum down to $10^{-10}$ Torr for chambers already preevacuated to within the $10^{-3}$ Torr range. The most commonly used momentum transfer pumps in SIMS instruments are *Turbomolecular pumps*. These are typically setup, along with a positive displacement pump, to evacuate an introduction chamber. Note: As an introduction chamber is routinely exposed to atmosphere, the best pressure that can be reached within an acceptable time is in the $10^{-7}$ Torr range. This, however, is sufficient to allow a sample to be transferred into

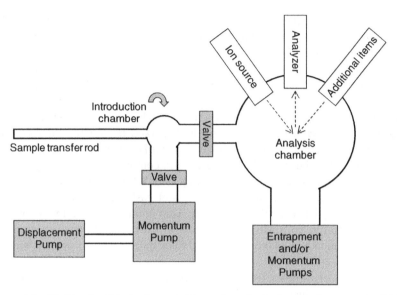

**Figure 4.2**   Highly simplified schematic illustration of a basic pumping setup for SIMS.

the analysis chamber held under UHV (once the valve between the two is opened), i.e. it does not allow significant contamination of the analysis chamber.

*Entrapment pumps* are self-contained units (these do not require any other pumps) capable of producing a vacuum down to $10^{-10}$ Torr. As the name suggests, entrapment pumps operate by trapping or condensing gas phase molecules into the solid state or by ionization and accelerating these molecules into a solid. *Ion Pumps* with *Titanium sublimation* pumps are more commonly included on state-of-the-art SIMS instruments. Ion pumps are operated continuously to sustain UHV conditions. Sublimation pumps, on the other hand, are operated in an intermittent manner to allow improvement in vacuum conditions. As such pumps should not be operated at pressures greater than $\sim 10^{-5}$ Torr, initial pumpdown of these chambers is carried out through a combination of displacement and momentum transfer pumps.

Pumping speed must also be considered when designing/constructing a chamber capable of supporting UHV conditions. The primary factors controlling pumping speed are the following:

1. The chamber volume (smaller chambers provide faster pump down rates)
2. The outgassing rates, which are a function of:
   a. the materials used
   b. the adsorbates present
   c. the surface area.
3. The ultimate pressure desired ($10^{-9}$ Torr or better).

Pumpdown time from Very High Vacuum (VHV) to UHV is the longest. This arises because in this regime, it is the removal rate of surface adsorbates, which controls the pressure. Adsorbates develop when any metal surface is exposed to atmospheric gases. Indeed, under ambient conditions, it can take weeks to pump a vacuum chamber from atmosphere to UHV, irrespective of the pumping system used. Note: This is also the reason why samples are not transferred directly from the atmosphere into analysis chambers.

To accelerate the removal of surface adsorbates, such chambers are baked whenever they are exposed to atmosphere. This entails heating the chamber to 100–200 °C for periods of between 12 and 48 h while being pumped. Extended cool-down times are also required. Under such operations, all temperature sensitive units and connections must be removed.

The effect of outgassing can be represented as

$$V\left(\frac{dP}{dt}\right) = G_r - P.S_p \tag{4.6}$$

where $V$ is the volume of the chamber in units of liter, $G_r$ the outgassing rate in units of liter. Torr/s, $P$ the pressure in units of Torr, and $S_p$ the pumping speed in units of liter/s. A good rule of thumb for clean stainless steel chambers is to provide at least 1 l/s pumping speed per 100 cm$^2$ of surface area. As typical SIMS instruments have a surface area equal to or greater than $\sim 2 \times 10^4$ cm$^2$, the pumping speed of the entrapment pump is typically around 200 l/s or more.

The ultimate pressure that can be reached can be derived from Relation 4.6 as:

$$P = \frac{G_r}{S_p} \tag{4.7}$$

The ability to measure the vacuum derived must also exist. In SIMS instruments, this is most commonly carried out using a combination of two types of gauges, these being the Pirani gauge (useful pressure range is $\sim 10$ Torr to $\sim 10^{-3}$ Torr) and the Ionization gauge (useful pressure range is $\sim 10^{-3}$ Torr to $\sim 10^{-10}$ Torr). These gauges are located within such instruments such that any emissions that may be produced from these detectors cannot influence the recorded spectra. This is accomplished by placing them "out of line of sight" of the sample surface and the detector/s. This is particularly important for retaining the high detection limits provided by Electron Multiplier detectors (these are covered in Section 4.2.4).

### 4.2.2 Primary Ion Columns

The *primary ion column* is the section in which the primary ions used to generate the secondary ions are formed, filtered, focused, and directed onto the sample. Cross sections of primary ion columns found on commercially available instruments are shown in Figure 4.3(a–c). The primary difference lies in the primary ion filtering

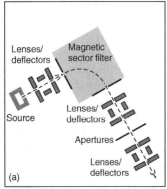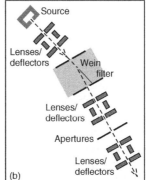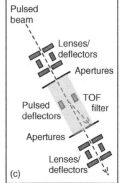

**Figure 4.3**   Simplified layouts of primary ion columns employing (a) Magnetic Sector filters, (b) Wien filters, and (c) Time Of Flight (TOF) mass filters. Operational aspects are discussed in the text. Additional lenses/deflectors/apertures are omitted for sake of clarity.

used. Filtering is required to remove ions of other $m/q$ ratio that are formed in the source region.

Mass filter types found on commercially available instruments include:

1. Magnetic sector mass filter
2. Wien mass filter
3. Time-of-Flight (TOF) mass filter

Magnetic sector mass filters comprise of a region in which an ion beam passes through a magnetic field. Wien mass filters utilize crossed electrostatic and magnetic fields. In both cases, the magnetic field induces filtering of ions of different $m/q$ ratio, a result of the fact that ions experience a change in their trajectories proportional to their $m/q$ ratio and the field applied. The electrostatic field used in Wien filters can correct for the primary ion beam deflection imposed while retaining the mass filtering capabilities. Note: Beam deflection is introduced in both Wien filters and Time-of-Flight filters to remove neutral species from the primary ion beam. If not removed, these will introduce signals from other regions of the sample, with one side effect being the loss of depth resolution (see Section 5.3.2.4). Note: Neutrals are formed in the ion source region and are not influenced by electromagnetic fields.

Time of Flight mass filters separate ions of different mass or energy as a result of their different velocities. This is understood from $E = \frac{1}{2}mv^2$, i.e. ions of greater mass will have lower velocities assuming that they all have the same energy. These, however, require pulsed ion beams, as each pulse defines a time zero from which the flight time (velocity) is recorded. The relation $E = \frac{1}{2}mv^2$ also allows separation of ions of the same mass but different charge states. This is realized as ions of different charge states are accelerated to different energies. This is needed as a Bismuth field ionization source, for example, produces $Bi_n^{q+}$ ions with $n$ extending from 1 to 7 and $q$ from 1 to 3.

The remainder of the primary ion column consists of a series of electrostatic lenses/deflectors and apertures. Additional sources, discussed further in Section 4.2.2.1, may also be present. Electrostatic lenses and deflectors are used to shape and transfer the ion beam to the region of interest. Apertures, along with focusing/defocusing of the beam, are used to control the current. As many analytical conditions are available (primary ion beam type, polarity, energy, and, in some cases, the angle of incidence), the primary ion column must be prealigned for the specific analysis on a daily basis.

Driven by the need for improved depth resolution, low-energy ion beams (denoted within this text as ion beam impact energies down to 500 eV impact energies) were introduced in the early 1990s in all instrument types. Ultra Low-Energy (ULE) ion beams (denoted within this text as down to 100 eV impact energies) were introduced in the late 1990s. This is accomplished by electrically floating either the primary ion source or the entire primary ion column. Of note is the fact that ion beam spot sizes decrease with decreasing primary ion current and increasing impact energy.

Ion beams used in SIMS may be rastered and/or pulsed as needed (the most commonly available Time of Flight instruments pulse the primary ion beam).

Electron sources are also commonly found on primary ion columns as these allow for the analysis of insulating samples (issues accompanying the analysis of insulating samples are covered in Section 5.2.1, with methods used to circumvent these issues covered in Section 5.2.1.1.1). These sources are focused/tuned to suite the analysis need.

### 4.2.2.1 Ion Sources

Almost all commercially available SIMS instruments contain more than one ion source and an electron source. The ion source types found in commercially available SIMS instruments fall into one of five groups, these being:

1. Electron Impact (EI) sources. These are used for producing positive inert gas ion beams ($Ar^+$ being one of the most common), as well as $O_2^+$ and $SF_5^+$. With modifications, $C_{60}^+, C_{84}^+, C_{24}H_{12}^+, Ar_n^+$, and so on, where $n$ extending to several thousand are also produced.

2. Duoplasmatron (Dou) sources. These are primarily used for producing $O_2^+$ and $O^-$ primary ion beams of greater efficiency than EI sources.

3. Radio Frequency (RF) sources. These are used for producing the same types of ions as Duoplasmatrons but to significantly greater efficiency.

4. Surface Ionization (SI) sources. These are used for producing alkali ion beams ($Cs^+$ is by far the most common in SIMS)

5. Field Ionization (FI) sources also referred to as Liquid Metal Ion Sources or Liquid Metal Ion Guns (LIMG). These are used for producing small spot $Ga^+, In^+, Au_n^+, Bi_n^{q+}$ ion beams, and so on, where $n$ and $q$ are integer values of one or greater ($n$ is typically limited to 1 through 7, whereas $q$ is limited to 1 through 3)

Each source has capabilities and limitations relative to each other, making each highly useful in specific areas. The capabilities/limitations of these sources, along with those noted when such sources are coupled with the appropriate primary ion columns, are outlined in Table 4.3. The beam energy and sizes (diameters) stated refer to general ranges of those recorded at the sample as opposed to that following extraction from the source. These parameters, primarily controlled by the remainder of the primary ion column, are listed as they are those of most interest during SIMS analysis. These sources are also used in many other technologies and, as a result, their operational parameters may be outside those listed in Table 4.3. Brightness values are those recorded at the source as these provide a measure of the maximum current that can be extracted. These are defined as a function of the angular dispersion and extraction voltage applied, i.e. in units of $Am^{-2}sr^{-1}V^{-1}$. Such values increase with increasing extraction voltage owing to the reduced space charge effects experienced. Aspects concerning charged particle beam transport are introduced in Appendix A.6. Further discussion on the ion sources most commonly used in SIMS are presented in Sections 4.2.2.1.1 through 4.2.2.1.4 (Table 4.4).

*4.2.2.1.1 Electron Impact Sources* EI sources are heavily used in SIMS to provide reliable beams of any inert gas ($Ar^+$ is most common) as well as other gases such as $O_2$ ($O_2^+$ ions), $SF_6$ (for $SF_5^+$ ions), $C_{60}$ (for $C_{60}^+$ ions), $C_{84}$ (for $C_{84}^+$ ions), Corenene (for $C_{24}H_{12}^+$), and Argon ($Ar_n^+$ where $n$ can extend to several thousand). Multiply charged ions are also produced, and indeed, $C_{60}^{2+}$ and $C_{60}^{3+}$ ions, for example, can be useful when higher impact energies/smaller probe diameters are required. EI sources are used in all forms of SIMS, i.e. Static, Dynamic, and Cluster ion SIMS (see Section 4.1.1 and sections within).

Inert gas ions introduce little in the way of chemical effects into the substrate under study. $O^-$, $O_2^+$, and $SF_5^+$, on the other hand, enhance positive secondary ion yields under Dynamic SIMS conditions, some by orders of magnitude (see Section 3.3).

**TABLE 4.4  Typical Properties of Ion Sources Used in SIMS within Optimized Primary Ion Columns along with the More Commonly Generated Primary Ions.**

| Source | Ions Used in SIMS | Energy (keV) | Brightness $(Am^{-2}sr^{-1}V^{-1})$ | Size $(nm)^a$ |
|--------|-------------------|--------------|----------------------|------------|
| EI | $Ar^+_n, O_2^+, SF_5^+, C_{60}^+$ | $< 1 \rightarrow 40$ | $\sim 10$ for $Ar^+$ | $\sim 5 \ \mu m$ |
| Dou | $Ar^+, O^-, O_2^+$, etc. | $< 1 \rightarrow 17$ | $\sim 500$ for $O_2^+$ | $< 0.2 \ \mu m$ |
| RF | $Ar^+, O^-, O_2^+$, etc. | $< 1 \rightarrow 17$ | $\sim 3000$ for $O_2^+$ | $< 0.1 \ \mu m^b$ |
| SI | $Cs^+$ | $< 1 \rightarrow 16$ | $\sim 100$ for $Cs^+$ | $< 50$ nm |
| FI | $Ga^+, In^+, Au_n^+, Bi_n^{q+}$ | $< 1 \rightarrow 30^c$ | $\sim 1 \times 10^6$ for $Ga^+$ | $< 10$ nm |

Note: Listed brightness and spot sizes are only noted at higher energies.

$^a$ Defines minimum possible spot size in optimized columns.

$^b$ Theoretical predicted value.

$^c$ Multiply value by the ion charge if using multiply charged ion populations.

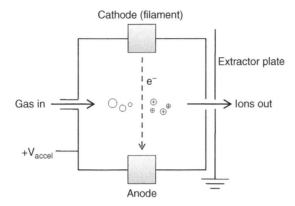

**Figure 4.4** Simplified schematic illustration of a typical crossed beam electron impact (IE) source. The energy of the electron beam is $\sim$70 eV.

These sources consist of a region in which the gas of interest (Ar, $O_2$, $SF_6$, etc.) is introduced and irradiated by electrons (Nier 1947). As the electrons are produced from a heated filament, these are sometimes referred to as *hot cathode* sources with a typical design depicted in Figure 4.4. Filaments are typically made out of Tungsten or Iridium. Energetic electrons interacting with gas atoms/molecules induce the emission of electrons from the respective gas atoms/molecules, thereby resulting in the formation of positive ions. The positive ions formed are then extracted through the orifice of a plate held at some negative potential relative to the ionization region. Beams with impact energies of <1–40 keV and up to $\sim$500 nA (noted at higher energies) are typically produced with brightness at the source of the order of 10 $Am^{-2}sr^{-1}V^{-1}$.

The production of large cluster ions ($C_{60}^{+}$, $C_{84}^{+}$, $C_{24}H_{12}^{+}$, $Ar_n^{+}$, etc.) requires additional hardware. For example, sublimation ovens are used to derive the organic vapors of interest, which are then irradiated by electrons. These sources are capable of producing low brightness beams of up to 40 keV, or multiples thereof for doubly and triply charged ions ($C_{60}^{+}$, $C_{60}^{2+}$, and $C_{60}^{3+}$ are some examples). Currents approaching 50 nA for singly charged ions are realized (the current decreases for doubly and triply charged ions). Inert gas cluster ions are produced in a Gas Cluster Ion Beam (GCIB) source by condensing the individual gas atoms into neutral clusters by cooling in a supersonic jet. These are then ionized via electron impact.

The popularity of these highly reliable and long-lived sources stems from the fact that these are relatively inexpensive and easy to work on. The lifetime of these sources is governed by the filament, which can be easily replaced. One drawback associated with such sources lies in their inability to be focused to much below $\sim$5 $\mu$m (a function of both the spherical and the chromatic aberrations suffered). That being said, a 1 $\mu$m beam has been demonstrated for a $C_{60}$ source in a purpose-built SIMS instrument (Fletcher et al. 2009).

*4.2.2.1.2 Duoplasmatron Sources* Duoplasmatron sources belong to the Plasma source or DC Glow Discharge family. Duoplasmatrons are heavily used in

SIMS for producing high current $O^-$, $Ar^+$ (or any other inert gas ion), $O_2^+$ ions, and so on. This, along with their brightness, makes these useful in Dynamic SIMS where high sputter rates and good depth resolution are required.

Inert gas ions introduce little in the way of chemical effects into the substrate, whereas Oxygen ions enhance positive secondary ion yields when examined under Dynamic SIMS conditions, some by orders of magnitude (see Section 3.3). Negatively charged primary ion beams can be useful for insulating samples (see Section 5.3.1.2.1).

Duoplasmatron sources essentially consist of a region in which the gas of interest (Ar, $O_2$, etc.) is introduced and allowed to form a plasma (Klemperer 1971). This plasma forms on generating an arc, as a result of the encapsulation of ions and electrons within a localized region by applied electrostatic and magnetic fields. The ions of interest (both positive and negative ions exist within the plasma) are extracted through the orifice held at the polarity needed. A typical source of this type is depicted in Figure 4.5.

Beams with impact energies from <1 to 17 keV and currents up to ~3 µA are attainable. Typical minimum spot sizes lie in the ~1 to 3 µm range, although 0.15 µm has been shown to be possible for a 16 keV $O^-$ beam within the nanoSIMS 50™-based instruments. The spot size is primarily limited by chromatic aberrations introduced as a result of the 5 → 15 eV energy spread of the ions making up the respective beam (aberrations are discussed in Appendix A.7.1). Typical brightness values for $O_2^+$ beams from a Duoplasmatron source are of the order of 500 $Am^{-2}sr^{-1}V^{-1}$.

Although highly effective and heavily used in Dynamic SIMS, the destructive nature of such plasmas requires that these sources be cleaned/replaced on a routine basis. The lifetime, between 50 and 500 h, is also unpredictable. Triplasmatrons are a related plasma source.

*4.2.2.1.3 Radio Frequency Sources* RF sources, also referred to as *RF antenna sources*, have found recent application in Dynamic SIMS (Smith et al.

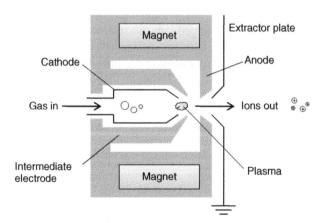

**Figure 4.5** Simplified schematic illustration of a typical Duoplasmatron source.

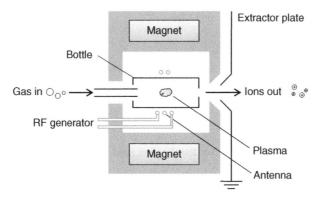

**Figure 4.6** Simplified schematic illustration of a typical Radio Frequency source.

2008). These sources, which belong to the plasma source family, have found attraction in Dynamic SIMS because of their capability to produce extremely bright $O_2^+$ ion beams with greater stability and much longer lifetimes than provided by Duoplasmatron sources.

Sources used in SIMS employ an electrodeless high-frequency discharge to induce ionization of gases present within a dielectric *bottle*. Ions are extracted as a result of the application of a DC electric field across this bottle (Newton and Unsworth 1976). A typical design is depicted in Figure 4.6. Internal degradation issues initially plaguing such sources have been circumvented through $TiO_2$ coating, and so on.

Beams with impact energies of <1–17 keV are readily attainable. Maximum currents in the mA range at the sample are possible when using impact higher energies. The beam spot size is expected to be below 0.1 μm (Smith et al. 2008). This is based on the low chromatic aberrations (aberrations are discussed in Appendix A.7.1) resulting from the narrow energy spread displayed by these sources (<5 eV). Note: Brightness values are in the 3000 $Am^{-2}sr^{-1}V^{-1}$ range.

Primary issues with these sources lie in their high cost relative to other sources used in SIMS and its complex maintenance (typically, customers are required to forward such sources to the respective vendor for annual refurbishment). These, however, show great promise in the area of ULE SIMS because of their high brightness. The same can be said for the stability and long lifetime between servicing.

*4.2.2.1.4 Surface Ionization Sources* These sources have found extensive use in Dynamic SIMS for producing high current density alkali ion beams, with $Cs^+$ ion beams by far the most common. Such sources also allow for $MCs^+$ or $MCs_2^+$ secondary ions (see Section 5.4.2) to be studied.

As Cesium reduces the work-function of the substrate under study, negative secondary ion yields can be enhanced by orders of magnitude (see Section 3.3).

The high current possibility arises from the propensity of alkali metals to form positive ions when present on a high work-function metal that is heated under high vacuum conditions (Alton 1988). The alkali metal is typically contained within

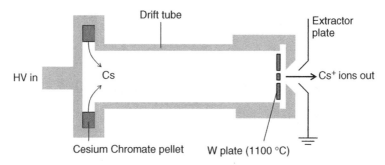

**Figure 4.7** Simplified schematic illustration of a typical Surface Ionization source.

a porous frit, termed the *reservoir*, in the form of Cesium carbonate ($CsCO_3$) or Cesium chromate ($Cs_2CrO_4$). Liquid Cesium is used in older systems. Cesium vapor is released on heating this *reservoir* to $\sim$400°C. $Cs^+$ ions are then formed when this vapor comes in contact with an *ionizer* (typically a Tantalum plate heated to $\sim$1100°C). The ions are extracted as a result of the potential field applied between the *ionizer* and the *extraction electrode*. A typical design is depicted in Figure 4.7.

Beams with impact energies of $<$1–16 keV and up to $\sim$10 μA (at higher energies) can be produced in these highly reliable long life sources. A minimum spot size just below 50 nm has also been demonstrated in the nanoSIMS 50™-based instruments. This small spot size stems primarily from the low initial energy spread ($<$0.5 eV) of the ions formed, as this reduces chromatic aberrations (aberrations are discussed in Appendix A.7.1). Brightness values from such sources are in the 100 $Am^{-2}sr^{-1}V^{-1}$ range.

The popularity of these sources lies in their relatively long lifetime. The lifetime is governed by the Cs content within the source frits, which themselves are easily replaceable. These are the most commonly found sources used in Dynamic SIMS analysis of electronegative elements in Dynamic SIMS.

*4.2.2.1.5 Field Ionization Sources* Those used in SIMS are liquid-based sources. These sources are referred to as an LMIG, a Liquid Metal Ion Source (LMIS), or electro-hydrodynamic ion source, with the former most commonly used in the SIMS community. These sources are used for producing small spot high current $Ga^+$, $In^+$, $Au_n{}^+$, $Bi_n{}^{q+}$, and so on ion beams, where $n$ and $q$ are integer values representing the number of atoms and charge, respectively. As covered in Section 4.2.2, filtering of $n$ and $q$ is carried out by the primary ion mass filter used.

Such ions tend to introduce minimal chemical effects into the substrate under study, hence have minimal effect on secondary ion yields.

Sources of this type operate by heating the metal/alloy to its melting point (27°C for Gallium) and allowing the liquid to be drawn up a high work-function metal needle (typically Tungsten) held under high vacuum conditions (Orloff 1993). This occurs via capillary action (Gilbert 1600).

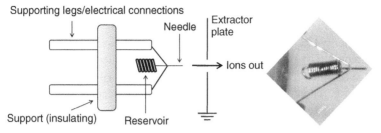

**Figure 4.8** Simplified schematic illustration of a typical liquid-based field ionization source (left) along with an optical image of the Needle and reservoir (right). This shows the needle to be ∼1 mm long. These are more commonly referred to as Liquid Metal Ion Guns (LMIGs).

In short, when the tip is held under a sufficiently high positive ion extraction field, the metal of interest is drawn off the tip apex as a positively charged ion. As this effectively depletes the tip region of the liquid metal, more metal is pulled to the tip resulting in further ion production, and so forth. A typical source design of this type is depicted in Figure 4.8. This region is commonly referred to as a *Taylor cone* in recognition of the mathematical description provided by Taylor (1964).

The small spot capability is a direct result of the area over which these beams are produced (tip of a refractory metal needle) as this reduces spherical aberrations suffered by other ion beam sources. These sources also display a relative narrow energy spread (∼5 eV) which in turn has the effect of reducing chromatic aberrations (aberrations are discussed in Appendix A.7.1).

Other ions can be produced by heating the source to the melting point of the respective alloy. An example is Manganese and Bismuth produced from BiMn alloys. Mass separation is then carried out within the primary ion column. Beams with impact energies of $<1-30$ keV for singly charged ions and up to ∼100 nA are typically produced in these reliable easy-to-use sources. Typical operating currents are, however, around 1 nA. Minimum spot sizes of less than 10 nm are achievable at low currents and high impact energies making these ideal for Focused Ion Beam (FIB) instrument applications. Brightness values for $Ga^+$ ions from such sources lie in the $1 \times 10^6$ $Am^{-2}sr^{-1}V^{-1}$ range.

This, along with their small spot capability, makes these sources ideal in imaging SIMS where high spatial resolution is required. Indeed, these are extensively used in Time of Flight based instruments in both Static and Dynamic SIMS applications. The downside of these sources is their relatively short lifetime. These can range from ∼400 to ∼1200 h

### 4.2.3 Secondary Ion Columns

The *secondary ion column* is the section of a SIMS instrument in which the ions emanating from the sample surface are collected, filtered, and focused onto the respective detector. Secondary ion columns typically contain:

1. Various electrostatic lenses and deflectors. These are used to efficiently collect and transport the secondary ions through the secondary ion column

2. An energy filter. This is used to limit the energy spread of the secondary ions as required by the analytical need and the mass filter used

3. A mass filter. This limits the $m/q$ ratio of secondary ions passing onto the detector, thereby allowing isotopic/elemental/molecular identification

4. Detection system. This records the pertinent secondary ion current and, in some cases, their spatial distribution relative to their point of emission

The layout of a typical secondary ion column is primarily dictated by the type of secondary ion mass filter used. In addition, the mass filter type essentially defines the capabilities and limitations of the respective SIMS instrument. Although other mass filter types exit, three distinct mass filter types are found in commercially available SIMS instruments. These are the:

1. Quadrupole mass filter

2. Magnetic sector mass filter

3. Time of Flight mass filter

Rudimentary secondary ion column designs based around these mass filter types are shown in Figure 4.9(a–c). More detailed illustrations are given in Appendix A.8. Note: The linear Time of Flight filter design shown in Figure 4.9(c), which appeared in early instruments has since been superseded by the Reflectron and TRIFT™ designs. Examples of commercially available SIMS instruments based around the Reflectron and TRIFT™ designs are presented in Appendix A.8.7 and A.8.8. As would be expected, all of these mass filters have advantages and disadvantages with respect to each other, all of which will be covered in Section 4.2.3.1.

There also exist variants of the layouts shown in Figure 4.9(a–c), with two examples being the reverse Magnetic Sector instrument, which, as the name suggests, places the Magnetic Sector before the energy filter and Time of Flight instruments in which the secondary ion beam is pulsed as opposed to the primary ion. These geometries have been commercialized in the form of the SHRIMP RG™ Magnetic Sector-based SIMS instrument (RG refers to Reverse Geometry as shown Appendix A.8.3), the MiniSIMS™ TOF, and J105 Time of Flight based SIMS instruments (see Appendix A.8.9). A list of SIMS vendors can be found in the Appendix. Additional variants of these geometries also exist within various research facilities.

For mass filters to work to their prescribed capabilities, the secondary ion beam must either be of a sufficiently narrow energy spread (as covered in Section 3.3.1.2, all secondary ion beams exhibit an inherent energy spread), or, the energy spread should be compensated for within the mass filter. Time of Flight mass filters of the Reflectron and TRIFT™ types can control this spread, hence do not require additional energy filters. Linear Time of Flight, Quadrupole, and Magnetic mass filters do not compensate for the energy spread. As a result, these mass filters require separate energy filters to be inserted within the respective secondary ion columns.

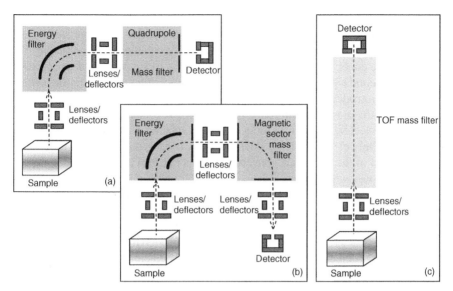

**Figure 4.9** Highly simplified illustrations of the layouts of secondary ion columns based around (a) Quadrupole mass filters, (b) Magnetic Sector mass filters, and (c) Time of Flight mass filters (a linear flight path is depicted for reasons of clarity). Additional lenses and deflector may also be present within these columns to further optimize the capabilities of the respective instrument.

Energy filters, covered in Section 4.2.3.2, also act to ensure that no line of sight exists between the sample and the detector. This is required to remove the secondary neutral emission also produced during the sputtering process. Note: As secondary neutrals in the secondary ion beam cannot be mass filtered (these are unaffected by electrostatic and magnetic fields), they will introduce a background signal if not removed.

In the case of Magnetic Sector mass filters, the optical properties of energy filters can be specifically matched such that the dispersion induced by one can be compensated for by the other. This property, referred to as *double focusing*, allows for improved ion transmission throughout these secondary ion columns and the ability of the respective secondary ion column to image the spatial location of secondary ion emissions via what is referred to as the *microscope mode*. Imaging in the *microscope mode* is also possible in specifically designed pulsed primary ion beam Time of Flight instruments, i.e. those based around the TRIFT geometry. Both are also capable of imaging in the *microprobe mode*. These imaging modes are covered in Section 5.3.2.2.

Extraction of the secondary ion beam from the sputtered surface is optimally carried out through the immersion of the sample in an appropriate electrostatic field. This is carried out by ensuring a potential difference exists between the sample and some metallic plate or cone bearing an orifice of $\sim 1$ mm in diameter. This plate/cone is typically situated in close proximity to the sample (in some instruments, this plate/cone is referred to as the immersion lens). This acts to accelerate the secondary ions from the sample toward the plate/cone, thereby forming a secondary ion beam on passing through the orifice in the plate/cone, and into the remainder of the secondary ion column. Accelerating the secondary ion beam to some energy also has the effect of reducing *space charge effects*, thereby improving secondary ion transmission. Aspects concerning charged particle beam transport are introduced in Appendix A.6.

To ensure the most efficient transmission of the secondary ions to the detector, lenses and deflectors are also introduced within specific regions of the secondary ion column. Only electrostatic lens/deflectors are used as the deflection imposed is not a function of the secondary ions mass (this is not the case in magnetic fields). The regions over which these lenses and deflectors are introduced are optimally defined at the secondary ion column design stage using ray tracing and/or phase space dynamics. As is covered in further detail in Appendix A.7.3, both describe theoretical means of defining the secondary ion beams transport characteristics.

Once the extracted secondary ion beam passes through the respective secondary ion column, it is detected by one of several different detector types. Detectors along with their capabilities and limitations are covered in Section 4.2.3.3.

The ability to extract and direct the secondary ion beam through the respective energy and mass filter and onto the detector is defined as the *transmission function*. As this is difficult to measure, and reported values tend not to relay the pertinent energy range, it must be used with caution when comparing such values. In any case, optimal reported transmission functions are typically less than 1% for Quadrupole-based secondary ion columns, as high as ∼50% for Magnetic Sector-based secondary ion columns and can theoretically approach 90–100% in Time of Flight based secondary ion columns. Analytically useful secondary ion transmission values tend to be less than those stated with the value dependent on the analytical conditions used.

### 4.2.3.1  Mass Filters   SIMS relies on the ability to resolve the $m/q$ ratio of secondary ions emitted from any solid, or made to be solid surface. Mass is derived once $q$ is known. As the vast majority of secondary ions emanate in the $+1$ or $-1$ charge state, it is often assumed that the mass is equal to the $m/q$ ratio. This explains why the abscissa in mass spectral plots is sometimes listed in units of mass rather than the more correct $m/q$ ratio.

The separation of secondary ions of different $m/q$ ratios is carried out in the mass filter region of a SIMS instrument. As listed in Section 4.2.3, mass filters come in several distinct forms, three of which are found in commercially available instruments (these being the Quadrupole, Magnetic Sector, and Time of Flight mass filters). As significant differences in the capabilities of the respective mass filters exist, the type of mass filter used defines the analytical capabilities of the respective SIMS instrument. This in turn defines the area that the respective instrument finds greatest usage in.

As an example, instruments utilizing Quadrupole mass filters are best suited to Dynamic SIMS applications on relatively simple substrates, i.e. those not displaying significant isobaric interferences (isobaric interferences are covered in Sections 5.1.1.1 and 5.3.1.3). Instruments using Magnetic Sector mass filters are best suited to high sensitivity Dynamic SIMS applications (shallow or deep) on any substrate, i.e. these can deal with most isobaric interferences. Instruments using Time of Flight mass filters are best suited to Static and shallow Dynamic SIMS applications on any substrate type, i.e. these too can deal with most isobaric interferences.

For completeness sake, the following discussion also covers Ion Trap mass filters (also referred to as a Paul trap or 3D ion Quadrupole trap) and Ion Cyclotron Resonance (ICR) mass filters (also referred to as a Penning trap). These mass filters are included as, although not used in commercially available SIMS instruments, they do appear in experimental facilities. Furthermore, Ion Trap mass filters are considered the workhorse in Mass Spectrometry, whereas the Fourier Transform (FT)-ICR mass filters are capable of displaying the highest mass resolution of all the mass filters presently available. At this point is also worthwhile mentioning another FT based mass filter that may see greater interest from the SIMS community moving forward. This is the Orbitrap mass filter which was developed from the Kingdon trap.

Typical values for the $m/q$ range, mass resolution ($m/\Delta m$), transmission, and the primary operation mode under which such mass filters are used are listed in Table 4.5. The mass resolution values stated pertain to those using the 10% peak width definition within the single ion mode (see Section 5.1.1.1.1). Of note is the fact that the mass resolution can vary with the $m/q$ of the secondary ions. This is noted for Time of Flight mass filters (these display decreasing values with increasing $m/q$) but not for Magnetic Sector mass filters. For Quadrupole mass filters, this is dependent on the operating conditions. Some of the parameters listed can extend beyond those stated. As an example, Quadrupole mass filters applied in Mass Spectrometry can scan an $m/q$ range extending to 4000. Likewise, the mass resolution of Time of Flight instruments is often stated as theoretically unlimited (significant practical limitations do, however, exist).

These mass filters are covered further in Sections 4.2.3.2.1 through 4.2.3.2.5.

**TABLE 4.5  Typical Properties of the Mass Filters that Have Been Incorporated in Both Commercial (Quadrupole, Magnetic Sector, and Time-of-Flight) and Experimental (Ion Trap and ICR) SIMS Instruments.**

| Mass Filter | $m/q$ Range | $m/\Delta m^a$ | Transmission | Primary Area of Use |
|---|---|---|---|---|
| Quadrupole | ~1 – 300 | Unit $m/\Delta m$ | < 1% | Dynamic and Static SIMS |
| Ion trap | ~1 – 4000$^b$ | NA-SIMS | NA-SIMS | NA-SIMS |
| MS | ~1 – 500 | > 25,000 | < 50% | Dynamic SIMS |
| ICR | ~1 – 10,000 | NA-SIMS | NA-SIMS | NA-SIMS |
| TOF | ~1 – 10,000$^c$ | > 10,000$^c$ | < 100% | Dynamic and Static SIMS |

NA-SIMS = Not Applied in commercially available SIMS instruments.

$^a$ Values derived using the 10% peak width definition within the single ion mode (see Section 5.1.1.1.1) for $m/q$ ratios less than 100.

$^b$ Values pertain to the use of these mass filters in Mass Spectrometry.

$^c$ Although these values are theoretically unlimited, practical limitations do exist.

*4.2.3.1.1 Quadrupole* Similar to most of the mass filters used in SIMS, the Quadrupole mass filters were conceived (Paul and Steinwedel 1953) well ahead of their introduction into SIMS (Benninghoven and Loebach 1971; Wittmaack 1975; Magee et al. 1978). These filters were extensively used in both Static and Dynamic SIMS from the mid-1970s through to the late 1990s. Since then, their popularity has been eroded as a result of improvements in both Magnetic Sector and Time of Flight based SIMS instruments, and the more stringent requirements for high mass resolution. An example of a commercially available Quadrupole SIMS instrument is presented in Appendix A.8.1.

SIMS instruments utilizing Quadrupole mass filters are highly effective when unambiguous elemental analysis from simple substrates is required, i.e. where isobaric interferences are minimal. An example would be in the depth profile analysis of Boron in Silicon where a detection limit to Boron to $\sim 1 \times 10^{16}$ atoms/cm$^3$ is possible. Quadrupole mass filters also provide for greater experimental flexibility owing to reduced extraction fields used as this allows for the sample surface angle to be varied with ease. Lastly, the relative simplicity of Quadrupole mass filters reduces purchase and ownership costs relative to Magnetic Sector or Time of Flight based instruments.

The major disadvantages associated with Quadrupole mass filters in SIMS lie in:

1. their limited $m/q$ ratio range ($1 - <500$ $m/q$ is typical)
2. their limited mass resolution (inability to deal with mass interferences)
3. their poorer detection limits and sensitivity relative to Magnetic Sector and Time of Flight instruments

The poorer detection limits and sensitivity primarily arise from the low transmission of these mass filters. The transmission of these filters is also dependent on the secondary ion $m/q$ ratio when set at constant mass resolution mode (the desired mode in SIMS), i.e. this decreases as the $m/q$ ratio increases. This poorer detection limits and sensitivity can, however, be partially compensated for using high primary ion currents (up to 10 µA can be used). Likewise, extraction fields of several hundred volts can be applied to capture a greater portion of the secondary ions produced, but only if they are subsequently decelerated before entering the Quadrupole mass filter region.

Quadrupole-based SIMS instruments only allow analysis of secondary ions of a particular mass at a time. Thus, if information on multiple secondary ions is required, the mass filter must be switched back and forth among the respective masses during analysis. All other secondary ions produced at the same time are thus discarded (filtered out). However, as the settling time of electrostatic units is much faster than that of electromagnetic units, Quadrupole mass filters do show the advantage of being able to switch to different masses at a much faster rate than Magnetic Sector mass filters. This can improve data acquisition times and can compensate, to some degree, for their low transmission.

A schematic of a Quadrupole mass filter is shown in Figure 4.10. In essence, this consists of four rods of radius $r$ whose axes lie parallel to the ions initial trajectory and are all an equal distance from each other, hence the name, Quadrupole, with

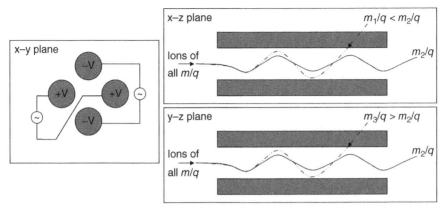

**Figure 4.10** Schematic of a Quadrupole mass filter as seen from the x–y, x–z, and y–z planes (ion beam is initially moving along the z-axis). This shows the higher $m/q$ ions being filtered out along one plane and the lower $m/q$ ions filtered out in the other plane according to the applied voltages (see text).

the rods separated from each other by a circle whose radius is 1.145 times the rod's radius.

A DC potential ($U$) of specific polarity is placed onto two opposing rods with the remaining rods experiencing the opposite polarity. An RF signal ($V \cos \omega t$) is then placed on all components. This generates a time varying field that will only allow ions of a specific $m/q$ ratio to pass through. The trajectories of all ions outside this $m/q$ range, as defined from the *Mathieu stability diagram* (Mathieu 1868), become unstable, hence are unable to reach the detector. The Mathieu stability diagram is derived by solving second-order differential equations pertaining to the RF and DC voltages, the physical dimensions of the Quadrupole ($r_0$ being the separation of opposing rods in units of the rod's radius), and the $m/q$ ratio of the ions. These provide the dimensionless relations:

$$a_Q = \left( \frac{8U}{r_0^2 \omega^2} \right) \left( \frac{q}{m} \right) \tag{4.8a}$$

$$q_Q = \left( \frac{4V}{r_0^2 \omega^2} \right) \left( \frac{q}{m} \right) \tag{4.8b}$$

Quadrupole mass filters can be scanned across the secondary ion $m/q$ range by adjusting the RF and DC components together. Two modes of operation are available. One allows for the transmission of secondary ions of different $m/q$ ratio with a constant mass resolution ($m/\Delta m$ as discussed in Section 5.1.1.1.1). This, however, results in a decrease in transmission with increasing mass (scales approximately as $m^{-1}$). The other mode allows for constant transmission but with a variable mass resolution with mass. Owing to the greater ambiguity introduced, the constant mass resolution mode is typically applied when these mass filters are used in SIMS.

*4.2.3.1.2 Ion Trap* Although Ion Trap mass filters are now the most heavily used mass filter types in Mass Spectrometry (they are considered the workhorse), they have not been applied in commercialized SIMS instrumentation. That being said, these have been applied in research environments (Todd et al. 2002). This lack of commercialization in SIMS is believed to be primarily due to:

1. the relatively recent developments of Ion Trap capabilities
2. the greater complexity associated with their modes of operation
3. their limited dynamic range
4. the longer acquisition times compared to that displayed by Quadrupole mass filters

Note: The mass range and resolution can be improved with the application of lower RF voltage scan rates.

Such mass filters do, however, display many attributes relative to other mass filters. As an example, they do not discard signals of different $m/q$ ratio as is done in other $m/q$ scanning-based mass filters such as the Quadrupole and Magnetic Sector mass filters. In addition, the transmission function of these mass filters does not decrease with increasing $m/q$ when operated in constant mass resolution mode as does in Quadrupole mass filters. And lastly, these mass filters are relatively small units compared to all but the Ion Cyclotron Resonance mass filter (commercial units range from 10 to 25 cm).

A schematic illustration of an Ion Trap mass filter is shown in Figure 4.11. As can be seen, this is composed of a ring electrode surrounded by two end cap electrodes. Both end cap electrodes have central orifices for ions to enter and exit the mass filter according to the fields applied.

The Ion Trap mass filter is closely related to Quadrupole mass filters in that the action of both is described using Mathieu stability diagrams (March 1997). This is realized in that a Quadrupole mass filter applies a two-dimensional time-dependent

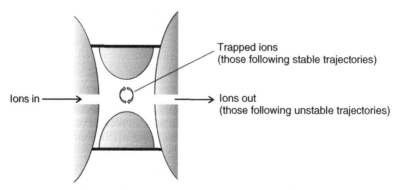

Ions in →

Trapped ions
(those following stable trajectories)

Ions out
(those following unstable trajectories)

**Figure 4.11** Schematic illustration of an Ion Trap mass filter. Signals are detected as the ions of specific $m/q$ ratio are allowed to escape the mass filter region.

electric field perpendicular to the axis of the ion beam entering the filter. An Ion Trap mass filter applies a three-dimensional time-dependent electric field, thereby trapping the ions entering the filter within a predefined spatial region.

Filtering is then carried out by sampling the ions exiting the Ion Trap as their trajectories are made unstable. This is done in a mass selective manner by adjusting the RF potential applied. This mass selective instability mode, developed in the 1980s, provides the advantage in that none of the signal is discarded (Glish and Vachet 2003). As a result, these are intrinsically more sensitive than Quadrupole mass filters, which do throw away un-measured signal. Note: Both only accept low-energy ions (<20 eV).

*4.2.3.1.3 Magnetic Sector* Magnetic Sector mass filters have been around since well before the advent of SIMS, with these still heavily used to this day in situations where the optimal detection limits, sensitivity, and dynamic range are required in Dynamic SIMS analysis.

As an example, depth profiling of Boron in Silicon can be carried out with Boron detectable down to $\sim 1 \times 10^{13}$ atoms/cm$^3$ (< 1 ppb). The sensitivity is derived in part because of the use of a single high current primary ion beam used in continuous mode (a primary beam of up to 10 µA can strike the sample) for both sputtering and analysis and the high transmission of the Magnetic Sector. This arises from the large energy acceptance (a few hundred electron volts) and high accelerating energies that can be accepted.

When these mass filters are well matched to the remainder of the ion column (in the form of double focusing secondary ion columns), these mass analyzers provide high mass resolution at high transmission, albeit at the cost of instrument size (these are the largest of the commercially available SIMS instruments). This matching also allows for imaging to be carried out in both the *microprobe* and the *microscope* modes (see Section 5.3.2.2). Examples of Magnetic Sector SIMS instruments can be found in Appendix A.7.1–4.

The major disadvantages associated with such mass filters lie in:

1. the relatively limited mass range available in commercially available SIMS instruments (those in Mass Spectrometry extend over a much greater range),
2. the degradation in detection limits when implementing high mass resolution above some limit (this arises from the need to reduce the energy and mass filter acceptances through the reduction in slits sizes),
3. the relative slow rate under which magnetic fields can be switched (this defines the $m/q$ ratio of the secondary ions reaching the detector),
4. the large size of these instruments (the largest are those applied in the geological fields that can span a room of 5 by 10 m in size), and
5. the high purchase and ownership costs relative to Quadrupole and Time of Flight based instruments (this is a result of their complexity and size).

Magnetic Sector-based SIMS instruments utilizing a single detector (most common) allow only the analysis of secondary ions of a particular $m/q$ at a time. Thus, if information on multiple secondary ions is required, the mass filter must be switched back and forth among the respective ions during analysis with all other ions discarded. This disadvantage can be partially circumvented through the introduction of multiple detectors, as is used in the nanoSIMS™, ims-1280™, and SHRIMP™ instruments.

Magnetic Sector mass filters are based around the fact that an ion passing through a magnetic field ($B$) applied perpendicular to the ions trajectory will experience a deflection such that the trajectory follows an arc of some radius ($r$) dependent on the ions $m/q$ ratio and velocity ($v$). This is indicated in Relation 4.9 as derived from the Lorentz force law. This is sometimes referred to as momentum filtering.

$$r = ((2\,mv)/(B^2 q))^{1/2} \tag{4.9}$$

As illustrated in Figure 4.12, only ions of a particular $m/q$ ratio can pass through the entrance and exit slits surrounding the Magnetic Sector and onto the detector region. All other ions are disregarded (deflected). The mass resolution of Magnetic Sector-based instruments is a property of the slit dimensions (several tens of microns), the radius of the arc defined by the Magnetic Sector (this can range from 1 to 2 m), and the energy of the incoming ions. For this reason, Magnetic Sector-based instruments accelerate secondary ions to $\sim$5 keV. Note: There are commercially available SIMS instruments with multiple detector arrays that can collect several prespecified ions of different $m/q$ ratio quasi-simultaneously.

Double focusing Magnetic Sector-based instruments, as the name suggests, refocus an ion beam of some specific $m/q$ ratio but exhibiting a finite energy spread back onto a single point (Castaing and Slodzian 1962, 1981; Dowsett 1992). This point is set to coincide with the exit slit to improve both the mass resolution and the transmission function of such instruments.

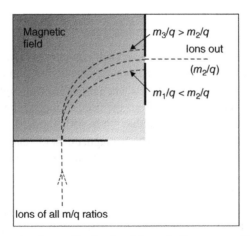

**Figure 4.12**    Schematic of mass filtering in a Magnetic Sector mass filter. Note: The magnetic field is applied perpendicular to the plane of the page through a curved drift tube in which the ions traverse.

Water-cooled laminated electromagnets are generally used because of their faster switching capabilities relative to traditional solid core magnets. Even so, a dwell time (waiting time) is still required to allow settling of the magnetic field before the secondary ion detection can be initiated (this can be as long as $1-2$ s). Note: Electrostatic fields can be adjusted faster than magnetic fields, hence the reason Quadrupole-based instruments can switch faster.

As revealed by Relation 4.9, $m/q$ filtering can also be carried out by adjusting the velocity of the secondary ions. This has been implemented in combination with the application of a magnetic field, by scanning the voltage applied to the flight tube in the Magnetic Sector. As this mode, referred to as *Electrostatic Peak Shifting* (EPS), allows for faster peak switching, it has been introduced into the latest generation Magnetic Sector SIMS instruments.

### *4.2.3.1.4    Ion Cyclotron Resonance Mass Filter*    Like Magnetic Sector mass filters, Fourier Transform Ion Cyclotron Resonance (FT-ICR) mass filters use magnetic fields to determine the $m/q$ ratio of ions. Although capable of extreme mass resolution ($m/\Delta m > 100,000$), these have only been recently introduced in the field of SIMS albeit in highly specialized research environments (Palmblad et al. 2000; Todd et al. 2002; Smith et al. 2011). These are also sometimes referred to as Fourier Transform Mass Spectrometers (FT-MS).

SIMS instrument designs based around FT-ICR have been able to replicate many of the advantages displayed by such mass filters when applied in mass spectrometry. As an example, mass resolution values of 385,000 have been demonstrated via the single ion method (see Section 5.1.1.1.1) albeit using the 50% definition (Smith et al. 2011). This was reported for molecular secondary ions produced via $C_{60}^{+}$ primary ion impact. Also demonstrated was the possibility of imaging the organic ions to unprecedented sensitivity and detection limits. This was carried out by synchronizing the pulsed $C_{60}^{+}$ beam raster with the FT-ICR mass filter detection electronics, i.e. the *microprobe* method (see Section 5.3.2.2).

These mass filters operate in a similar manner to Ion Trap mass filters, in that both of these filters trap ions of interest. Furthermore, both require the injection of low-energy secondary ions, similar to Quadrupole mass filters. The difference lies in the fact that FT-ICR mass filters utilize strong magnetic fields for trapping the respective ions and for recording their masses. This possibility is realized as the resonance frequencies at which the ions trapped within the FT-ICR cavity is a function of their $m/q$ ratio. The oscillation frequency, termed the *characteristic orbital frequency*, also known as the *ion cyclotron frequency* ($\omega_c$) induced by the applied magnetic field ($B_0$) relates to the $m/q$ ratio via:

$$\omega_c = \frac{q \cdot B_0}{m} \qquad (4.10)$$

The fact that these mass filters also act as the detector presents itself as a key difference between these and the other mass filters discussed (all other mass filters use pulse counting detectors as discussed in Section 4.2.3.3). As a result, the dynamic range is now a function of these mass filters. For these filters, this is of

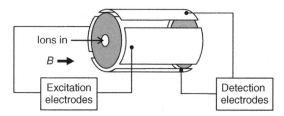

**Figure 4.13** Schematic of mass filtering in a Fourier Transform Ion Cyclotron Resonance mass filter. Note: The front and rear plates (grayed) are earthed. The entire unit is immersed in a strong magnetic field ($B$) pointed in the direction of the incoming ions.

the order of $10^5$, i.e. less than that for instruments employing external detectors. In addition, these mass filters display the lowest throughput of all the mass filters, thereby increasing analysis times.

FT-ICR mass filter cavities are $\sim$2 to 5 cm in size. The associated hardware, which includes a super conducting magnet required to produce fields of $7-12$ Tesla, and so on, significantly increases the size of such instruments such that they will fill a small room. A schematic diagram of a typical FT-ICR mass filter cavity is depicted in Figure 4.13.

*4.2.3.1.5 Time of Flight* Although Time of Flight mass filters were introduced shortly after the release of fully functional Quadrupole and Magnetic Sector-based SIMS instruments, their application and development in SIMS have since resulted in this being the fastest growing sector within the SIMS community. Instruments utilizing Time of Flight mass filters are most effective in Static SIMS studies and shallow Dynamic SIMS studies. Indeed, their detection limits when operated in the Static mode can approach $1 \times 10^7$ atoms/cm$^2$ (for Iron on Silicon), whereas in the Dynamic mode, this can approach $\sim$1 $\times 10^{15}$ atoms/cm$^3$ (for Boron in Silicon). The strength of these mass filters lies in their ability to simultaneously detect all secondary ions within some $m/q$ range (this data can be stored for later use if needed). Likewise, a high $m/q$ range and the high mass resolution accessible with minimal loss in sensitivity/detection limits (the transmission remains unaffected when using high mass resolution) lend well to their application fields.

Disadvantages associated with instruments utilizing these mass filters lie in:

1. The loss of detection limits/sensitivity and increased analysis times when operated in the Dynamic mode relative to Magnetic Sector-based instruments
2. The relatively large data files produced per analysis (these can fill terabyte drives relatively quickly when extensive depth profile analysis is required)
3. The dependence of mass resolution on $m/q$ and surface quality (smoothness)

The degradation in detection limits/sensitivity along with increased analysis times applies only to instruments utilizing pulsed primary ion beams when operated in the Dynamic SIMS mode. Reasons for this are discussed in a latter part of this section.

The sizable data files arise from the fact that these instruments generate mass spectra for every point the primary ion beam is scanned over when operated in the *microprobe* mode (this imaging mode is discussed in Section 5.3.2.2). As data is typically collected over $128 \times 128$ or $256 \times 256$ points, this translates to 16,384–65,536 mass spectra collected per layer. The number of data points per mass spectra can also be significant (this is a function of the mass resolution and mass range). This extensive information content often translates into increased data processing complexity.

The dependence of mass resolution on $m/z$ as well as surface quality (smoothness) can be understood on the basis that such filters separate the different $m/q$ ions based on their average velocity ($v$), i.e. light ion such as $H^+$ has a much higher velocity than a heavy ion such as $U^+$. This is possible as all secondary ions are emitted with a peak energy of $1-10$ eV, some with high energy tails that can extend to several hundred electron volts (Figure 3.27). Thus, by recording the time ($t$) it takes for a select pulse of ions to travel over some path length ($L$), the time spectrum recorded can be converted into a mass spectra via the relation:

$$t = L(m/2qv)^{1/2} \qquad (4.11)$$

The trick is to induce the emission of secondary ions from a smooth sample surface within some time interval so that the flight time can be measured. This can be accomplished one of two ways, i.e. by pulsing of the primary ion beam or by pulsing the secondary ion beam. These ions are then accelerated to between 2 and 8 keV before entering the Time of Flight mass filter.

Time of Flight based instruments thus provide for the quasi-simultaneous collection of all secondary ions, with those of interest selected following analysis rather than before analysis. The fact that all data is saved also presents this as a unique advantage in the area of failure analysis where the signals of interest may not be apparent during the analysis. This is in contrast to Magnetic Sector and Quadrupole mass filter-based SIMS instruments that require that the signals of interest be prespecified before the analysis being carried out, with all other secondary ions discarded.

Most commercially available SIMS instruments pulse the primary ion beam at frequencies between 10 and 50 kHz. This is achieved by deflecting the primary ion beam across a small aperture in the primary ion column. Pulse durations last from a few nanoseconds to several hundred nanoseconds with the time between individual pulses ranging from tens to hundreds of microseconds. Combining this with a secondary ion flight path of 1 to 2 m and time-sensitive detection electronics allows for the flight time of secondary ions produced per pulse to be measured. The primary ion beam described earlier is referred to as the *analysis beam*. Pictorial examples of a primary ion pulse-based Time of Flight mass filters based around the Reflectron and TRIple Focusing Time of Flight (TRIFT™) geometries are shown in Figure 4.14(a) and 4.14(b). The two most common geometries produced through Ion Time of Flight and Physical Electronics are shown in Appendix A.8.7 and A.8.8.

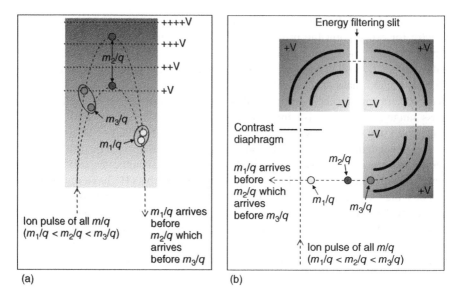

**Figure 4.14** Schematic of mass filtering of positive ions (snapshot in time) in (a) a Reflectron-based Time of Flight mass filter and (b) a TRIFT-based Time of Flight mass filter. Note: The pulse of ions entering mass filter arrive at the detector at different times according to their $m/q$ ratio.

Pulsing the primary ion beam, henceforth referred to as the *analysis beam*, also allows for the greater mass resolution to be attained, particularly when beam bunching is applied. Beam bunching describes, as the name suggests, the bunching of a specific beam pulse along the axis the beam is traveling in. This mode, carried out using specific electrostatic lenses, shortens the pulse width at the surface, thereby shortening the pulse duration. This allows the mass filter to work closer to its full potential. The use of field ionization sources (LMIGs) also helps because of the narrower energy spread exhibited (see Section 4.2.2.1.5).

Under optimal conditions, pulse widths of less than 1 ns are routinely achievable. This allows for mass resolution values in excess of 10,000 to be attained with relative ease (recall, mass resolution scales with $m/q$ for these mass filters). In the microprobe more, this comes at the cost of spatial resolution. The loss of spatial resolution stems from the increased space charge effects experienced by the ions within the bunched pulse, which effectively spreads out the pulse in the spatial dimension (think of a balloon that when compressed in one direction expands in the other). Thus, the decision must be made before analysis as to whether high mass or high spatial resolution is needed. Analytical considerations and alternative methodologies are covered further in Section 5.3.2.1.2.

At this point, it is worth noting that a 50 kHz analysis beam cannot contain more than $\sim$1000 ions per pulse (the remainder is discarded). This translates into a current of $\sim$80 pA striking the sample (1000 ions times $50 \times 10^3$ ions/s all divided by $6.28 \times 10^6$ cps, where the latter equates to 1 pA of current) when using an initial beam current of 800 nA (80 pA times 100 μs between pulses divided by a 10 ns pulse duration).

Although 80 pA appears ideal for Static SIMS applications, this is much too low to satisfy the sputter rates required in Dynamic SIMS. To counter this, a second primary ion beam of much higher current is used in such instruments to etch some volume of the sample's surface between *analysis beam* cycles. This beam, commonly referred to as the *sputter beam*, is pulsed at the same frequency but out of phase with the *analysis beam*. $O_2^+$ or $Cs^+$ beams are most commonly used for this purpose, as these can also induce a significant enhancement of respective secondary ion yields (see Section 3.3.2) owing to the small fraction implanted into the substrates surface. The downside of this approach is that the sensitivity and detection limit gains resulting from simultaneous ion detection are lost. This is due to the fact that the sputtered population cannot be recorded (only ions produced by the analysis beam are recorded).

Time of Flight mass filtering through the act of pulsing the secondary ion beam can also be carried out. Advantages of this approach lie in the fact that the loss of sensitivity/detection limits associated with the use of the dual beam Dynamic SIMS approach described earlier are circumvented, i.e. all the secondary ion emissions are collected. The pulsing sequence used is applied to the secondary ion beam shortly after their emission through electrostatic deflection, with beam bunching providing additional benefits. Mass separation of ions within an individual secondary ion pulse then ensues as this pulse travels toward the detector.

Highly simplified SIMS instruments based around this concept are commercially available in the form of the miniSIMS instrument developed by Millbrook Instruments Ltd. (now available from Scientific Analysis Instruments Ltd.). Although effective, these suffer greater constraints in the analytical conditions that can be applied (this is discussed further in Section 5.3.2.1.2). On the plus side, such instruments are available at a significantly lower cost than pulsed primary ion beam instruments. Indeed, these have been introduced in the hope of introducing SIMS to a greater audience.

An example of a research-based instrument in which mass filtering is carried out through pulsing the secondary ion beam is shown in Appendix A.vii.h. This has been developed specifically for molecular depth profiling and three-dimensional imaging of organic materials. The use of a continuous primary ion beam also allows for improved spatial resolution (1 μm is reported for $C_{60}$ ions (Fletcher et al. 2009)). Note: Such values cannot be reached using pulsed $C_{60}^+$ primary ion beams. The increased energy spread exhibited by cluster ion beams does, however, result in increased pulse times to values approximately 10 times greater than that possible using field ionization sources.

Both geometries, however, suffer in that pulsing either the primary ion or the secondary ion beam limits the number of secondary ions of a specific $m/q$ that can be recorded per analytical beam pulse. This is a direct result of the dead time effects/pulse pair resolution exhibited by detectors (see Section 4.2.4.2). In the case of Channel Plates (those most commonly used in TOF SIMS instruments), the pulse pair resolution approaches ∼10 ns, which is still greater than the primary ion pulse width. The maximum count rate ($\sim 10^5$) is thus a function of the primary ion pulse width and the primary ion pulse frequency (10 and 50 kHz). Methodologies for extending this range are discussed in Section 4.2.3.3.4.3.

**4.2.3.2   Energy Filters**   Energy filtering of the secondary ion beam, particularly atomic secondary ions, is required in all SIMS instruments if optimal data quality is to be attained. This can be carried out internally in instruments using Time of Flight mass filters using the Reflectron or TRIFT™ geometry or through the addition of an energy filter in instruments using Magnetic Sector, Quadrupole, or linear Time of Flight mass filters.

In the case of Quadrupole-based instruments, energy filtering is needed to constrain the secondary ion energy spread to within some relatively narrow range as is required by the respective mass filters (< 20 eV). This is also required in Ion Trap and FT-ICR mass filters. Note: The atomic secondary ions emitted from the sample can span several hundred electron volts, whereas the molecular secondary ions span only several tens of electron volts (see Section 3.3). These maybe accelerated to several hundred electron volts to improve their initial collection efficiency. As a result, the ions passing through the energy filter must be decelerated before entering the mass filter. The low energy acceptance of these mass filters is one of the key elements describing the low transmission of Quadrupole mass filter-based SIMS instruments (see Table 4.5).

As for Magnetic Sector-based instruments, energy filtering is required to reduce chromatic aberrations experienced by the secondary ions. Chromatic aberrations describe the loss of beam focus owing to some spread in the energy of the ions of interest (the focal point of electrostatic lenses is ion energy dependent), which in turn affects the mass resolution as well as the spatial resolution if using the *microscope* imaging mode (see Section 5.2.3.2). Note: These instruments can also handle a greater secondary ion energy spread than Quadrupole instruments, hence their greater transmission. In an effort to reduce both Chromatic and Spherical aberrations, while also improving the secondary ion collection efficiency, the secondary ions are generally accelerated to around 5 keV in such instruments. Indeed, as discussed in Section 4.2.3.1.3, instruments utilizing Magnetic Sector mass filters are designed around the double focusing geometry. This describes the refocusing of secondary ions of a finite energy spread.

Energy filtering can also be employed to reduce secondary ion intensities to values within the respective detectors range and remove molecular interferences from elemental signals (see Section 5.1.2.3). In research applications, this also allows greater insight into the mechanisms of secondary ion formation/survival (see Section 3.3).

The energy filters found in commercial SIMS instruments employing Quadrupole or Magnetic Sector mass filters are independent units referred to as Electro Static Analyzers (ESA). These are typically placed before the respective mass filter. In the case of Magnetic Sector-based instruments, the energy filter can also be placed after the mass filter. In this case, the instrument geometry is referred to as a reverse Magnetic Sector with a commercial example being the SHRIMP

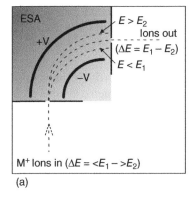

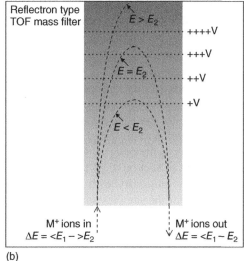

**Figure 4.15** Pictorial representations of the energy filtering experienced by positive ions passing through (a) Electro Static Analyzers (ESA) used in Quadrupole and Magnetic Sector instruments and TRIFT type Time of Flight analyzers and (b) reflectron type Time of Flight analyzers. For negative secondary ions, the polarities applied are simply reversed.

RG (RG refers to Reverse Geometry). A schematic of this instrument is presented in Appendix A.7.1.

As illustrated in Figure 4.15(a), these energy filters essentially consist of a pair of curved metallic plates between which the secondary ion beam passes. By applying the respective potentials to these plates, the electrostatic field produced deflects the secondary ion beam according to their energy (secondary ions of greater energy require a greater deflection force). The energy spread is then controlled by the width of the slit through which the secondary ions must pass after exiting the electrostatic field imposed by these plates. This must be set to a value of < 20 eV for Quadrupole mass filters but can be up to 300 eV for Magnetic Sector mass filters (continuously tunable over any energy range). Slits can also be placed at the entrance of these energy filters to further constrain the energy range of ions departing the filter.

Time of Flight mass filters based around the Reflectron or TRIFT™ designs do not require an additional energy filter. This is realized as in these instruments, ions of different energies but the same $m/q$ ratio can be made to arrive at the detector within a very narrow time interval (this capability is illustrated in Figure 4.14(a). Even so, energy filtering can still be carried out. In the case of the TRIFT™ configuration, this is accomplished by narrowing the energy slits before and following the electrostatic sectors in the same manner as illustrated in Figure 4.15(a). Note: The TRIFT™ design encapsulates three such sectors as depicted in Figure 4.14(b). Mass filters based around the Reflectron type configuration can only filter out higher energy ions. This is accomplished by adjusting the potentials of the mass filter such that only a specific population reaches the detector as is illustrated in Figure 4.15(b). All other ions can then be refocused onto the detector as a result of the electrostatic fields applied.

***4.2.3.3   Detectors***   In the case of SIMS, the detector records the number of secondary ions passing some point in space per unit time. As depicted in Figure 1.2, this point in space follows the mass filter such that it is the number of secondary ions of some specific charge to mass ratio that is recorded. FT-ICR mass filters provide the one exception to this rule because, as covered in Section 4.2.3.1.4, these mass filters also act as the detector. Further discussion on the mass filters used in SIMS can be found in Section 4.2.3.1.

Secondary ions carry with them a charge. This charge allows the ions to be extracted from the samples surface and passed through the respective secondary ion column with greatest efficiency. Detector efficiencies improve with the impact energy of the respective secondary ions, at least up to some limit ($\sim$4 keV). Current is defined as the passage of charge per unit time. The passage of a single unit of charge (ion or electron) per second equates to $1.602 \times 10^{-19}$ A.

The detector acts to convert the number of ions into a quantity that is easily relayed, i.e. a current of electrons that is then amplified before being converted into some frequency value, i.e. counts per second (cps). One of the most important characteristics of a detector is the linearity of the conversion process.

There exist a range of detectors applied in SIMS. The prime difference lies in:

1. the responsiveness of the detector (this defines the detection limit),
2. the number of ions per second measurable (this defines the dynamic range), and
3. the ability to provide information on the spatial distribution of the ions impacting the detector.

In general, more than one type of detector is found in commercially available SIMS instruments. Although all operate through the conversion of the energy released on ion impact into electrical current, the way this is done defines the capabilities/limitations of the respective detector. This in turn defines their use, i.e. pulse counting, imaging, or both (depends on whether imaging is carried out via the *microprobe* or *microscope* modes as discussed in Section 5.3.2.2). If spatial imaging is carried out using the *microscope* mode, the detector must be capable of the spatial recognition. This requirement is not necessary in the *microprobe* mode.

The detectors most commonly found in commercial SIMS instruments include:

1. Faraday Cup (FC)
2. Discrete Dynode Electron Multiplier (DD-EM)
3. Micro-Chanel Plate Electron Multiplier (MCP-EM)

The detection characteristics of these are listed in Table 4.6.

**TABLE 4.6   Detectors and the Typical Range Over which These Are Used within Commercially Available SIMS Instruments.**

| Mass filter | Min/Max Count Rates (cps) | Single ion detection | Dynamic Range | Spatial Recognition |
|---|---|---|---|---|
| FC | $\sim 5 \times 10^4$ to $5 \times 10^{9a}$ | No | $10^{5a}$ | No |
| DD-EM | $\sim 1$ to $2 \times 10^{6b}$ | Yes | $10^6$ | No |
| MCP-EM | Variable[c] | Yes | $10^4$ | Yes |

[a] The linear range is a function of the counting electronics used, as covered in Section 4.2.3.3.1.

[b] These will suffer increased dead time effects on approaching the specified count rates (counting speed is limited).

[c] This depends on amplification used (freely adjusted in these detectors).

Note: Channel Plates belong to the family of Electron Multipliers. The Electron Multiplier moniker has, however, been applied in SIMS to refer to the discrete dynode design (another design within the Electron Multiplier family). Faraday Cups and Electron Multipliers are discussed further in Sections 4.2.3.3.1 and 4.2.3.3.1.

*4.2.3.3.1   Faraday Cups*   The FC, as named after Michael Faraday, is the oldest and simplest of all the current sensing devices. This essentially consists of a metallic cup into which the secondary ions are directed. On impacting the interior base of this cup, the charge is deposited (one unit for singly charged ions and multiple units for multiply charged ions). By electrically isolating the FC, the deposited charge can be converted from a current to a voltage using a high impedance amplifier. This voltage is then passed onto a voltage/frequency converter such that the signal can be relayed in cps. A schematic example of an FC is shown in Figure 4.16.

The maximum count rate stated for an FC is actually a limitation imposed by the counting electronics of the SIMS instrument. In short, this is a function of the maximum input voltage of the voltage/frequency converter divided by the charge of an electron ($1.602 \times 10^{-19}$ C) times the value of the resistor placed across the input and output thermals of the impedance amplifier. Assuming for the sake of argument, values of 10 V and 10 MΩ are applied, which results in an upper count rate of $6 \times 10^9$ cps.

The accuracy of the signal recorded by an FC is primarily dependent on the conversion efficiency occurring on ion impact and the operating temperature of the

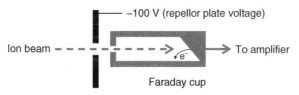

**Figure 4.16**   Schematic of a Faraday Cup detector along with the incoming ion beam and the secondary electrons ($e^-$) produced/trapped by the repellor plate.

associated electronics. The effect of the operating temperature is typically not considered as SIMS instruments are typically located in well-controlled environments.

The conversion efficiency is ensured by a metallic plate with a small orifice and a low negative potential placed at the mouth of the cup (this is sometimes referred to as the *repellor* plate). This plate allows the secondary ions to pass through such that it can impact with the detector walls, while preventing the escape of electrons produced as a result of this impact (a multitude of secondary electrons can be produced). If these were allowed to escape, this would result in a net removal of current from the detector.

The primary advantage associated with FCs lies in their robustness and the fact that these detectors have essentially no upper counting limit (the limit typically stipulated is a result of the associated electronics). As a result, these are the only detectors used for both primary and secondary ion detection in SIMS instruments. As amplification is not carried out within these units, a current of $> \sim 1000$ ions per second must strike the detector if a signal is to be recorded. Note: 626 cps equates to $\sim 1 \times 10^{-16}$ A and the lower detection limit of the most sensitive electrometers is around a factor of two times this value. The practical lower limit of these detectors is, however, signal to noise dependent, hence, will also be a function of, among other things, the dwell time applied.

*4.2.3.3.2 Electron Multipliers* An *Electron Multiplier* refers to any type of photon, electron, or ion detector in which the signal to be recorded generates many electrons following impact. The electrode at which the initial impact takes place is called the *conversion electrode*. The probability that an ion results in electron emission is termed the *detection quantum efficiency*. Electrons emitted from this electrode are then accelerated onto subsequent electrode surfaces whereupon more electrons are produced. In the case of Discrete Dynode Electron Multipliers, the subsequent electrode surfaces are called *acceleration electrodes*. This multiplication process can result in the overall emission of between $10^4$ and $10^9$ electrons per single ion impact, with the conversion factor (multiplication factor) being a direct function of the type of Electron Multiplier, the impact energy of the signal to be recorded, and the acceleration voltage used.

Electron multipliers come in several forms as illustrated in Figure 4.17.

The DD-EM is a unit in which the conversion electrode and the acceleration electrodes are all individual electrically isolated metallic electrodes. These electrodes are referred to *dynodes*. The primary advantage of these detectors lies in their high accuracy and sensitivity, i.e. a conversion factor (multiplication factor) of $10^9$ is reached for ions impacting at energies of greater than $\sim 4$ keV (secondary ion post acceleration of may be needed to reach these impact energies) and an acceleration voltage between each dynode fixed at several hundred volts. This makes these the most sensitive of the detectors used in SIMS as well as being one of the more robust within the EM family. Disadvantages lie in the saturation effects above $\sim 1 \times 10^6$ cps (a result of the recovering time required from each electron cascade) and the drift in gain even below $\sim 1 \times 10^6$ cps (introduces mass fractionation variability).

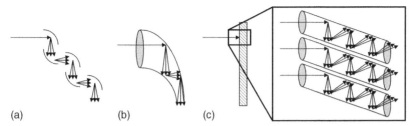

**Figure 4.17** Schematic illustrations of (a) a Discrete Dynode Electron Multiplier, (b) a Channeltron, and (c) a Micro-Channel Plate, along with the electron cascades generated when these detectors are biased. Reprinted with permission from van der Heide (2012) Copyright 2012 John Wiley and Sons.

A Channeltron is a horn-shaped continuous dynode structure that is coated on the inside with an electron-emissive material such as PbO. An ion striking the opening of the Channeltron creates a secondary electron cascade in the same manner as seen in DD-EMs. Although exhibiting many similar characteristics to DD-EMs, Channeltrons are more fragile and hence less heavily used in SIMS. In addition, both are highly effective for pulse counting, but without the ability of providing information on the spatial distribution of the incoming ions.

A Micro-Channel Plate (MCP) consists of a 2D array of parallel glass capillaries whose inner diameter ranges from 10 to 25 $\mu$m, and whose axis lies $\sim 7°$ off that of the trajectory of the signal to be recorded. As with Channeltrons, these capillaries are internally coated with an electron-emissive material allowing ions or electrons striking the inside wall to create an avalanche of secondary electrons once biased. This cascade creates a conversion factor (multiplication factor) of up to $10^4$. The fact that MCPs consist of many detectors ($\sim 2000$ across the diameter of a 25 mm circular array) essentially reduces the apparent dead time such that the pulse pair resolution approaches $\sim 10$ ns (Note: the pulse pair resolution of the electronics associated with pulse detection is of the order of $\sim 100$ ps). Use of two MCP arrays back to back with one rotated $180°$ to the other can result in a gain of $10^6$. These are sometimes referred to as a *chevron MCP*. The resulting current pulse is then passed onto an anode collector (for pulse counting) or phosphor screen or resistive anode encoder (for spatially resolved signal detection). One downside of chevron MCPs lies in the short lifetime of the second plate (a result of the increased count rates detected). Moreover, there are also a range of hybrid detectors such as MCP's coupled with scintillator Photo Multiplier (PM) tubes that have found extensive use in Time of Flight instrumentation. As further amplification ensures (more than one photon is produced per electron on phosphor screens), an overall gain approaching $10^8$ can be realized. Although capable of single ion detection, their dynamic range is, however, more limited when compared to DD-EMs.

As with FCs, the current recorded by the respective EM must first be converted to a voltage. This is carried out using a high impedance amplifier. This voltage is then passed onto a voltage/frequency converter to allow the signal to be relayed in units of cps. Once the noise associated with these types of detectors is removed (see Section 4.2.3.3.2.1), EMs are easily capable of recording individual ion counts per second. This stems for the fact that their background signals are well below 1 cps when operated under optimal conditions.

In the case of ion Magnetic Sector-based instruments operated in the *microscope* mode, the information on the point of secondary ion ejection is retained throughout the secondary ion column. As a result, both pulse counting and spatial resolving detectors are commonly found in such instruments. DD-EMs are used when maximum sensitivity is required (these also allow for imaging in the *microprobe* mode), FCs are used when increased dynamic range is needed, and MCPs combined with a phosphor screen or reactive anode encoder when imaging in the *microscope* mode.

As secondary ions passing through Quadrupole mass filters lose all information concerning their point of emission (see Section 4.2.3.2.1), only pulse counting detectors are needed. As a result, DD-EMs and FCs detectors are common with the former ensuring optimal sensitivity and the latter extending the dynamic range. This also allows imaging in the *microprobe* mode (see Section 5.2.3.1).

As for Time of Flight mass filter-based SIMS instruments, MCPs are preferred over DD-EMs, as the latter constitutes a single detector, whereas the former many detectors (needed to contend with the extremely fast detection required).

4.2.3.3.2.1   EM NOISE   Although Electron Multipliers are a type of pulse counting device, the signal measured from many pulses (that resulting from many secondary ion impacts per second) displays a distribution in current (the number of electrons produced per ion impact). This distribution arises from the statistical nature of the ion to electron conversion process, as well as the processes responsible for the additional electrons formed in subsequent electron-surface collisions.

This output signal distribution is generally represented by what is referred to as a *Pulse Height Distribution* (PHD) curve. These plot the distribution of the charge (from the electrons produced per ion impact) versus the number of ions resulting in a specific charge. The latter can be defined by adjusting the discriminator voltage (applied in millivolts) and measuring the number of pulses. Typical PHD curves are shown in Figure 4.18.

Of note is the fact that at low Discriminator Voltage values, the number of ion impacts (pulses) resulting in a specific number of electrons (as converted to millivolts) deviates from that expected from a statistical distribution and rapidly rises in value. This deviation is due to what is referred to as *noise*. This noise is associated with both the Electron Multiplier itself and the associated counting electronics. As this noise becomes more severe at lower frequencies, it is akin to $1/f$ *noise* or even *flicker noise* (Hung et al. 1990).

This contribution can be filtered out by setting the lower range over which such signals are collected, i.e. by setting the discriminator level to the position of the trough. This typically equates to a Discriminator Voltage of around 5–10 mV. This, however, changes with EM type, age, applied voltage (see Figure 4.18), and secondary ion momentum. As a result, this should be checked and recalibrated periodically.

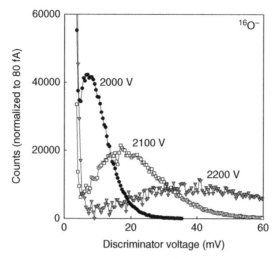

**Figure 4.18** Schematic example of Pulse Height Distribution analysis carried out for 4500 eV $^{16}O^-$ secondary ions impinging on an ETP Discrete Dynode Electron Multiplier operated at the listed voltages. Pulse counting was carried out using custom built ECL logic pre-amplifier/discriminator units. Discriminator voltage in this case should be set at ∼5 mV. Reproduced with permission from van der Heide and Fichter (1998) Copyright 1998 John Wiley and Sons.

4.2.3.3.2.2  EM Dead Time   As mentioned in Section 4.2.3.3.2, all Electron Multipliers exhibit an effect referred to as *dead time*. This describes the interval of time after an ion impact has occurred over which the detector is not able to record any signal, i.e. appears *dead*. This stems from the fact that each electron cascade generated from the respective ion impact on the respective detector spans some length of time.

Although the dead time is extremely short (typically of the order of 1 μs), its effects will be noted when the frequency of ion impact exceeds this time. For DD-EMs, this frequency is of the order of 1 MHz, or in units of ion impacts per second, of the order of $1 \times 10^6$ cps. An example of the effect the dead time can have on a depth profile is illustrated in Figure 4.19. As can be seen, as the count exceeds some value (1 MHz for a DD-EM), the detector no longer can record the secondary ions generated. As a result, the profile is distorted from its actual shape.

Such dead time effects can be corrected if the dead time of the counting system is known ($\tau$). This is described via the relation:

$$N' = \frac{N_C}{(1 - N_C \tau)} \tag{4.12}$$

where $N'$ is the corrected signal and $N_C$ the recorded signal.

Also of consideration is the *instantaneous dead* time. This is an effect that results when the area analyzed via a rastered primary ion beam is gated, whether electronically or optically (gating and its effects are discussed further in Section 5.3.4.2). This will induce a more severe effect than described by Relation 4.12 as the signal

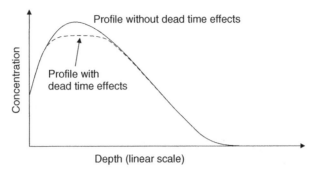

**Figure 4.19** Schematic example of the effect of dead time effects over the peak intensities recorded from an ion implanted substrate.

is now recorded for a fraction of the frame time as opposed to the total frame time. Note: Recorded count rates do not account for this effect as these are automatically averaged over the entire frame in the form of cps. This effect can be corrected by utilizing the relation

$$N = \left(\frac{1}{Y_{EM}}\right) \cdot \left(\frac{N_C}{(1 - (N_C/\eta_{GATE}) \cdot \tau)}\right) \tag{4.13}$$

where $Y_{EM}$ is the Electron Multiplier Yield (output pulse rate/input ion rate) and $\eta_{GATE}$ the fraction of time the signal is collected versus the frame time, i.e. the analyzed area/raster area. Note: $N_C/\eta_{GATE}$ is the instantaneous count rate.

4.2.3.3.2.3    MEASUREMENT OF INTENSE SIGNALS    When the signal count rate (frequency) approaches the dead time of the Electron Multiplier, there are several options that can be implemented to avoid the introduction of dead time effects. Such count rates are commonly noted when examining multiple secondary ion signals of very different intensities.

In Magnetic Sector mass filter-based SIMS instruments, the first option is to switch to another detector able to measure the more intense signal/s and then switch back to the more sensitive detector for other signals of interest. As a result, such instruments are commonly fitted with both Faraday Cups and Electron Multipliers. An alternative approach is to bias the sample during the acquisition of the intense signal, or to adjust the energy band pass of the energy filter, such that only the higher energy population of the more intense signal is recorded. The sample bias or energy filter is then reset during the acquisition of the less intense signal/s of interest.

Owing to the added constraints of Time of Flight mass filter-based SIMS instruments (these only use MCP detectors), additional options have had to be developed. The two most common approaches include the use of the Poisson statistic correction approach and the introduction of the Extended Dynamic Range (EDR) filter. The Poisson correction approach, applied during data collection, is a statistic method that attempts to correct for the dead time effect suffered. Although effective, this method is applicable only over a limited intensity range, i.e. the maximum correction extends only over a factor of two or three times the maximum count rate the MCP can handle. EDR filters are applicable over a much greater intensity range as they can attenuate the intense signal by a factor of $10-100$ (dependent on the filter used) before it reaching the detector. This is made possible by redirecting the portion of the secondary ion pulse containing the intense signal through one of these filters while ensuring the remainder of the secondary ion pulse carries on directly to the detector (the latter is to retain detection limits/sensitivity to the less intense signals). Filtering of each secondary ion pulse in this manner requires not only extremely fast electronics but also the pre-selection of the signal to be filtered. Successful application of this approach can increase the dynamic range of such instruments by up to two orders of magnitude.

## 4.3  SUMMARY

SIMS has experienced extensive growth and sophistication within many divergent fields over the past few decades. With its commercialization starting in the late 1960s, there now exist numerous instrument types and geometries (price tags ranging from ~1 to several million dollars US). As a result, SIMS is now considered the most heavily used of the ion spectrometries (these are techniques that measure ion emissions) for examining submicron scale regions within solids to high sensitivity, dynamic range, and detection limits, all with relative ease (minimal sample preparation, etc. is needed).

The application of SIMS can broadly be subdivided into two modes, that being Static SIMS and Dynamic SIMS. Static SIMS is a methodology used to examine the elemental and molecular distributions present over the outermost monolayer present on any solid. Dynamic SIMS examines the elemental and molecular distributions as a function of depth. Note: The ability to examine molecular distributions is a relatively recent capability introduced through the implementation of large cluster primary ion beams. This has also opened up the possibility of three-dimensional imaging of molecular distributions.

SIMS derives compositional information from the solid under study by directing a focused energetic ion beam (these are termed *primary ions*) at the surface of interest. This induces the emission of atoms and molecules from the solid's surface, a small percentage of which exist in the ionized state (these ions are termed *secondary ions*). These secondary ions are then collected and focused into a mass spectrometer, hence the name secondary ion mass spectrometry, before detection.

Instrument designs are defined by the type of mass filter used within the secondary ion column. Indeed, there are three types that have been heavily commercialized in the field of SIMS, with more available in the parent field of Mass Spectrometry. Those commercialized are based around:

1. The Quadrupole mass filter
2. The Magnetic Sector mass filter
3. The Time of Flight mass filter

Along with these, the Ion Trap and Fourier Transform Ion Cyclotron Resonance have been examined. All provide capabilities and limitations specific to the mass filter. The Quadrupole mass filter, first introduced to SIMS in the 1970s, uses a two-dimensional oscillating electric field to influence the trajectories of ions. As the trajectories are depended on the ion's $m/q$ ratio (defined by the *Mathieu stability diagram*), specific ions can be selected by adjusting the RF and DC voltages, with all other ions discarded. The popularity of this type of mass filter stemmed from their small size, ease of use, and fast switch characteristics provided. However, owing to the poor transmission ($< 1\%$) and the relatively poor mass resolution available in SIMS instruments, Quadrupole mass filters have steadily been displaced by Magnetic Sector and Time of Flight mass filters.

The Magnetic Sector mass filter was the first mass filter used in both Mass Spectrometry and SIMS. As recognized in the early twentieth century, placing a magnetic field perpendicular to the direction of an ion beam induces a displacement in the respective trajectories that is proportional to the field strength and the ion's $m/q$ ratio. Thus, by scanning this field, mass spectra can be acquired. Although stabilization times are the longest of the mass filters, and the fact that these are the largest in size, these mass filters gained popularity, particularly in the area of Dynamic SIMS. This is primarily due to the high transmission ($10-50\%$) and high mass resolution possible (up to and above 25,000 in commercially available SIMS instruments).

The Time of Flight mass filter, first appeared in SIMS in the 1980s, is conceptually the simplest. This is realized as mass filtering is carried out by recording the time for an ion to travel some distance. Mass separation then occurs as heavier ions of the same charge travel more slowly. The relatively late appearance of these mass filters is, however, attributed to the timing/pulsing electronics required. Pulsing can be carried out on the primary beam (best for Static SIMS applications) or the secondary beam (best for Dynamic SIMS applications), with most instruments utilizing the former. Such instruments have gained significant popularity in both Static and Dynamic SIMS owing to their high transmission (up to 100%) and high mass resolution (10,000 is typical in commercially available instruments, although this is theoretically unlimited).

To retain mass resolution, energy filtering of the secondary ions is also needed. Time of Flight mass filters of the Reflectron or TRIFT designs provide this within the respective mass filter. In all other cases, the separate energy filter is placed before the respective mass filter. The one exception lies in the case of Reverse Geometry Magnetic Sector-based instruments as these situate the energy filter after the mass filter. Energy filters induce an energy-dependent deflection using an electrostatic field.

Electrostatic deflectors, lenses and secondary ion extraction fields make up the remainder of the ion optical elements within the secondary ion column. These are utilized to ensure that the highest possible fraction of the secondary ions reach the detector. Detectors used in SIMS comprise of Faraday Cups for the measurement of intense signals and Electron Multipliers for pulse counting of less intense signals. Electron Multipliers come in several different designs referred to as Discrete

Dynode Electron Multipliers, Channeltrons, and MCPs. Which one is used depends on their relative capabilities and limitations.

Primary ion columns contain one or more ion sources, along with the associated electrostatic lenses, deflectors, and a mass filter for producing an isotopically pure focused beam with impact energies that are generally within the 0.1–50 keV range. The associated electrostatic lenses, deflectors, and mass filter are similar to those used in the secondary ion column. The one exception is the Wien mass filter. This uses opposing electrostatic and magnetic fields to ensure the purity of the resulting primary ion beam. Wein filters are only found on primary ion columns in SIMS instruments, owing to their poorer mass resolution capabilities.

The ion sources used in SIMS fall into five different groups these being:

1. EI sources
2. Duoplasmatron sources
3. RF plasma sources
4. Surface ionization sources
5. Field ionization sources

EI sources are used for producing positive inert gas ion beams with some examples being $Ar^+$, $O_2^+$, and $SF_5^+$. These sources are also used in generating large cluster ions. Although not producing the highest currents or smallest spot sizes, these highly reliable sources are commonly found in many SIMS instruments.

The Duoplasmatron and RF are both plasma sources used for producing the same ions as in an EI source, as well as $O^-$, but to greater efficiency (the RF source displays by far the greatest efficiency). These are of interest because of the extensive secondary ion yield enhancement of positive secondary ions induced by Oxygen. The Duoplasmatron is used extensively in Dynamic SIMS instruments for the analysis of mostly electronegative elements.

Surface Ionization sources are used for producing alkali ion beams ($Cs^+$ is by far the most common in SIMS). These are of interest because of the extensive secondary ion yield enhancement of negative secondary ions induced by Cesium. As a result, Surface Ionization sources are used heavily in Dynamic SIMS instruments for the analysis of mostly electropositive elements.

Field Ionization sources, also referred to as Liquid Metal Ion Guns (LMIGs), are used for producing small spot $Ga^+$, $In^+$, $Au_n^+$, $Bi_n^{q+}$ ion beams, and so on where $n$ and $q$ are integer values of one or greater. These are used extensively in pulsed primary ion Time of Flight instruments for both Static and Dynamic SIMS analyses. Owing to their low currents, sputtering is typically carried out using an EI or Surface Ionization source with both pulsed in an interleaved manner to each other.

As with most surface analytical techniques, the ability to produce a vacuum played a pivotal role in defining the introduction timeline of SIMS. The importance of vacuum is realized as this allows the passage of charged particles over the distances needed (typical flight path is one to several meters) and provides for a contamination free surface (required in the analysis of monolayer composition through to in-depth distributions). Such requirements can be understood within the context of the kinetic theory of gases. The production of vacuum requires that specific materials be used in instrument construction, along with specialized pumping systems to produce ultimate pressures of $10^{-9}$ Torr or better.

As in any of the physical sciences, understanding the capabilities and limitations of the instrumentation applied not only provides for the maximization of the potential of the respective technique, but also allows for a fuller understanding of the results obtained (this includes understanding whether the result obtained is an effective representation of the substrate analyzed or a distortion introduced by the measurement process).

This understanding has also provided for the further extension of the capabilities of the SIMS technique, with the introduction of large cluster ion sources being a prime example. Indeed, such sources have not only resulted in a significant resurgence in research (this started around the beginning of the twenty-first century) but has opened up new fields of application not previously thought possible, i.e. molecular depth profiling.

In closing, it seems pertinent to state the quote by the nineteenth century chemist Sir Humphry Davy "*Nothing tends so much to the advancement of knowledge as the application of a new instrument. The native intellectual powers of men in different times are not so much the causes of the different success of their labors, as the peculiar nature of the means and artificial resources in their possession.*"

# CHAPTER 5

# Data Collection and Processing

## 5.1 THE ART OF MEASUREMENT

If correctly applied, Secondary Ion Mass Spectrometry (SIMS) has the ability to provide a wealth of information on the elemental and/or molecular distributions on or within the near surface region of any solid substrate of interest. There are, however, a myriad of conditions that can be applied before, during, and after data collection. As a result (and as stated in Section 4.1), understanding the ideal conditions required for a specific type of analysis is often considered an *art*. Gaining this understanding requires, at the very least, knowledge of:

1. the optimal sample handling and preparation procedures (should they apply) required for specific substrate types,
2. the data collection methodologies available, inclusive of primary ion choice, analytical conditions applied (inclusive of all primary and secondary ion column parameters), the signal measured, and so on,
3. the degree and type of damage that can be suffered by the sample during analysis, i.e. understanding what is real versus an analysis-induced distortion, and
4. the reference procedures, and when the need arises, the quantification methodologies that can be applied to the collected data.

Indeed, it is often stated that *to be comfortable with SIMS is to be unaware of the potential errors that can be introduced during and/or following analysis*. As an example, if a $Cs^+$ primary ion beam of too low an energy or too high an impact angle (relative to the sample normal) is used in the depth profile analysis of dopants in a Silicon substrate, crater base roughening in the form of ripple topography will ensue after some sputtering time. Such topography will induce variations in all secondary ion signals, including that of major elements not exhibiting compositional variations. The loss of depth resolution also noted may be misinterpreted as arising from some Diffusion or Segregation process.

In addition, SIMS is a highly diverse analytical technique. This diversity pertains to both the capabilities supplied and the instrumentation available. As a result,

*Secondary Ion Mass Spectrometry: An Introduction to Principles and Practices*, First Edition.
Paul van der Heide.
© 2014 John Wiley & Sons, Inc. Published 2014 by John Wiley & Sons, Inc.

there exists extensive freedom in the choice and conditions under which the instrumentation can be used. For example, commercially available SIMS platforms all consist of several primary ion sources whose operational parameters can be freely adjusted over a wide range depending on the analytical needs. These platforms also provide the possibility of routinely switching from the collection of positive secondary ions to negative secondary ions. Most other analytical techniques generally use a single source operated under fixed conditions to produce the specific signal of interest. For example, the extremely effective surface analytical technique of X-ray Photo-electron Spectroscopy (XPS) uses photons to produce and collect electrons. This reduced instrumental complexity removes the requirement for routine beam realignment, which is a standard prerequisite for effective SIMS analysis.

This chapter aims to bring insight to the common methodologies for reducing various distortions prevalent in SIMS. From these methodologies, sample-specific Standard Operating Procedures (SOPs) or Best-Known Methods (BKMs) can be developed.

This insight is launched with a discussion on the data formats available to SIMS and their definitions in Section 5.1.1. Sample handling procedures are then introduced in Section 5.2. After this, data collection procedures, along with commonly encountered problems, are presented in Section 5.3. For completeness, the primary data processing procedures, inclusive of quantification, are covered in Section 5.4. Examples from the highly diverse areas of Materials Science, The Earth Sciences, and the Biosciences are presented.

### 5.1.1 Data Formats and Definitions

As illustrated in Figure 1.1, data can be recorded in three formats, these being:

1. Mass spectra (this option is available to all forms of SIMS)
2. Depth profiles (this option is not available in Static SIMS)
3. Imaging, whether in two or three dimensions (only spatial two-dimensional imaging is available in Static SIMS)

These formats are covered in the following subsections along with the definitions of:

1. Mass resolution
2. Depth resolution
3. Spatial resolution

Note: Resolution defines the ability to separate a signal into its constituent parts, whether as a function of its mass, the depth from which it originates, its spatial location, etc.

**5.1.1.1 Mass Spectra** A mass spectrum (singular for spectra) is a plot of the intensity versus the mass, or more correctly, the mass to charge ratio ($m/q$) of the recorded secondary ions. The $m/q$ scale is often plotted in mass units alone as most secondary ions possess a single unit of charge. This is plotted on the abscissa. Intensity, plotted along the ordinate, is most commonly in units of counts per second (cps). This is typically plotted on a logarithmic scale because of the large dynamic range available.

A mass spectrum, derived by scanning the mass filter over some predefined $m/q$ ratio range, constitutes all the secondary ions of the polarity of interest (only one polarity can be collected at a time). Such spectra are of interest when the sample type is unknown (this allows for the identification of the elemental and/or molecular constituents) and/or when information on the optimal signals for the acquisition of images and/or depth profiles is required. This option is available and commonly used in all applications of SIMS, whether in Static or Dynamic modes.

An example of a mass spectrum collected on a Quadrupole-based SIMS instrument operated at nominal mass resolution is shown in Figure 5.1. This negative secondary ion spectrum, collected from a Silicon surface bearing a native oxide, is a result of energetic $Cs^+$ primary ion impact applied under Dynamic SIMS conditions. As discussed in Section 3.3.2, $Cs^+$ primary ions enhance the yield of negative secondary ions from electronegative elements while suppressing those from electropositive elements. Such conditions also introduce significant damage such as to remove any molecular information that may originally have been present. As a result, the most intense peaks, noted over the mass region shown, are from the major isotopes of Oxygen (nominal mass of major isotope is 16 u) and Silicon (nominal mass of major isotope is 28 u).

Along with these signals are signals from all of the other isotopes of Oxygen and Silicon (relative abundances of the isotopes are defined in Appendix A.2). As the secondary ion yield for different isotopes of the same element is not

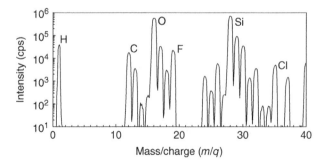

**Figure 5.1** A mass spectrum of the negative secondary ions emanating from a Silicon substrate collected on a Quadrupole SIMS instrument as a result of $Cs^+$ primary ion impact operated under Dynamic SIMS conditions. The minimal mass resolution available on these types of instruments allows only for ions of nominal mass difference to be separated.

strongly affected in the secondary ion formation process (the effects that are noted, referred to as *isotope* or *mass fractionation*, are quantifiable as discussed in Section 3.3.2.2.3), the intensities of all isotopes follow actual isotope abundances to within a small fraction of a percent.

In addition to the signals from all the above-listed elements are signals from various other elements. These elements, which may be intentionally introduced (examples in the semiconductor industry include Boron, Phosphorus, or Arsenic) or unintentionally introduced (these arise from exposure to some prior environment) include the following: Hydrogen (nominal mass of the major isotope is 1 $u$), Carbon (nominal mass of the major isotope is 12 $u$), Fluorine (nominal mass of the major isotope is 19 $u$), Chlorine (nominal mass of the major isotope is 35 $u$).

More complex examples of mass spectra common to the area of Materials Sciences, the Earth Sciences, and the Biosciences are shown in Figure 5.2(a–c). These were collected on Time-of-Flight (Figure 5.2a and 5.2c), Magnetic Sector (Figure 5.2b), and Fourier Transform-Ion Cyclotron Resonance (Figure 5.2c)-based SIMS instruments. In all cases, improved mass resolution was required to separate the increased prevalence of spectral overlaps or *isobaric interferences*, also referred to as *mass interferences*.

Such interferences arise when there exist secondary ions of different elements and/or molecules that are of the same *nominal m/q* ratio (*nominal* values are those rounded to the closest whole mass number). These become more prevalent:

1. At higher $m/q$ (owing to the greater likelihood of molecular combinations of the lighter elements).
2. As the spread in the $m/q$ ratio of the secondary ion signals increases (this increases the overlap with other secondary ion signals of similar $m/q$ values).
3. When Static SIMS conditions are employed (owing to the greater prevalence of molecular ions present in the respective mass spectra).

The spread in the $m/q$ ratio for any specific secondary ion signal is a distortion introduced by the instrument optics and hence is a function of the analysis conditions applied. This is realized as isotopic mass is a highly discrete value that remains fixed under nonrelativistic conditions (values are listed in Appendix A.2). Examples of the spread in $m/q$ ratios can be seen in Figure 5.2(a–c). Note: All secondary ions suffer the same spread under the same analytical conditions.

Improving the mass resolution (definition is covered in Section 5.1.1.1.1) reduces this $m/q$ spread, thereby allowing for the separation of secondary ion signals of similar (nominal) $m/q$ ratios. Minimal spread is noted in the FT-ICR mass spectra shown in Figure 5.2(c) where an extremely High Mass Resolution (HMR) was used (Smith et al. 2013).

In addition to this, secondary ion peaks can exhibit a round or a flat peak top shape (c.f. Figure 5.2a and b). This too is an instrumental manifestation introduced and controlled by the analytical conditions applied. For example, the round top

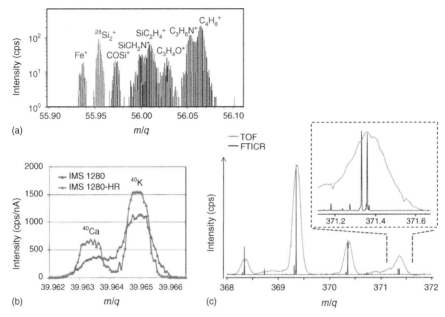

**Figure 5.2**  Examples of mass spectra collected under high-mass resolution conditions over: (a) the 55.90–56.10 *m/q* range from a Silicon wafer on a Time-of-Flight (TOF)-based SIMS instrument (Supplied by Physical Electronics), (b) the 39.962–39.966 *m/q* range from an Adularia mineral sample on two different IMS-1280 Magnetic Sector-based SIMS instruments (care of Cameca), and (c) the 368–372 *m/q* range from a rat brain section on a TOF-based SIMS instrument in imaging mode (this affects mass resolution as covered in Section 5.3.2.1.2) overlaid with that from a Fourier Transform-Ion Cyclotron Resonance (FT-ICR)-based SIMS instrument. Reproduced with permission from Smith et al. (2013) Copyright 2013 Springer.

arises from the Gaussian broadening suffered by the secondary ion beam as it passes through the respective ion columns. This occurs in all SIMS instrument types, albeit minimized in well-designed systems when operated under HMR conditions. Likewise, flat-top peaks arise from the refocusing action experienced by secondary ions in double focusing Magnetic Sector-based instruments. These are useful in studies requiring high sensitivity (examples include isotope ratio and dopant measurements) as this increases the insensitivity of the secondary ion signal to slight variations in the various electrostatic or magnetic fields applied in focusing and filtering of these secondary ions. Note: Such measurements maybe derived through peak switching (the act of switching the electro-magnet such that the secondary ions of the desired mass to charge ratio are collected in an iterative manner). Alternatively a multi-detection system may be used. To re-iterate, peak shapes and peak widths are a function of the instrumentation and the analysis conditions applied. Instrumentation is covered in Chapter 4.

Lastly, mass spectra can be displayed as line spectra. This is an artificial display mode introduced in the data processing stage in which each line represents the averaged mass of the respective secondary ion signal or signals of interest.

As described in Section 3.3.3.4, multiply charged secondary ions can also be present. Owing to the increased charge, these appear in mass spectra at a lower $m/q$ ratio than their singly charged counterpart ions. For example, the $^{28}Si^{2+}$ ion is situated at a nominal $m/q$ ratio of 14 (this can be misinterpreted as that of the extremely low yielding $^{14}N^+$ ion), whereas the $^{28}Si^{3+}$ ion is situated at an $m/q$ of 9.3333. Note: Multiply charged negative secondary ions are rarer than their positively charged counterparts.

### 5.1.1.1.1  *Mass Resolution*  Mass resolution describes the ability of an instrument to separate ions of different masses (Note: The $m/q$ ratio is more correct). This is defined as:

$$R = \frac{m}{\Delta m} \tag{5.1}$$

where $m$ is the average mass of the two signals of interest and $\Delta m$ their mass difference between the two signals of interest.

In most instruments, $R$ can be varied from a value slightly better than unity (that needed to separate signals that are one mass unit apart) to many thousands (that needed to separate signals that are a small fraction of a mass unit apart). The former is referred to as *nominal mass resolution,* whereas the latter is referred to as *high mass resolution.* HMR is covered further in Section 5.3.1.3.2.

Examples of similar mass ions commonly measured in Dynamic SIMS include the following:

1. $^{31}P^-$ from $^{30}Si^1H^-$. As their masses are 30.973762 $u$ and 30.981595 $u$, respectively, the mass resolution required to separate these is at least 3955. This is routinely attainable in Magnetic Sector and Time-of-Flight-based SIMS instruments (see Section 4.2.3.1)
2. $^{238}U^+$ and $^{238}Pu^+$. As their masses are 238.050788 $u$ and 238.049560 $u$, respectively, the mass resolution required to separate these is at least 193,804. This is out of reach of all but Fourier Transform Ion Cyclotron Resonance (FT-ICR)-based SIMS instruments (see Section 4.2.3.1.4).

An instrument's mass resolution, under the conditions applied, can be defined from the peak shape of the respective ions via one of two methods. These will otherwise be referred to as:

1. The single-ion method
2. The double-ion method

In the single-ion method, $R$ can be defined from the respective isobaric interference-free peak, as equal to the average value of the peak mass ($m$) divided by the peak width ($\Delta m$) at some percentage of the maximum peak height. Although

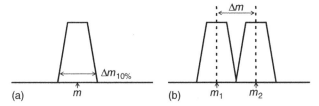

**Figure 5.3**    Mass resolution definitions referred to as the (a) single-ion method and (b) the double-ion method for flat-top peaks. The same definitions apply to round topped peaks.

the peak width at 10% of the peak height is commonly used, the 1% and 50% values, with the latter being the Full Width at Half Maximum (FWHM), have been applied.

In the double-ion method, $R$ is defined as the average mass of the two peaks of interest divided by their mass difference (the respective masses are taken as those noted at the respective peak maxima). These concepts are illustrated in Figure 5.3 using linear intensity scales for flat-top peaks (linear scales are used as this provides for improved visual sensitivity). The same concepts apply to round-top peaks.

### 5.1.1.2    Depth Profiling

Depth profiling provides SIMS with the ability to probe the substrate of interest to different depths. This can be carried out to a depth of a few nanometers to many microns over the spatial region of interest. In order to initiate depth profiling, Dynamic SIMS conditions must be applied. Recall that Dynamic SIMS describes the removal of one or more atomic/molecular layers per analytical cycle (see Section 4.1.2). Each data point in a depth profile thus represents an analytical cycle. The total depth accessed is a function of the total sputtering time applied, the sputtering conditions used, and the substrate type examined (see sputtering properties covered in Section 3.2).

A depth profile is a plot of the intensities or concentrations of the specific secondary ion or ions of interest (plotted on the ordinate) versus the sputtering time or depth (plotted on the abscissa). Intensities and sputtering time are the values recorded, whereas concentrations and depth are those derived via the procedures discussed in Section 5.4. As in the collection of mass spectra, a logarithmic scale is typically used in plotting the intensity/concentration as this most effectively relays trends over the large dynamic range, as well as the high detection limits made available by SIMS. Improved visual sensitivity is noted when using a linear intensity/concentration scale. An example of a fully quantified elemental depth profile is shown in Figure 5.4(a), along with a partially quantified (abscissa only) molecular depth in Figure 5.4(b).

A depth profile, in units of intensity versus sputtering time, is most typically constructed in real time (in Time-of-Flight-based SIMS instruments, these can also be constructed retrospectively for ions that were not initially thought of interest). This constitutes plotting the intensities of the secondary ions of interest per analytical cycle, with each cycle taking some predefined time. The overall time applied is otherwise referred to as the *sputtering time*. Only those secondary ions of a specific

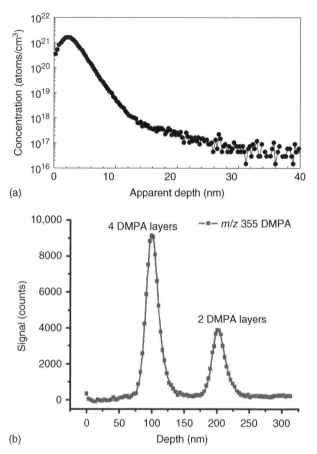

**Figure 5.4** SIMS depth profiles resulting from (a) 1 keV $Cs^+$ impact on an As-doped Silicon wafer (the quantified Arsenic distribution was defined from the $AsSi^-$ secondary ion signal once normalizing to the $Si_2^-$ secondary ion signal followed by the application of the required intensity to concentration and sputter time to depth scale conversions as outlined in Section 5.3) and (b) alternating Langmuir–Blodgett multilayer films of Barium Arachidate and Barium Dimyristoyl Phosphatidate (the Protonated Dimyristoyl Phosphatidate signal arising from 40 keV $C_{60}$ primary ion impact at low temperatures is the signal measured) resulting in a molecular delta-layered structure. Reprinted with permission from Lu, Wucher, and Winograd (2011). Copyright 2011 American Chemical Society.

polarity can be collected at one time. If secondary ions from the opposite polarity are also needed, an additional depth profile must be collected from a representative (typically adjacent) area.

To ensure the best possible precision, the signal collected should remain spatially homogeneous within the volume probed, or at the very least be representative of any larger scale variations. If there exists doubt, spatial images should be examined (either before or during depth profiling) to identify an area satisfying these prerequisites.

The signals recorded are typically those displaying the highest intensities along with minimal isobaric (mass) interferences. Isobaric interferences and methods to reduce these are discussed in Section 5.3.1.3. In some cases, less intense signals may need to be collected to avoid detector saturation issues (see Section 4.2.3.3.2.2) and/or excessive isobaric interferences (see Section 5.1.1.1). When isobaric interferences cannot be removed through the selection of other isotopes and so on, HMR conditions can be implemented. The sputtering conditions used will depend on:

1. The depth region of interest, i.e. to elucidate thin films or signal variations occurring over some small depth interval, will require a slower sputtering rate along with conditions resulting in improved depth resolution. The opposite conditions will be needed if larger volumes and/or improved detection limits are of interest.
2. The signals of interest, i.e. elemental versus molecular secondary ions, and so on.

Quantification of the intensity scale can be carried out once secondary ion matrix effects are removed. This is possible through the corresponding analysis of substrate-specific reference materials. This along with the conversion of sputtering time scale typically is carried out retrospectively. In specific cases, secondary ion matrix effects can be reduced, thereby easing quantification of the intensity scale. Specific examples of this can be noted in the collection of Cesium cluster ions ($MCs^+$ and $MCs_2^+$, where M is an isotope of the element of interest) and in the analysis of near homogeneous organic substrates or ultra thin homogeneous layers deposited on metallic surfaces. Note: Cesium cluster ion analysis requires the use of a Cesium primary ion beam and comes at the cost of reduced detection limits. Quantification is covered in Sections 5.4.2 and 5.4.3.

*5.1.1.2.1 Depth Resolution* Depth resolution describes the ability to separate two or more regions exhibiting compositional variations as a function of depth. SIMS is intrinsically capable of providing exceedingly high-depth resolution values, i.e. values to less than 1 nm. The instrumental factors affecting depth resolution and methods for reducing these effects are discussed in Section 5.3.2.4. The sputter-induced factors affecting depth resolution are covered in detail in Section 3.2.

Depth resolution is most commonly defined as the depth over which a signal from some abruptly appearing layer climbs from 16% of its maximum intensity to 84% when plotted on a linear scale as this represents two standard deviations ($\pm 1\sigma$) of the convolution of a Gaussian function with a step function. This is illustrated in Figure 5.5. This definition is also applied to decaying signals from abruptly terminated layers. Caution must, however, be exercised when matrix effects and/or radiation-enhanced segregation are active, as these can modify the value derived relative to the absolute sputter-induced depth resolution.

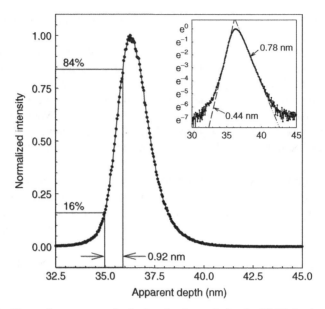

**Figure 5.5**    The various means of relaying depth resolution in SIMS depth profile analysis. Note: Intensities are plotted on a linear scale. In the inset is the same data with the intensity plotted on a natural logarithmic scale (decay lengths are described using natural logarithmic arguments). The data is of the sixth Boron delta layer from the Boron delta layer Silicon structure (a structure with atomically abrupt interfaces with the composition following a delta-like function with depth) examined under 0.5 keV $O_2^+$ primary ion conditions at normal incidence, i.e. from that shown in Figure 5.25.

Growth or decay lengths are also commonly used in SIMS to provide a measure of the depth resolution. This is derived from the slope, or inverse of the slope, of the profile during the onset or decay of the respective signal if plotted on a log or natural log scale as illustrated in the inset of Figure 5.5. As both definitions are in common usage, caution must be exercised when comparing values. The linear trend noted (see dashed lines) arises from the fact that the atomic mixing induced by the primary ion beam (that responsible for the loss in depth resolution) is an exponential function of the energy deposited.

Only after all potential distortions are removed or accounted for, can the fundamental depth resolution limit be defined. As covered in Sections 3.2.3 and 5.3.2.4.1, this limit is defined by sputter-induced atomic mixing.

**5.1.1.3  *Imaging***    Imaging can be used to reveal the spatial and/or volumetric distributions of elements or molecules represented by their respective secondary ions. Note: Although all elements/molecules can be detected in SIMS, the conditions used will enhance/quench specific signals. Imaging can be carried out if the position from which the secondary ions emanated from can be relayed pictorially, with the relative intensities illustrated using some gray or color scale. Although a logarithmic intensity scale can be applied, linear intensity scales are preferred because of improved visual sensitivity.

Two-dimensional images are acquired in SIMS via either the:

1. *Microprobe* mode
2. *Microscope* mode

The *microprobe* mode can be applied in most any instrument, whereas the *microscope* mode is restricted to specifically designed instruments (see Section 5.3.2.2).

Three-dimensional images are constructed by overlaying two-dimensional images collected as a function of sputtering time in much the same way a depth profile is constructed. An example of this stacking approach is illustrated in Figure 1.5. The volume becomes apparent when these two-dimensional images, which are stacked over each other, are shifted along the sputtering time/depth axis. Most three-dimensional images are, however, heavily distorted along the depth axis because:

1. The depth scale is typically a small fraction of the analyzed area (the former is typically one to two orders less than the latter) with the two not usually scaled.
2. The presence of any surface topography/structure is typically imaged as a flat surface with a uniform sputter rate then assumed in all reconstructions.
3. Different crystal planes tend to sputter at different rates (this arises from differences in surface atomic densities as discussed in Section 2.1.1.2).

As covered in Section 5.3.2.4.5, sample rotation (also referred to as Zalar rotation) during image depth profiling can minimize the latter. Alternatively, surface topography measurements carried out before and after image depth profiling can account for the latter two. Indeed, the effectiveness of this approach has been demonstrated via *ex-situ* Atomic Force Microscope (AFM) measurements (van der Heide 1998; Robinson et al. 2012). Based off of the success of this approach, Ion-Tof has designed a hybrid AFM/TOF-SIMS instrument.

As in depth profiling, the signals recorded are typically those displaying the highest intensities and lowest isobaric interferences. Two examples of three-dimensional imaging are shown in Figures 5.6 and 5.7. Note: Quantification of such images can be extremely difficult because of the inability to account for site-specific matrix effects (see Section 3.3.1.2) and/or the limitation in effective reference materials/procedures.

In the case of elemental distributions, semi-quantitative imaging can be aided through the collection of Cesium cluster ions ($MCs^+$ and $MCs_2^+$, where M is an isotope of the element of interest). This approach, however, requires the use of a $Cs^+$ primary ion beam and yields poor sensitivity and detection limits relative to their atomic secondary ion emissions. When using this approach, the $Cs^+$ and/or $Cs_2^+$ signals should also be collected as these are needed in normalizing the respective

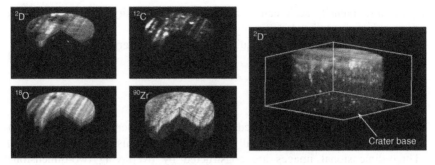

**Figure 5.6**  Three-dimensional images of the $^2D^-$, $^{12}C^-$, $^{18}O^-$ and $^{90}Zr^-$ secondary ions emanating from a $ZrO_2$ film on a Zirconium-based alloy (from a section of tube used in the nuclear industry) as a result of 14.5 keV $Cs^+$ primary ion impact. Of particular interest was the Deuterium ingress as this provides information on the corrosion processes suffered. Two visualization methodologies were applied. One is a "pie" type display with a slice taken out to reveal internal faces of the volume examined (any section can be cut). A three-dimensional rendition was then generated to reveal the overall internal distribution of the $^2D^-$ distribution (appears as light gray) with the remaining volume made transparent. The arrow in the rendition, which can be freely rotated along any axis, defines the crater base. The field of view is 150 µm in diameter, with the depth sputtered being $\sim$5 µm. These were collected using the microscope mode to a spatial resolution approaching $\sim$1 µm on a Magnetic Sector instrument. Reproduced with permission from van der Heide and Fichter (1998) Copyright 1998 John Wiley and Sons.

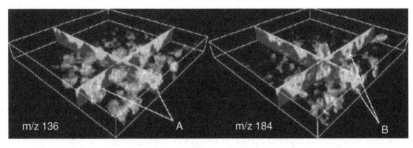

**Figure 5.7**  The three-dimensional renditions of 136 and 184 $m/q$ secondary ions from a Silver foil-supported cell membrane arising from 40 keV $C_{60}$ cluster ion impact. The 136 $m/q$ signal is representative of the adenine distribution, whereas the 184 $m/q$ signal is representative of the phospholipids distribution (the respective protonated signals were imaged). Orthogonal slices are shown (signified as: A and B) to facilitate visualization of the internal structure. All data was collected using the microprobe mode to a spatial resolution approaching $\sim$1 µm on a Time-of-Flight-based SIMS instrument operated in the continuous primary ion beam/pulsed secondary ion beam mode. The field of view is 180 × 180 µm square. Images reproduced with permission from Vickerman et al. (2009) Copyright 2009 Elsevier.

$MCs^+$ or $MCs_2^+$ signals. This approach is discussed further in Sections 5.1.1.3.1 and 5.4.3.

As for molecular distributions, reduced matrix effects can be experienced. This is particularly evident for specific molecular secondary ion signals from organic substrates examined under large cluster primary ion impact. As covered in Sections 3.3.4.1 and 5.4.3, this results from the reduced variations in the primary components making up the matrix and the reduced influence of the respective primary ions used (these are not chemically active). An example is shown in Figure 1.5.

*5.1.1.3.1 Spatial Resolution*  Spatial (lateral) resolution describes the ability to separate signals from neighboring regions (those in the spatial plane) that exhibit compositional and/or sputter rate variations. Like many other imaging techniques, spatial resolution is defined as the distance over which some signal from the edge of some localized region climbs from 16% of its maximum intensity to 84% when plotted on a linear scale. Other definitions, such as the 20%–80% intensity values can, however, also be found. These are generally applied owing to their ease of use and the artificially improved numbers provided. The 16%–84% definition is the preferred method because, as in depth resolution, this represents two standard deviations ($\pm 1\sigma$) of the convolution of a Gaussian function with a step function.

An example of the application of this method can be seen when examining Figure 5.8(a–c). These images were collected by different surface imaging techniques (Scanning Electron Microscopy (SEM), Energy Dispersive X-ray (EDX),

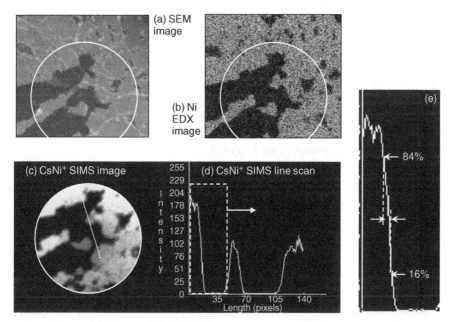

**Figure 5.8**  Chalcopyrite grains imbedded in a prepolished conductive epoxy substrate as imaged by (a) SEM, (b) EDX (the Ni distribution shown), and (c) SIMS operated in the microscope mode (the CsNi$^+$ secondary ions are displayed). The SIMS image resolution is defined via line scan analysis (see white line in SIMS image) as illustrated in (d) and in (e). Raw MCs$^+$ signals, where M is an isotope of Nickel, are shown.

and SIMS) from Chalcopyrite grains embedded in an epoxy support. Line scan analysis of the SIMS image is shown in Figure 5.8(d) and 5.8(e) with the 16%–84% spatial resolution definition shown in the latter. The $CsNi^+$ image was used in this case as this population suffers reduced matrix effects. Although the spatial resolution from such cluster ions tends to be slightly poorer than that obtained from their monoatomic counterparts (consistent with the belief that such clusters are formed at a short distance above the surface as discussed in Section 3.3.3.1), the often significant matrix effects exhibited by monatomic secondary ions can (and often does) distort the apparent spatial resolution from that actually supplied by the instrument.

The spatial (lateral) resolution is a function of a number of parameters inclusive of:

1. The imaging mode used (see Section 5.3.2.2.)
2. The aberrations present (see Appendix A.6.2.1)
3. The secondary ion yield (high yields allow for optimal resolution)
4. The pixilation applied (pixels should be smaller than required resolution)
5. The matrix effects active (these can distort the actual spatial resolution)

Only once all possible instrumental contributions and potential distortions are removed or accounted for, can the fundamental spatial depth resolution limit be defined. This limit is defined by the collision cascade volume (Collision cascades are discussed in Section 3.2.1), which typically extends over a surface diameter of $\sim 10$ nm. This is assuming additional atomic mixing occurring during the sputtering process is minimized (atomic mixing and its effects are discussed in Section 3.2.3).

Optimally, spatial resolution values should be derived from well-defined reference materials such as the BAM-L002 or BAM-L200 series (BAM is short for *Bundesanstalt für Materialforschung und -prüfung*), which comprise of robust nano-scale strip patterns.

## 5.2 SAMPLE PREPARATION AND HANDLING

Unlike most analytical techniques, SIMS generally requires little in the way of sample preparation. Indeed, all that is generally required is:

1. The sample is Ultra High Vacuum (UHV) compatible, i.e. does not evaporate
2. The sample is of a size amenable to that of the instrument
3. The substrate's surface remains clean before and during analysis
4. The substrate's surface is flat and smooth over the area of interest

Smooth surfaces are required as this reduces topographically induced secondary ion yield variations and is important when HMR in Time-of-Flight instruments using pulsed primary ion beams is required.

As far as dimensions are concerned, samples should optimally be the size and geometry of a small coin. This is suggested as it allows for ease of handling and is of a sufficiently small surface area to allow for acceptable pump-down speeds. Note: Access ports on commercially available instruments are also of limited dimensions, as this keeps their cost of purchase and ownership within acceptable levels (larger samples will require larger introduction chambers, larger analysis chambers, greater pumping speeds, etc.). There are specialized instruments that can accept entire wafers. However, because pump-down times scale with sample size, there must be a good reason for introducing such large samples. Note: The same information can often be obtained through cleaving carefully selected areas.

One of the reasons why UHV conditions are required in SIMS is that this controls the re-adsorption of gas phase contaminants onto the surface of interest during analysis. Note: Such adsorption issues are accelerated over regions experiencing electron (for charge neutralization) or ion impact, particularly if freshly sputtered (a pure Silicon surface, for example, is highly reactive, hence the reason for the native oxide formation that occurs within seconds on air exposure). A particular instance where this is noted is in the analysis of elements present in the gas phase, i.e. those of Hydrogen, Carbon, Nitrogen, and Oxygen. Such analysis can be facilitated through:

1. Further improvements in the vacuum whether resulting from increased pump-down times and/or the use of Titanium sublimation pumping
2. Use of a Liquid Nitrogen Cryo-shroud surrounding the sample as this enhances the removal of condensables from the sample region
3. Presputtering of some effective gettering material such as Titanium or Silicon before the analysis of the material of interest

The requirement for UHV also places limitations on the sample size (larger samples require longer pump-down times) and the types of samples that can be examined, i.e. the sample must be UHV compatible. In other words, the sample cannot de-gas or deteriorate excessively during analysis.

There are, however, specific cases in which more extensive sample preparation is required. This is most commonly noted in the application of SIMS to the Biosciences and the Earth Sciences. Specific protocols found in the primary three areas in which SIMS is applied are covered in Sections 5.2.1–5.2.3.

### 5.2.1   Preparation in the Materials Sciences

Although samples from the Material Sciences exhibit the greatest diversity in morphology, they tend to be easiest to handle. Indeed, all that is generally required is that the sample be cut/cleaved such that it can be affixed to the respective sample holder. This ease arises from the fact that most are in the form of solids of sizes amenable to handling with some examples including Silicon wafers, Metallic coupons, or oxide fragments. Additional care must, however, be taken when working with vacuum-sensitive materials and/or hazardous materials. In the case of radioactive materials, specifically designed shielded SIMS instruments should be applied.

If the samples are of a stable solid form and have spatial dimensions greater than a few millimeters, they can be directly inserted into or affixed (clipped or glued) onto the respective sample stub/platform with no special preparation required. Care must, however, be taken to ensure that any clips used are not in the immediate vicinity of the area to be analyzed. Aside from potentially blocking the primary or secondary ion beams, this can distort the secondary ion extraction field thereby influencing the secondary ion signal intensities recorded. For the same reason, analysis toward the very edge of a sample or around topographical features should be avoided where possible.

If the samples are of a stable solid form but have spatial dimensions smaller than a few millimeters or are in the form of powders, some alternative support is needed. One practice is to use Indium foil strips into which the sample is pressed using another piece of Indium foil and/or a precleaned spatula. The Indium strips are then affixed into/onto the respective sample holder. Any remaining loose powder should be removed before inserting the sample into the introduction chamber by inverting and tapping the sample holder and/or directing a pressurized $N_2$ flow via an $N_2$ gun over the surface.

If the samples are of a stable solid form but exist as nanosized powders, they can be dispersed in some inert solvent and then pipetted onto a substrate whereupon it is allowed to dry. Once dried, they can be inserted into the introduction chamber and pumped down accordingly. Samples in the form of Langmuir Blodgett films deposited on some solid support such as Silver or Silicon can also be examined. Additional care must however be taken to ensure that out-gassing rates are not excessive, and that a suitable sample volume remains during analysis. Owing to the highly specific nature of such samples, this may require a trial-and-error approach.

If the sample is of a liquid form under standard temperature and pressure conditions, they must be converted into a solid form that will be vacuum stable. This is generally accomplished through freeze-drying procedures covered in Section 5.2.3.

If thermal treatment is required before or during analysis, care must be taken with the materials used. Indeed, this negates the possible use of the solvent or Indium support approaches described earlier as well as the use of frozen samples. Thermal studies can be aided using specialized vendor-specific sample supports. Note, however, that no standardized sample holder presently exists, i.e. all are vendor specific, hence cannot be interchanged between instruments from different vendors.

### 5.2.2   Preparation in the Earth Sciences

Samples of interest to the Earth Sciences tend to be highly stable but insulating in nature. In addition, these rarely exhibit smooth flat surfaces unless polished. Samples may also be small in size, i.e. may take the form of millimeter- to micron-sized grains. As all of the above can pose problems in carrying out the highly precise and distortion-free SIMS measurements, specific sample preparation procedures have been developed.

In the case of small grains, sample preparation generally takes the form of:

1. Preparing a mold to the size amenable to the sample holder to be used (a shallow cylindrical mold with internal dimensions of either 10 mm or 25 mm in diameter and 5–10 mm deep is common in many SIMS instruments)

2. Mixing an epoxy resin in this mold. Note: The material used must be vacuum compatible. Specific resins that have been used include:

   a. Buehler No# 20-8130 resin with No# 20-8132 hardener
   Applying the 20-8135 release agent to the mold before pouring the epoxy helps in puck removal

   b. Aradite No# 502 along with Dodecenul Succinic Anhydride (DDSA) and Bensyldimethylamine (BDMA)

   c. KÖRAPOX No# 439
   On the other hand, Woods alloy or Indium once heated to its liquid form can be poured into the cavity of a sample holder.

3. Imbedding the grains of interest along with a suitable reference sample on the surface of the support used (epoxy, Woods alloy, Indium, etc.) before it cures/cools. Note: These should all be in close proximity to each other with their locations well characterized.

4. Curing for a sufficient time (overnight) such that these become sufficiently vacuum stable

5. Polishing down the front side of the cured puck until the surface roughness is less than 1 μm. This also exposes the internal regions of the respective grains. Over-polishing should be avoided as this will result in rounding of grain surfaces and the possibility of crevice formation. As Silica, Alumina, and/or diamond polishing media tend to be used, these will introduce Carbon, Silicon, Aluminum, and other signals into the respective mass spectra. Which signals are introduced will depend on the polishing solutions and epoxy resin used.

6. Ultrasonic cleaning of the pucks in soapy water for 5 to 10 minutes and then rinse with Deionized water. For Zircon U-Pb analysis, Ultrasonic cleaning in One-Molar Hydrochloric (HCl) acid solution is preferred as this reduces Lead contamination issues. Ultrasonic cleaning using other solvents such as ethanol may also prove useful.

7. Drying in a vacuum oven set between 50 and 80°C for an extended period of time (days). This helps reduce out-gassing when in the vacuum environment.

8. Applying a thin conductive coating to the surface of interest. Carbon or Gold can be used, with the latter being preferred because of its faster sputtering rate and the reduced isobaric interferences introduced (Isobaric interferences below 197 u for singly charged secondary ions (multiply charged ions noted in the positive secondary ion spectra are exceedingly weak) are not noted when Gold is used). This is particularly problematic in U-Th analysis as the

use of Carbon will introduce significant isobaric interferences. The thickness of the Coating should be sufficient to result in a Resistivity in the 10 Ω range. This is generally in the range 10–20 nm. Note: Trace impurities inclusive of Carbon, Fluorine, and so on are introduced in all forms of coating.

For samples of a few millimeters in diameter, thinned sections can be glued to a polished metallic cylinder using one of the epoxy resins described above or afixed using double sided tape. The latter will, however, de-gas more heavily. For even larger sized samples, a Petro-graphic cross-sectional disc should be cut with the face of interest polished in the same manner as mentioned earlier. This can be a stand-alone disc or may be supported in an epoxy resin puck of dimensions amenable to the sample holder to be used. On the other hand, a thinned section can be supported on a glass slide or amenable dimensions along with a suitable reference material. The epoxy preparation will be similar to that above.

Insulators can be examined using procedures discussed in Section 5.3.1.2.1.

### 5.2.3 Preparation in the Biosciences

Samples of interest in the Biosciences that pose the greatest problem are Biological cells. This is realized since these tend to be insulating and usually contain water. As water in its liquid form exhibits a significant increase in vapor pressure when placed under vacuum conditions, all water present must be either removed or frozen. Both approaches can, however, also result in significant damage to the sample of interest if carried out in an uncontrolled manner (a fact realized as water tends to be contained within various vessels, cells, etc.). Indeed, freeze drying can induce cell rupture if the crystallization rate is not well controlled.

Sample preparation also requires careful fixation to avoid redistribution of the inorganic macronutrient elements including Potassium, Calcium, and Magnesium (these exist in their ionized states).

Commonly applied sample preparation approaches in these instances include (Chandra and Morrison 1992):

1. Chemical fixation
2. Cryo-fixation
3. Freeze fracturing

Chemical fixation can be carried out using the following step-by-step approach:

1. Cutting the tissue section of interest
2. Applying Glutaraldehyde, Osmium Tetra-oxide, or some other fixative
3. Dehydrating the solution through application of Acetone or Ethanol
4. Impregnation of the resulting tissue section with a Plastic Resin
5. Heating of the plasticized tissue section to ~60 °C (this enhances curing)

The end result is a water-free solid block of tissue that is UHV compatible. Although heavily used in Scanning Electron Microscopy and Transmission

Electron Microscopy (TEM) studies, this approach is not extensively used in SIMS. This stems from the fact that such preparation procedures tend to result in modification of the chemical composition, particularly for inorganic macronutrient elements, on or within the respective tissue sections. This approach can, however, be applied when nondiffusible ions or elements bound to large molecules are of interest.

Cryo-fixation describes a rapid freezing approach carried out using Liquid Nitrogen, Propane, or Iso-Pentene. When carried out rapidly enough, this limits the size of the crystals formed, thereby controlling any damage common to the more standard freezing approaches. This is further facilitated by reducing the time between excision and freezing. The resultant solid can then be analyzed as is if a cryogenic stage exists in the respective SIMS instrument or this can be analyzed after the ice has been removed through either freeze substitution or freeze drying followed by resin embedding.

Freeze substitution, as the name suggests, describes the substitution of ice present within the interior of the respective cells with Ether, Acetone, or Acrolein. This is carried out at $-80\,°C$ in a dry environment. Freeze drying, on the other hand, is carried out by sublimating the water off the frozen sample through vacuum exposure in a controlled manner. This can take several days. In some cases, resin impregnation may then be applied as this allows for more effective control of surface topography (Recall: Surface roughness can affect secondary ion signal intensities). This step, however, may take months. Once cured, ultramicrotome sections from predefined regions can then be prepared and mounted on a TEM grid or some other support. Careful preparation can result in near complete localization of highly diffusible ions.

Freeze fracturing is an alternative methodology that involves depositing the tissue cells of interest onto a piece of Silicon wafer. A second piece of Silicon wafer is then placed on top with the sandwich then frozen rapidly using the same procedures as mentioned earlier. After this is completed, the two pieces of Silicon wafer are pulled apart leaving some fraction of the cells fractured. This is then freeze dried. Although this can prove effective, this has been plagued by topographical issues. The effects of surface topography, however, can be minimized by normalizing the signal of interest of some element or molecule whose concentration is uniform over the region of interest. And there are other methodologies.

Insulators can be examined using procedures discussed in Section 5.3.1.2.1.

## 5.2.4 Sample Handling

Concomitant to UHV compatibility is the requirement for sample cleanliness. This is realized as any undesired high vapor pressure contaminant introduced onto the area to be analyzed will cause vacuum deterioration and will introduce additional signals into the resulting mass spectra. Furthermore, some fraction of the desorbed species will adsorb on to the analysis chamber walls. This will then affect subsequent analysis because, as the vacuum improves, any species adsorbed on the chamber walls will desorb and re-adsorb onto the "fresh" samples being analyzed.

Cleanliness during sample preparation can be assured using gloves and the appropriate precleaned tools (tweezers, screw drivers, spatulas, sample support holder, etc.). Although Latex gloves can be used, polyethylene gloves are preferred (some Latex gloves contain Silicones, which can segregate to a sample's surface if unintentionally handled). Cleaning is usually carried out using acetone, methanol, or isopropyl alcohol. Ultrasonic cleaning can be of use if more aggressive methods are required. The tools to mount the samples onto the respective sample stubs or platforms must also be kept clean. Likewise, for tools needed to ensure that the sample is of a size that can fit into the instrument. Under no circumstance should samples be handled with bare hands.

In specific cases, prior removal of additional surface layers may be required. This can be carried out through:

1. Rinsing the sample in specific solvents/reagents before placement of the sample in the introduction chamber
2. Presputtering the area of interest with large cluster ions. This option allows the removal of surface organic layers while not introducing damage to the underlying regions. For example, see Figures 3.13 and 3.14
3. Heating thermally stable samples while in the introduction chamber.

The requirement of a flat surface is realized as topography can alter sputter yields, and thus the recorded secondary ion intensities. In addition, mass resolution in Time-of-Flight-based instruments using pulsed primary ions can be affected. Removal of surface topography can be carried out using various polishing and/or cleaving techniques. These, however, tend to be sample specific, as the various methods are prone to introducing contaminants and their associated signals.

Once mounted (samples are usually held down by clips or inserted into spring loaded sample holders), the samples can be introduced into an introduction chamber. These tend to be of a minimal size to allow for quick pump down, with pump-down times typically in the $10-15\,min$ range for clean samples of low porosity. In the case of highly porous samples, it is advisable to pump these down in the introduction chamber for extended periods of time (overnight before analysis may be required).

Samples are transferred to the analysis chamber once a pressure in the $10^{-7}$ Torr range is reached. Introduction chambers thus allow the fast transfer of samples from atmosphere to the analysis chamber with minimal loss of the UHV conditions required during analysis (days would be required if no introduction chamber existed). Introduction chambers can also allow for storage of multiple samples using carousel-like mechanisms. Likewise, multiple samples can be mounted on specific sample holders, with an extreme example being the capability of loading up to fifty $1 \times 1\,cm$-sized samples at one time (this capability exists on Cameca WF instruments).

## 5.3    DATA COLLECTION

As mentioned in Section 5.1, SIMS is a highly diverse micro-analytical technique, possibly the most diverse. This diversity, even noted when restricting the discussion to commercially available instruments, covers the capabilities provided and the freedom in instrumental parameters available. As an example, almost all commercially available SIMS instruments contain two or more primary ion guns and are all able to collect secondary ion spectra in either the positive or the negative polarities. In addition, there are almost always multiple signals from which to choose from for following a specific element or molecule. In some cases, these may be summed. In short, one or more of the following may be collected:

1. Atomic positive or negative secondary ions. These are collected when best sensitivity/detection limits are required. Note: There also exist element-specific methodologies with some examples including the collection of dimer ions, i.e. $SiN^-$ when Nitrogen in Silicon is of interest, the use of $O_2$ leak conditions for increasing the sensitivity/detection limits for metal ions, etc.

2. $Cs^+$ cluster secondary ions ($MCs^+$ where M represents some electropositive element, or $MCs_2^+$ where M represents some electronegative element). This is used when the matrix effects present need to be minimized, albeit at the cost of sensitivity/detection limits. Note: $Cs^+$ primary ion impact is required

3. Molecular positive or negative secondary ions representative of the molecular distributions of interest. Cationized positive secondary ions (the cation can be Hydrogen, a specific 1$^{st}$ row transition metal, or some alkali metal) may also be recorded as these can result in improved sensitivity/detection limits

In some cases, there does not appear any clear systematics to the choice of the conditions or the secondary ions used. This is partially a result of the diverse community that utilizes SIMS and the channel by which the respective methodologies were developed. In any case, much of this information is available in journal publications and/or in texts specific to the area of application. Some texts from the various areas in which SIMS is commonly applied include Wilson et al. 1986; Riviere and Myhra 2009; Fayek et al. 2009; Mahoney 2013. Note: As the application of SIMS to the Biosciences is rapidly expanding/changing, the reader is also advised to access the most recent literature for the latest methodologies, many of which are sample specific.

The section presented henceforth introduces the reader to the fundamentals associated with any of the data collection protocols described earlier. This starts with an overview of the various referencing procedures that should be applied along with the issues noted when dealing with insulators. This is covered in Section 5.3.1 along with the commonly applied methodologies. As isobaric interferences and methodologies for dealing with them can also affect mass calibration, these are also covered in Section 5.3.1. Following this, Section 5.3.2 covers the different data collection modes that can be applied within the various different instrument geometries. Data conversion methodologies describing how the sputter time scale is converted into depth and how the intensity scale is converted into concentration are covered in Section 5.4.

### 5.3.1 Secondary Ion Mass, Energy, and Intensity Scales

Once the sample of interest is located within the analysis chamber and the desired vacuum level is attained, sample-specific prealignment procedures must be carried out on the respective SIMS instrument before data collection can proceed. As these tend to be analysis specific (dependent on the sample type and the information of interest), these should be carried out before each analysis, i.e. on a daily basis. These prealignment procedures must be carried out to ensure that:

1. The primary and secondary ion beams are well focused with all aberrations minimized according to the conditions used, i.e. the mass resolution, energy, and/or spatial resolution required (aberrations are covered in Appendix A.6.2.1)
2. The region of interest on the samples surface is within the instrument's ion optical field of view for both the primary and the secondary ion beams. This will also depend on the axes of both ion beams as these must intersect at the sample's surface.
3. The analytical conditions must be optimized according to the sample type, the potential issues noted, the information required, and the instrument type used (Quadrupole, Time-of-Flight, or Magnetic Sector-based SIMS). This will include optimizing:
   a. Secondary ion intensities (these should be stable, of sufficient intensity to provide good statistics, and must be within the linear range of the respective detector used)
   b. Sputter rates when operating in the Dynamic SIMS mode (this should be optimized to cover the depth scale of interest within an acceptable time while still providing acceptable depth resolution)
   c. Mass resolution, energy resolution, and/or spatial resolution according to the information required, the sample type, and the potential analytical issues that may be prevalent

Such procedures assume that the secondary ion energy, mass, and intensity scales of the instrument are calibrated. Calibration procedures are discussed henceforth.

#### 5.3.1.1 *Referencing the Mass, Energy, and Intensity Scales* Calibration of the mass scale is required as this represents one part to the data recorded in SIMS, and this is used to elucidate the identity of the element or molecule from the

secondary ions of the isotopes measured (isotope mass is defined in Section 2.1.1.1, whereas the mass filters used to separate the secondary ions are discussed in Section 4.2.3.1). Note: Although it is more correct to use the $m/q$ ratio (SIMS separates secondary ions based on this ratio), mass alone is sometimes used in plotting mass spectra as the vast majority of secondary ions exist with a unit charge.

Calibration of the mass scale is carried out by recording the secondary ions from some known substrate whose secondary ions exhibit significant intensities with their masses easily identified. Common substrate examples include Silicon, Molybdenum, and Indium oxide. Secondary ion emissions that span a sizable fraction of the mass range should be used as this ensures effective mass calibration throughout the entire mass range accessed. Those listed earlier allow calibration from 1 to $\sim$400 $m/q$. Conductive substrates are preferred as charge buildup occurring during insulating substrate analysis can affect the energy scale, which, in turn, can affect the mass scale if not accounted for. If nonconductive substrates are examined, charge neutralization should be implemented. Charge neutralization is covered in Section 5.3.1.2.

Calibration of the energy scale is required as secondary ion intensities vary significantly with their emission energy. Indeed, the ability of an instrument to collect and separate secondary ions as a function of their $m/q$ ratio scales with the secondary ion extraction energy (higher extraction potentials improve both). Note: Singly charged secondary ions passing through an electrostatic field with the gradient placed normal to the direction of travel are accelerated/decelerated (which of the two depends on the polarities of the respective secondary ions and applied potentials) by 1 eV for every 1 V applied (2 eV for doubly charged ions, 3 eV for triply charged ions, and so forth). This charge-dependent variation in acceleration energy explains why doubly charged ions appear at half the mass of the respective element, i.e. $^{28}Si^{2+}$ appears at 14 $m/q$. As all secondary ions are emitted with similar energies (energy distributions peak between 1 and 10 eV as discussed in Section 3.3), a common energy scale is used for all secondary ions of unit charge. Multiply charged secondary ions are generally of little interest.

Alignment of the secondary ion emission energy scale from that recorded (derived by accounting for any acceleration/deceleration experienced by the respective secondary ions within the SIMS instrument) can be carried out by ensuring that the leading edge of a secondary ion energy distribution (collected by scanning the electrostatic field used to efficiently extract the secondary ions from the sample surface) lies at 0 eV. The leading edge is typically defined as the position at which the secondary ion intensity drops by 50% relative to the peak intensity. Owing to the intrinsic energy spread of secondary ions and the energy acceptance imposed, effective calibration is best carried out using large molecular secondary ions (these have the narrowest energy spread) and a narrow energy acceptance (a few electron volts) if needed. As with mass calibration, energy calibration should be carried out using conductive substrates as this avoids the potential effects of sample charging (see Section 5.3.1.2).

Calibration of the intensity scale describes the optimization of the secondary ion detectors such that their sensitivities are maximized (detectors are covered in

Section 4.2.3.3). Detector optimization relies on complete capturing and conversion of the signal of interest into an electrical signal, which is relayed in intensity units. This requires optimization of their respective operation voltages, which, in the case of Electron Multipliers, requires routine adjustments to account for variations in gain owing to aging effects, and so on. For low-energy secondary ion beams, complete conversion also requires post acceleration of the secondary ion beam before striking the detector. Electron Multiplier noise must be minimized. This is ensured by correct adjustment of the discriminator level as covered in Section 4.2.3.3.2.1. When more than one detector type is used, i.e. Faraday Cups and Electron Multipliers, their output signals should be aligned with each other over their shared linear ranges.

Alignment of the intensity scale with the concentration of the signals measured is otherwise referred to as quantification. Procedures in common use (see Sections 5.4.3) assume a linear dependence between the number of secondary ion emitted and the intensity recorded under the specific set of data collection procedures used.

### 5.3.1.2  *Charge Buildup*

If the solid being analyzed is insulating, charge will accumulate over the area sputtered. This results from the uncompensated removal of charge from the emission of secondary electrons and ions relative to charge deposited from impacting primary ions. This is illustrated in Figure 5.9(a). Positive charge buildup invariably occurs because of high secondary electron yields. This is not seen in conductors as electrons from a grounded terminal compensate for any charge loss. This is illustrated in Figure 5.9(b).

Charge buildup is detrimental in all types of SIMS instruments as this imposes an additional potential field around the sample over which the incoming primary ions and emitting secondary ions have to deal with. As an example, positive charge buildup will further accelerate positive secondary ions/decelerate negative

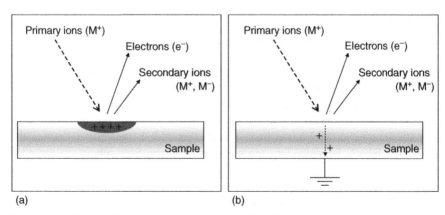

(a)                                          (b)

**Figure 5.9**  Pictorial illustrations of (a) charge buildup that occurs during sputtering on an insulating substrate or a substrate that is not in electrical contact with the instrument and (b) a conductive sample in electrical contact with the instrument. Although positive charge buildup invariably occurs, other situations are possible depending on the applied fields, and so on.

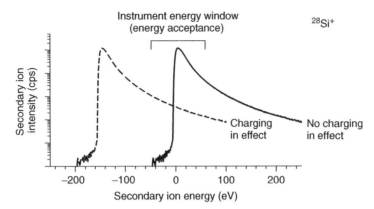

**Figure 5.10** Example of the effect of positive charge buildup on the positive secondary ion energy distribution after accounting for the applied accelerating voltage. This was collected from a Silicon surface with the energy window set at ~5 eV on a Magnetic Sector SIMS instrument.

secondary ions thereby moving the respective emission energy distributions as is illustrated in Figure 5.10. As the energy acceptance of all instruments is fixed, this will alter all recorded secondary ion intensities, and in some cases, the $m/q$ ratio at which these are noted.

If uncompensated, charge buildup will increase over time with the recorded secondary ion signals steadily decreasing. In extreme cases, complete electrical break down will occur. The break down may result from the decomposition of an insulating film on the sample analyzed or may result from arcing to conductive elements in close proximity to the area experiencing charge buildup.

In the case of Time-of-Flight-based SIMS instruments, charge buildup can also lead to the introduction of additional *"fictitious"* peaks in the respective mass spectra when depth profiling from a grounded conductive layer to an insulating layer. This issue is noted as these instruments apply extraction fields to the sample and assume that these extraction fields remain unaffected. When the electrical potential of the surface changes, so too does the velocity of the departing secondary ions. This change in velocity is recorded as a change in the ions $m/q$ value, hence the introduction of the fictitious peaks.

Charge buildup also affects primary ion trajectories, irrespective of the instrument type. This is particularly evident when low-energy primary ions are used. That being said, primary ions are less susceptible to charge buildup owing to their greater energies with respect to secondary ions.

*5.3.1.2.1 Charge Compensation* Charge compensation describes the act of ensuring that no charge buildup occurs on the surface of the sample being analyzed. This typically occurs on insulating samples and samples not in electrical contact with the instrument during sputtering. If charge compensation is not ensured, the loss of secondary ion signal along with the other deleterious effects described in Section 5.3.1.2 will occur.

Charge compensation in SIMS is most commonly carried out by coirradiating the affected sample surface region with electrons. High-energy electrons (typically 0.1–1 keV) can be used but only if the electron current matches the charge buildup and the beam is directed at the area affected (this can be difficult). Use of low-energy electrons (< 10 eV) provides the additional possibility of self-compensation, thereby easing the charge compensation criteria (charge matching). This is realized as charge buildup on the sample surface will deflect the incoming low-energy electrons to the affected area. This self-compensation mechanism will ensue on the assumption that the initial incoming electron beam current density exceeds that of the charge buildup.

Other methods that can be used in conjunction with electron beam irradiation include:

1. Coating the sample with a conductive layer, e.g. Carbon, Gold, and Platinum, before analysis. Note: This is only effective under Dynamic SIMS conditions because, as the term implies, the surface is coated. Also of note is the fact that coating will introduce additional signals (those from Carbon, Gold, Platinum, etc. and any impurities that may be introduced during the application of the respective conductive layer) into the respective mass spectra

2. Placing an earthed conductive grid (Tungsten grids are common) or conductive tape (Aluminum or Copper tape) in close proximity to the area analyzed. Note: Care must be taken to ensure that additional signals are not introduced. Directing an electron beam onto these surfaces can also facilitate charge compensation owing to the additional secondary electrons produced

3. Pressing the sample into Indium foil. This applies to powdered samples and when combined with one of the other methods described earlier. Note: If care is not taken in setting up the analysis area, this can introduce additional signals

4. Heating the sample during analysis. This can promote electrons into the conduction band of large band gap materials when temperatures greater than ~400°C are used. This method should only be applied to thermally stable precleaned samples. As a result, Indium should not be used in such cases

5. Use of a negative polarity primary ion beam, i.e. $O^-$. This method, commonly used in the Earth Sciences, is effective, in that, the primary beam introduces negative charge onto the samples surface and in the area being sputtered. This complements electron beam charge neutralization

6. Use of $O_2$ leak or Ar leak during analysis. It is presumed that when combined with electron beam irradiation, this yields more low-energy electrons.

7. In the case of Time-of-Flight instruments using pulsed primary ion beams, the effects of charge buildup during dual beam depth profile analysis (analysis beam and sputter beam) can be reduced by modifying the primary ion beam pulsing sequence. For example, a noninterlaced (phase) pulsing sequence as opposed to interlaced (interleaved) sequence provides additional time for charge equilibration (pulse sequencing is discussed in Section 5.3.2.1.2)

Although more rarely implemented, there are other methods that can be applied for charge compensation. One such method is the *Specimen Isolation* method (Metson et al. 1983). This method, applied in Magnetic Sector-based SIMS instruments, places an electrically isolated sample, hence the name, under a conductive aperture earthed through the sample holder. During primary ion impact, charge builds up on the sample's surface and increases until a potential difference of several hundred volts relative to the conductive aperture/sample holder is reached. Equilibration of the charge then occurs as a result of the additional secondary electrons produced from the conductive aperture. These electrons are believed to be produced as a result of spurious secondary ions/electrons striking this aperture and the potential field present. By adjusting the sample voltage, the offset can then be compensated so as to adjust the energy of the secondary ions entering the mass spectrometer. As a result, this method can be used to facilitate Kinetic energy (KE) filtering (see Section 5.3.1.3.3). Through careful optimization, imaging of insulators is also possible (van der Heide and McIntyre 1993). To date, this method has only found application to samples of interest to the Earth Sciences.

### 5.3.1.3 *Isobaric Interferences*

A common issue encountered in SIMS is the presence of secondary ions from different elements and/or molecules that exhibit the same nominal $m/q$ ratio. Such overlapping signals arise from what are otherwise referred to as *isobaric interferences*. More commonly, these are referred to as *mass interferences*.

An example of different secondary ions of the same nominal mass are noted during the analysis of a Phosphorus-doped Silicon wafer are $^{30}Si^1H^-$ and $^{31}P^-$. These are commonly noted when Phosphorus is less than $\sim 0.1$ atomic%. SiH formation arises from $H_2O$ adsorption on a freshly sputtered Silicon surface. Although, a reduction in the $^{30}Si^1H^-$ signal intensity relative to that of $^{31}P^-$ is noted on improving vacuum, such hydride mass interferences are still evident when analysis is carried out under a vacuum in the $10^{-10}$ Torr range. Additional examples of potential isobaric interferences are illustrated in Figure 5.2(a) and 5.2(b).

As noted in Section 5.1.1.1, isobaric interferences become more prevalent:

1. At higher $m/q$ ratios (the number of possibilities increases quadratically as the secondary ion $m/q$ ratio increases owing to the greater likelihood of molecular combinations of the lighter elements).
2. When the spread in the $m/q$ ratio of secondary ions increases (this arises from the fact that an increase in the spread increases the chance of overlap).
3. When Static SIMS conditions are employed (this is due to the greater prevalence of molecular ions in Static SIMS mass spectra with respect to Dynamic SIMS mass spectra).

The prevalence of isobaric interferences stems from the spread introduced into the $m/q$ ratio for all secondary ions. As also introduced in Section 5.1.1.1, this spread is a distortion introduced by the instrument, i.e. is a function of the analytical conditions applied during data collection. Under the same analytical conditions, all secondary ions experience the same spread.

Options to remove or at least reduce these interferences employ the use of:

1. Peak stripping
2. High Mass Resolution (HMR)
3. Kinetic Energy (KE) filtering

These approaches are discussed henceforth. Owing to the improved ease of implementing HMR on commercially available SIMS instruments, this has become the most heavily used approach. HMR is, however, not possible on commercially available Quadrupole-based SIMS instruments.

*5.3.1.3.1  Peak Stripping*    As the name suggests, peak stripping is a retrospective subtractive approach in which some fraction of the secondary ion signal suffering the interference is removed such that the resulting peak represents the intensity from an interference-free signal.

An illustrative example of this is shown in Figure 5.11 in which hydride interferences experienced by the $^{29}Si^-$ and $^{30}Si^-$ secondary ions from a Phosphorus-doped Silicon wafer are derived as the difference in the signals recorded relative to the natural abundance of the isotopes. From this, an expected value for the $^{30}Si^1H^-$ interference can be estimated. This can then be used to derive an interference-free signal from $^{31}P^-$. The fraction subtracted from the intensity at 31 $m/q$ would represent the interference introduced by the $^{30}Si^1H^-$ ions. This fraction would either be defined by the relative intensity of some other secondary ion related to the secondary ions responsible for the interference, or would be derived to produce the expected result. Note: Caution must be exercised when using this approach to ensure the validity of the faction subtracted. Indeed, this approach should only be used as a last resort.

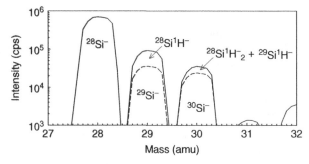

**Figure 5.11**    A mass spectrum of the negative secondary ion emission from a Silicon wafer. The solid line represents the collected signal and the dashed lines represent the peak stripped spectra, i.e. the spectra not suffering the hydride interferences. This mass spectrum was collected on a Quadrupole-based SIMS instrument at unit mass resolution (from Figure 5.1).

*5.3.1.3.2    High Mass Resolution*    HMR can be used during analysis to separate out signals of the same nominal mass as that of the secondary ion of interest. This does so by reducing the $m/q$ ratio spread exhibited by all secondary ions (all ions exhibit the same spread when analyzed under the same conditions), thereby allowing for the separation of secondary ion signals exhibiting similar (nominal) $m/q$ ratios. An example of the use of HMR in the analysis of a metallic alloy reference material is shown in Figure 5.12. The definition of mass resolution is covered in Section 5.1.1.1.1.

This separation in $m/q$ of different secondary ions is possible as the exact mass of every element, with the exception of $^{12}C$, deviates slightly from their nominal mass (reasons for this are outlined in Section 2.1.1). Indeed, all ions with $Z < {}^{12}C$ have a slightly greater mass than their nominal mass, whereas those with $Z > {}^{12}C$ but $< {}^{220}Rn$ (mass increases again for $Z > {}^{220}Rn$) have a slightly lesser mass than their nominal mass, i.e. the masses of $^{1}H$ and $^{30}Si$ are 1.007825 and 29.973770 u, respectively. Separating the $^{30}Si^{1}H^{-}$ interference (mass equals $1.007825 + 29.973770$ or 30.981595 u) from the $^{31}P$ signal (mass equates to 30.973762 u) thus requires a mass resolution ($m/\Delta m$) of 30.973762/0.007833 or

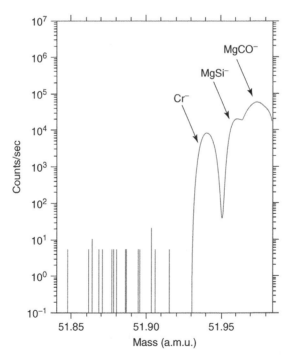

**Figure 5.12**    Use of high mass resolution spectra reveals that three separate signals are present at the nominal mass to charge ratio of 52 $m/q$ when examining the secondary ions from an NBS 649 reference material (NBS is now NIST). This spectrum was collected on a Magnetic Sector-based SIMS instrument.

~3955. As noted from Table 4.4, this is easily accessible in most SIMS instruments (the exception being those using Quadrupole or Ion Trap mass filters).

### 5.3.1.3.3 Kinetic Energy Filtering

Like HMR, KE filtering is an alternative method that can be used during analysis to separate out signals of the same nominal mass as that of the secondary ion of interest. KE filtering is, however, more limited in that it only removes molecular interferences from monatomic secondary signals, i.e. one cannot apply this to any signal (elemental or molecular) as can be done when using HMR conditions. Loss of signal intensity is also noted.

KE filtering is carried out by adjusting the secondary ion energy acceptance of the instrument such that only ions of higher energy are recorded. As illustrated in Figure 5.13, this filters out almost all molecular signals as their energy distributions are almost always narrower than monatomic secondary ion energy distributions.

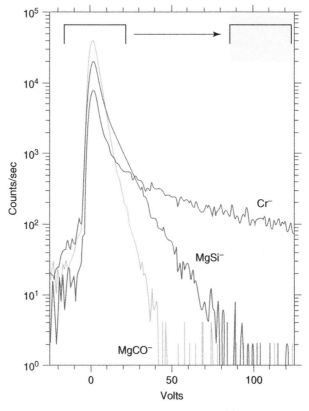

**Figure 5.13** Overlay of emission energy distributions collected from the sample/species illustrated in Figure 5.12 under high mass resolution conditions. This reveals that by adjusting the instruments energy acceptance, the molecular secondary ion signals of MgCO⁻ and MgSi⁻ can be filtered out from the Cr⁻ signal. This, however, results in poorer sensitivity and detection limits. These energy distributions were collected on a Magnetic Sector-based SIMS instrument with the energy filter set to ~10 eV.

Hydrides are an exception as they display similar energy distributions as their parent elemental signals.

### 5.3.2 Instrument Operation Modes

The objective in SIMS is to record secondary ion intensities over time (this is relayed as secondary ion current over time) and/or to record secondary ion intensities over space (this is relayed as the secondary ion current emanating from some specific location on the samples surface) as this allows for insight into the relative concentration and/or concentration gradients of the respective isotopes, elements, or molecules of interest on or within the solid of interest.

There are, however, a number of modes under which a SIMS instrument can be operated, with each optimized for a particular type of analysis, whether pertaining to the collection of mass spectra, depth profiles, or images in two or three dimensions (these formats are illustrated in Figure 1.2 and discussed in Section 5.1.1). This is realized as each mode can affect one or more of the following:

1. Detection limit (the ability to detect the signal of interest if present at low concentrations)
2. Mass resolution (the ability to separate signals in close proximity to each other in mass)
3. Energy resolution (the ability to separate signals in close proximity of each other in energy)
4. Spatial resolution (the ability to separate signals in close proximity of each other in space)
5. Depth resolution (the ability to separate signals in close proximity of each other in depth)

The different modes exist as it is impossible to attain data with all of the earlier mentioned factors being maximized at the same time, i.e. optimizing one tends to result in the deterioration of one or more of the others. As a result, the prior choice between the different modes of operation must be made according to the parameter/s of greatest importance. The different modes, some of which are instrument type specific, entail not only control of the primary and/or secondary ion optics in a highly specific manner but also the vacuum conditions present during analysis. All are dependent on the specific secondary ion signal used. These are discussed henceforth for commercially available SIMS instruments.

#### 5.3.2.1 *Primary Ion Beam Operation Modes* Aside from the plethora of primary ion types, there also exists a range in primary ion impact energies, primary ion incidence angles, and primary ion currents available.

The choice of primary ion beam type is dependent on the elements/molecules of interest, the damage that can be accepted, and the sensitivity/detection limits required.

The impact energy controls:

1. The sputter rate (sputter rates increase with increasing energy)
2. The depth resolution (depth resolution degrades with increasing energy)
3. The beam spot size (the beam diameter decreases with increasing energy)

The impact angle controls the sputter rate (sputter rates increase at more glancing impact angles) and the depth resolution (depth resolution increases at more glancing impact angles). Likewise, the primary ion current controls the sputter rate (these increase with each other) and the beam spot size (these increase with each other).

Aside from the freedom to choose the primary ion beam type, impact energy, impact angle, and impact current, there exist numerous modes under which these can be operated. For starters, the primary ion beam, or beams, can be directed at the sample's surface over some predefined area ranging from $< 1 \, \mu m^2$ to $> 10,000 \, \mu m^2$. This can be directed in a *stationary mode* or the *rastered mode*.

*Stationary mode* describes the use of an un-scanned beam. Advantages of this mode lie in the fact that instantaneous dead-time effects are removed (see Section 4.2.3.3.3.2). Note: Although this is also referred to as *static mode*, this definition will not be used in this text so as to avoid confusion with Static SIMS).

*Rastered mode* describes scanning of a focused primary ion beam over some predefined area over the sample. In this mode, the beam spot size must be substantially smaller than the analyzed area and instantaneous dead-time effects associated with primary beam rastering must be considered (see Section 4.2.3.3.3.2).

In addition, the primary ion beam can be directed at the sample's surface either in a *continuous manner* or in a *pulsed manner*.

*Continuous primary ion beams* are applied in Quadrupole-based SIMS instruments, Magnetic Sector-based SIMS instruments, and Time-of-Flight-based SIMS instruments in which secondary ion beam pulsing is induced through bunching of the secondary ion beam alone (see Section 4.2.3.1.5).

*Pulsed primary ion beams* are used in Time-of-Flight-based SIMS instruments in which secondary ion beam pulsing is derived from primary ion beam pulsing (see Section 4.2.3.1.5). Pulsing can also be carried out using different sequences.

These modes are discussed separately in the following subsections.

### 5.3.2.1.1 Continuous Primary Ion Beams
As the name suggests, a continuous primary ion beam is one in which a continuous flow of ions reaches the sample over the entire duration of analysis, i.e. this beam is not interrupted between analysis cycles. Continuous primary ion beams are preferred when a significant amount of the sample is to be probed (as in Dynamic SIMS analysis to depths greater than $\sim 0.1 \, \mu m$) and/or when the smallest primary ion beam diameter is required (as in imaging in the *microprobe* mode (see Section 5.3.2.2)). Continuous primary ion beams are used in Magnetic Sector-based instruments, Quadrupole-based instruments, and specific Time-of-Flight-based instruments, i.e. those in which the secondary ion beam is pulsed through bunching of the secondary ion beam. Such primary beams are focused in either the *Gaussian mode* or the *Shaped mode*.

The *Gaussian mode* consists of forming an image of the source, or near source region, on the sample's surface. Under this condition, a small spot size is formed, which scales with the primary ion current. An ultimate spot size of ~10 nm is possible as is demonstrated in FIB instruments using field emission sources (McPhail et al. 2011). As this is of the same size, or smaller, than the volume over which a collision cascade extends, the spatial resolution, which is only available when imaging in the *microprobe mode* (see Section 5.3.2.2), then becomes limited by the collision cascade dimensions. Such beams should be operated in a rastered manner, with the rastered area being significantly larger than the beam spot size at the sample. Rastering of the beam is stipulated because such beams exhibit a Gaussian beam density across the beam cross section. Imaging in the *microscope mode* (see Section 5.3.2.2) can also be carried out when operating a rastered primary ion beam in the *Gaussian mode*. These are the modes used in the Cameca IMS™ series of instruments with a resultant spatial resolution of ~1 μm. The *Gaussian mode* is used on all the commercial instruments, with the exceptions being the ASI SHRIMP™ and Cameca WF/SC™ series (these use the *shaped mode*).

The *shaped mode* consists of forming an image of a primary beam aperture on the sample's surface (one form is referred to as *Kohler illumination*). Under this condition, the beam spot size remains independent of the primary ion current (the spot size is larger than under *Gaussian mode*). As the current density across the spot remains constant, rastering of the beam is then not necessary, although sometimes still applied. This mode of operation can be used when imaging in the *microscope mode* (see Section 5.3.2.2) and/or when carryout analysis where detector dead-time effects should be minimized (see Section 4.2.3.3.3.2).

*5.3.2.1.2  Pulsed Primary Ion Beams*  Pulsed primary ion beams are required in Time-of-Flight-based instruments in which the secondary ion column is not used to generate the pulse (as covered Section 4.2.3.1.5, Time-of-Flight instruments can operate by either pulsing the primary ion beam or secondary ion beam). Pulsed primary ion beams are generally preferred in situations in which shallow regions of the sample are to be probed. This generally extends to Static SIMS or Dynamic SIMS to depths less than ~0.1 μm. Most commercially available Time-of-Flight instruments pulse the primary ion beam/s.

Pulsed primary ion beams used for generating the secondary ion signal (this beam is henceforth referred to as the *analysis beam*) are pulsed at frequencies of between 10 and 50 kHz. Each pulse then lasts from a few nanoseconds to several hundred nanoseconds with the time between pulses ranging from tens to hundreds of microseconds. These times are adjusted to control:

1. The mass range. This is determined by the time between analysis beam pulses with longer times translating to larger mass ranges
2. The mass resolution. This is determined by the analysis beam pulse width, i.e. a smaller width translates to higher mass resolution
3. The secondary ion signal intensity. This can also be controlled by adjusting the analysis beam pulse duration

In addition, the primary ion analysis beam focusing/pulse sequencing can be tailored to optimize the specific information required. As an example, the pulse width, which defines the mass resolution, can be adjusted by pulsing the electrodynamic fields within the primary ion column such as to accelerate slower ions/decelerate faster ions within a specific pulse, such that they all arrive at the sample surface at the same time. This is referred to as *beam bunching*. The downside is that spatial resolution is lost when operating in the micro-probe mode.

Specific modes of operation commonly used in such instruments include the:

1. *Imaging mode* also referred to as *Unbunched mode*. This is used when a finely focused beam is needed. As this requires the *Gaussian mode* of focusing, the spot size scales with the primary ion beam current.

2. *Spectroscopic mode* also referred to as *Bunched mode*. This is used when optimal secondary ion mass resolution is needed. Note: This comes at the cost of image spatial resolution.

3. *Burst Mode*™, patented by Ion-Tof, is used when either:
   a. HMR is required under conditions requiring high spatial resolution (imaging) or
   b. Improved sensitivity/detection limits are required under conditions requiring HMR.

4. $HR^2$ *mode*™, developed by Physical Electronics, provides improved high mass and high spatial resolution at the same time. This is made possible through the re-design of the bunching region within their primary ion columns. Note: TRIFT-based instruments are also able to image via the *microscope* mode.

These modes were introduced as the laws defining ion beam transport do not allow for a finely focused primary ion beam to display a short pulse width. One or the other must be preselected. For an analogy, a pulse containing a select number of ions can be compared to a balloon containing a finite amount of water; compress this in one direction and it will expand in the other. The *Burst* and $HR^2$ modes attempt to reconcile the limitations imposed with the former most effective at low $m/q$ and the latter all $m/q$. The *Imaging*, *Spectroscopic*, and *Burst* modes are illustrated in Figure 5.14, with the resulting analytical interdependencies illustrated in Figure 5.15.

*Imaging mode* also referred to as *Unbunched mode* describes the use of a primary beam of the smallest possible spot size (beam diameter) when arriving at the samples surface. This small spot size is required when employing the *microprobe* mode of imaging (see Section 5.3.2.2). The minimum spot size is attained by accelerating the analysis beam to as high an energy as possible (the use of doubly charged ions increases the energy by the factor of two) and with as low a beam current as feasible. Recall: The beam spot size decreases with current as this reduces the space charge effects. Imaging is carried out using a beam operated in *Gaussian mode* (either extreme cross-over or a collimated approach). Focusing and accelerating a pulsed primary ion beam, however, stretches the beam pulse out over time,

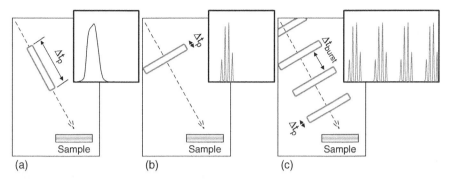

**Figure 5.14** Pictorial illustration (not to scale) of (a) *Imaging mode*, (b) *Spectroscopic mode*, and (c) the use of *Burst mode*™ while in *Spectroscopic mode* for improved signal intensity. In the respective insets are shown the resulting mass spectra for the same secondary ions. The parameters $\Delta t_p$ and $\Delta t_i$ refer to pulse width and time between pulses, respectively.

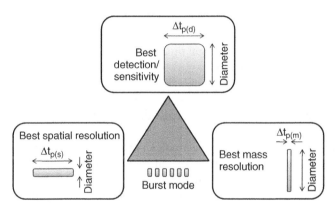

**Figure 5.15** Analytical interdependence signifying the beam conditions required for best detection limits (pulse width = $\Delta t_{p(d)}$), best spatial resolution (pulse width = $\Delta t_{p(s)}$), or best mass resolution (pulse width = $\Delta t_{p(m)}$). The analysis beam pulse shape is illustrated in the three respective boxes, with the beam traveling across the page, i.e. from left to right. Implementation of *Burst Mode*™ allows for high mass resolution with high spatial resolution to be attained, i.e. this represents a cross-over between the *Imaging mode* and the *Spectroscopic mode* of operation.

thereby increasing the pulse width (this stems from the fact that ions further from the central axis travel a greater distance when convergence/divergence is induced, thereby spreading out the pulse width over time). Typical pulse width values after collimation of pulsed primary ion beams are thus in the 20–100 ns range. A longer beam pulse width translates to poorer mass resolution but improved detection limits and thus sensitivity. Pulsing also introduces beam distortions, thereby causing deterioration in the minimum beam spot size.

*Spectroscopic mode*, also referred to as *Bunched mode*, brings together the ions within a particular primary ion pulse such that they arrive at the sample's surface over the shortest possible time interval. This bunching is desired when HMR is required (recall: mass resolution in Time-of-Flight instruments is a function of the pulse width). Pulse widths of less than 1 nanosecond can be routinely attained using this approach. This can result in mass resolution values of 10,000 or greater without any loss of sensitivity. The only major downside of this approach is that this induces a spreading of the primary ion beam in space, thereby increasing the minimum spot size attainable to between 2 and 5 μm. Note: HMR will, however, only be retained if the surface to be examined exhibits minimal surface topography. This is understood as any surface topography will extend the apparent time, a primary ion beam pulse strikes the surface.

*Burst mode™* only available on Ion-Tof instruments allows for improved detection limits/sensitivity while in the imaging mode for signals requiring HMR. As is shown in Figure 5.14(c), this accomplished by inserting additional analysis beam pulses such that additional spectra from the signal/s of interest appear over mass spectral regions that are void of any secondary ion signals. As an example, there are no signals between the masses of $\sim 28.04$ ($^{28}$Si presents a signal at 27.977 $m/q$, which will be accompanied by various other signals such as $C_2H_4$ at 28.032 $m/q$) and 28.96 ($^{29}$Si has a signal at 29.976 $m/q$). Thus, by inserting additional primary ion analysis beam pulses with a slight time shift, additional $^{28}$Si secondary ion signals can be recorded over the mass spectral region typically void of any signals, i.e. the region within 28.04–29.96 $m/q$. These can then all be summed to provide increased $^{28}$Si secondary ion intensities. This approach can result in intensities that can be as much as 10 times that of the individual signals. This mode is best suited to low $m/q$ secondary ions.

As indicated earlier, Physical Electronics introduced the $HR^2$ *mode™*, to provide improved high mass and high spatial resolution at the same time. This is implemented while in the Spectroscopic or Bunched mode through re-design of the LMIG region so as to minimize the energy spread of the primary ion beam pulses. It is quoted that 500 nm spatial resolution at moderate beam currents with sub nanosecond pulse widths is attainable.

As a lower current density of the analysis beam reaches the sample surface per unit time (relative to instruments utilizing continuous primary ion beams), increased sputter rates can only be realized through the irradiation of the analyzed area of the sample by a second pulsed primary ion beam. These beams are thus referred to as the *sputter beam*. Note: Both beams must be operated in an interleaved manner with respect to each other, i.e. only one can be sticking the sample at a time.

To ensure effective sputtering, the sputter beam is operated such that the dose is significantly higher (tens of nanoampere) than that of the analysis beam (< 1 picoampere). The sputter rate is then controlled through the adjustment of the sputter beam pulse width, between tens to hundreds of microseconds, as opposed to the adjustment of the ion optics as used in continuous primary ion beams. The extraction ion optics is also switched off over the interval the sputter beam is directed at the sample (this also aids in providing for additional charge compensation). Depth resolution will then become a function of the sputter beam conditions. For high-depth resolution, these can be operated at 250 eV and below impact energy, whereas the analysis beam remains at its standard impact energy (25–30 keV for singly charged ions). Depth resolution is also retained by ensuring that the sputter beam raster pattern covers a significantly larger area than that of the analysis beam (this is to remove crater edge effects as described in Section 5.3.2.4.2).

There also exist different pulse sequences for the analysis and sputter beams during depth profiling. These are referred to as the *interlaced or interleaved* and the *noninterlaced or Phase modes*. Both essentially produce the same result on conductive samples when species present in the gas phase are not of interest, i.e. Hydrogen, Carbon, Oxygen, and Nitrogen. When these are of interest, the interlaced mode is preferred.

The *interlaced or interleaved mode* describes a situation in which the sputter beam is switched on during the flight time of the secondary ions produced by the analysis beam. To avoid interferences between the analysis and sputter beams, a delay and lead-off time of several μs is implemented. This represents the faster and hence, more commonly used mode for depth profiling. This mode is also better for interface analysis and allows for reduced adsorption of gas phase species.

The *noninterlaced* or *phase mode* describes the situation in which the sputtering and analysis beams are treated separately. In this mode, many analysis cycles are typically implemented before initiating the sputter beam. If carried out on insulating samples, this can provide improved charge neutralization relative to the interlaced mode. This is realized as the sputter beam imparts a much higher dose than the analysis beam. The downside of the *noninterlaced mode* is that there exists the possibility of missing the interface while depth profiling, and a greater time for adsorption of gas phase species onto the sample surface, and analysis times are longer.

### 5.3.2.2  *Secondary Ion Imaging Modes*    Imaging in SIMS refers to the reproduction of the distribution of any element or molecule on or within the volume of interest. Reproduction of the spatial distribution is the most common form of imaging. This can be carried out via one of two modes, these being the *Microprobe mode* or the *Microscope mode*.

The *Microscope mode* forms spatial two-dimensional maps by ensuring that the spatial information of the point of secondary ion emission is retained throughout the secondary ion column. In other words, the image of the sputtered surface is formed by projecting the secondary ion signals onto a position-sensitive detector in much the same way an optical *microscope* works. As illustrated in Figure 5.16(a), the entire spatial region of interest can be imaged simultaneously without focusing or rastering the primary ion beam (both are required when imaging in the *microprobe mode*). Primary ion beams focused via the Gaussian mode (see Section 5.3.2.1.1) should, however, be rastered in order to remove intensity variations resulting from the primary ion beam profile (this is Gaussian shaped, hence the name). Advantages of this approach lie in the fact that the image resolution remains independent of the primary ion focusing characteristics. As a result, much higher primary ion currents

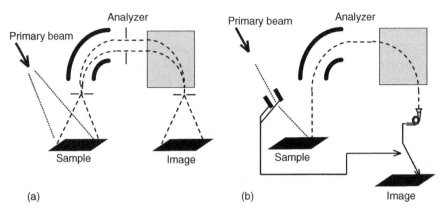

**Figure 5.16** Pictorial illustrations (not to scale) of SIMS imaging via (a) the *microscope mode* and (b) the *microprobe mode*. Reproduced with permission from van der Heide and Fichter (1998) Copyright 1998 John Wiley and Sons.

can be used during image acquisition relative to the *microprobe mode*. This results as image resolution remains dependent only on the secondary ion beam alignment and filtering. Indeed, a heavily filtered secondary ion beam can result in image resolution values in the range of 1 μm. This mode is best suited to Dynamic SIMS studies requiring analysis to significant depths. Indeed, as illustrated in Figure 5.6, this can be carried out to several μm in depth within acceptable time periods. Imaging via the *microscope* mode is only available on specifically designed double focusing Magnetic Sector and Time-of-Flight (TRIFT geometry)-based SIMS instruments.

The *Microprobe mode* probes individual regions and then stitches these together to form a spatial two-dimensional map of the region of interest. In SIMS, this is carried out by scanning the primary beam over the surface (any raster pattern can be used). As is illustrated in Figure 5.16(b), this requires synchronizing the detection electronics with the primary ion scan rate (raster) such that the counts recorded at some specific time can be related to some specific region of the surface. As a result, the image resolution becomes a direct function of the primary beam (probe) spot size. As the beam spot size scales with beam current (this is a result of the repulsion induced by space charge effects between charged particles within the respective beam) primary beams of low current must be used. This limits this imaging mode to surface and shallow volume studies. Advantages of this approach lie in the improved spatial resolution when compared to the *microscope* mode (although values to approaching collision cascade dimensions are possible (McPhail et al. 2011; Wirtz et al. 2012), 50–100 nm is more commonplace) and the fact that the secondary ion beam does not need to be heavily filtered (this improves secondary ion transmission). This type of imaging can be carried out on any instrument type, including those capable of imaging in the *microscope* mode. Note: The *microprobe* mode is the more heavily used of the two modes of imaging.

Image depth profiling is used to derive three-dimensional images of some specific volume. Such images are constructed by collecting and overlaying

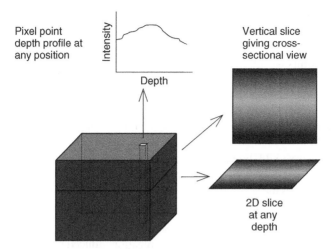

Pixel point depth profile at any position

Intensity

Depth

Vertical slice giving cross-sectional view

2D slice at any depth

**Figure 5.17** Retrospective evaluation possibilities of the volume probed in Dynamic SIMS whether in the *microprobe mode* or in the *microscope mode*. Reproduced with permission from van der Heide and Fichter (1998) Copyright 1998 John Wiley and Sons.

two-dimensional spatial images (an array of pixels) from the same location as a function of sputtering time, assuming the sputter rate is raised to a sufficient level to remove more than one atomic layer per cycle. These reconstructions can then be viewed in a number of different ways to most effectively relay the apparent intensity variations (see Figures 5.6 and 5.7 for two such examples). Such volumes can also be retrospectively chopped up along any plane as illustrated in Figure 5.17. Likewise, localized depth profiles can be derived by plotting the intensities within a particular array of voxels. For effective reproduction of any such distributions, matrix effects should be minimized.

### 5.3.2.3 The $O_2$ Leak Methodology

As covered in Section 4.2.1.1, SIMS requires a vacuum of better than $\sim 1 \times 10^{-4}$ Torr (that at which the mean free path of an ion equates to $\sim 1$ m, which equates to the typical path length of a primary or secondary ion column). More typically, a vacuum of better than $1 \times 10^{-8}$ Torr is used to control the adsorption of gas phase molecules. As an example, a monolayer of Oxygen will form on Silicon in $\sim 1$s when in the $10^{-6}$ Torr range.

There are, however, instances where the presence of Oxygen or some other highly electronegative element, on a sample's surface can be advantageous, i.e. when improved sensitivity and/or detection limits are required. This is realized in that, as indicated in Table 3.3, the presence of Oxygen enhances the positive secondary ion yield of specific elements (metals in particular) by as much as several orders of magnitude. Oxygen-induced enhancement effects are typically not noted in the negative atomic or the molecular secondary ion emissions. Note: There are exceptions such as in the production of $Si^-$ secondary ions from Silicon (this is covered in Section 3.3.2.2) and, of course, in the production of oxide-related molecular secondary ion emissions.

The introduction of $O_2$ close to a sample's surface during sputtering (usually via a thin tube extending to within a few millimeters of the region being sputtered) has thus become a relatively common methodology when extreme sensitivity or detection limits for specific elements are required. Note that although this also results in an increase in the analysis pressure (best results are noted at $\sim 1 \times 10^{-6}$ to $1 \times 10^{-7}$ torr), the use of high purity $O_2$ gas and the effective use of differential pumping of the primary and secondary ion columns, both act to minimize the detrimental effects commonly associated with carrying out analysis under such pressures.

Such analysis should, however, be carried out with extreme care, because:

1. Not only does the secondary ion yield vary with the $O_2$ leak pressure used, but so too does the sputter rate (a decrease in sputter rates is commonly observed)
2. The oxidation of the surface being analyzed
   a. Can alter the radiation-induced segregation induced by the sputtering process (this phenomenon is covered in Section 3.2.3.1)
   b. Can modify the formation or otherwise of sputter-induced surface topography (covered in Section 3.2.3.3)
   c. Will result in the introduction of additional oxide-related secondary ion signals in the mass spectra that may result in additional isobaric interferences (see Section 5.3.1.3)
3. Higher chamber pressures will result in increased neutralization of the primary ion beam. This, in turn, can result in a loss of depth resolution (as is discussed in Section 5.3.2.4, this will occur if the neutral population were to strike an area away from that of the primary ion beam)
4. The adsorption of Oxygen onto the internal surfaces of the analysis chamber that also occurs during the implementation of $O_2$ leak conditions will require extended pump-down times if UHV conditions are required in subsequent analysis. Such analysis should, therefore, be carefully planned in accordance with any other analysis required.

### 5.3.2.4 Depth Profiling and Related Aspects
In the area of Dynamic SIMS, depth profiling is the most heavily used of the three data collection modes. As an example, this mode is used for over 90% of the semiconductor research and development work. The collection of mass spectra before depth profiling or imaging tends to be used to reveal the most useful analytical signals. Imaging tends to find heavier usage in areas such as the Biosciences.

Depth profiling refers to the reproduction of any secondary ion intensity variation as a function of depth recorded from spatially homogeneous substrates, i.e. variations in one dimension (analysis of secondary ion intensity variations in two or three dimensions is referred to as *imaging*). Depths ranging from $\sim 10$ nm to greater than 10 μm can be routinely accessed, with examples shown in Figure 5.3.

The primary parameters of importance in depth profiling are:

1. The *sputter rate*
2. The *depth resolution*
3. The *dynamic range*

The sputter rate, as mentioned in Section 3.2.2, defines the rate at which the material is being removed from the sample. For a specific substrate, this is controlled by the primary ion energy, angle of incidence, current and the area sputtered. These are preset before analysis to match the volume of interest. Methods to derive sputter rates are discussed in Section 5.4.2. Uniform sputter rates over substrate regions not exhibiting any chemical/structural variations are typical. Sputter rate variations are, however, noted during the onset of sputtering (see Figure 3.20) and over interfacial regions (see Figure 3.26) of the substrate. This stems from ion beam-induced segregation of elements within the substrate of interest (this is discussed in Section 3.2.3.1) and the nonsteady-state implantation of some fraction of the primary ion beam into substrate (see Section 3.2.2.2.2)

Nonsteady-state sputtering during the onset of depth profiling arises from the loading of the near surface region of the substrate by primary ions. This loading occurs through implantation of some fraction of the primary ions into the substrate. An example of this loading is depicted for $Cs^+$ primary ions in Silicon in Figure 3.19. The re-sputtering rate of implanted primary ions will, however, eventually catch up to the implantation rate. Once equilibrium between the primary ion implantation and the removal rates is reached, no further sputter rate variations are noted. This is referred to as *steady-state sputtering*. The region suffering sputter rate variations typically extends from the topmost surface to $\sim 2.5$Rp of the respective primary ion. Assuming surface topography growth is controlled (see Section 3.2.3.3) and the sputter rate is sufficiently high such as to overcome the effect of surface swelling, Rp values can be approximated using Relation 3.6a–d.

As discussed in section 3.3.2, reactive primary ion beams also introduce significant variations in secondary ion yields and hence the measured secondary ion signals. Thus, any variation in the amount of the implanted primary ion's population will induce significant variations into the measured secondary ion signals. Examples of this are shown in Figures 3.21, 3.22, 3.29 and 3.31. These effects are typically referred to as: *Transient effects*, with the region referred to as the *Transient region*.

As transient effects are difficult to remove, the transient region should *in general* not be used unless well understood or methods effective in reducing or removing these effects are available. Indeed, some success can be derived when comparing samples of the same matrix, as these will exhibit the same transient effects. Likewise, point-by-point data correction algorithms, with one example being the PCOR™ method (see Appendix A.9.4), can be devised. Another approach lies in the capping of the substrate with a thin layer of the same material as the substrate's matrix. An example commonly used in the semiconductor industry is to use a thin Silicon cap. This effectively moves the transient region outside the region of interest.

The depth resolution defines the minimum depth over which two signals can be separated. This parameter, discussed in Section 5.1.1.2.1, depends on:

1. The primary ion beam conditions, inclusive of the resulting data point density
2. The secondary ion spatial acceptance applied
3. The sample quality

Ultimate depth resolution can only be attained if:

1. The sample is spatially homogeneous over the area analyzed
2. The sample's surface is free of foreign objects (dust, particles, fibers, etc.)
3. The sample's surface is free of scratches

This is realized as the sputtering front follows the initial surface topography. If this is not parallel to any subsurface interface or feature/s of interest, the apparent sharpness of the interface/feature/s will be a convolution of both the surface topography and the abruptness of the interface/feature. These effects, referred to as *crater base effects* within this text, are covered in Section 5.3.2.4.3.

Optimal primary ion beam conditions also assure that sputtering proceeds at a uniform rate over the region of interest, i.e. that the crater base being formed runs parallel to the surface and any subsurface interface and/or feature of interest. This is ensured by rastering the primary ion beam in a uniform manner over a predefined region, with the raster pattern (usually square) covering a much larger area than the depth being sputtered to (typical factors are 100× or more). This requires effective focusing of the primary ion beam at the substrate's surface along with alignment of the primary and secondary ion beams such that their axes coincide at the substrate's surface.

Any neutrals present in the primary ion beam must also be removed because this can result in nonsymmetrical sputter crater formation. Recall: The trajectories of neutrals are not influenced by electrostatic or magnetic fields, and as a result, they may impinge on areas not specified by the primary ion beam optics. Such neutrals arise from primary ion interaction with apertures present within the primary ion column (typically the last set of apertures) and/or collisions with residual gas molecules. A vacuum of at least $10^{-7}$ Pa should be used to ensure a dynamic range of $10^6$ or better (Wittmaack 1982).

In the case of molecular depth profiling using large cluster ion beams, the substrate temperature can also play a part. Indeed, lowering the temperature to below 100 K can provide optimal results, with an example shown in Figure 5.4(b) (Lu et al. 2011). Owing to the relatively recent introduction of molecular depth profiling in the field of SIMS, this area is experiencing extensive developments. Hence, the reader is advised to refer to the latest literature in this area.

The loss of depth resolution can also result in decreased peak intensities of the element/s and/or molecule/s of interest, and/or deterioration in dynamic range.

Decreased peak intensities result from the dispersion of the elemental and/or molecular signals of interest over a greater depth, whereas deterioration in dynamic range stems from the longer time for the secondary ion signal/s representative of the element/s and/or molecules of interest would reach background levels following the layer of interest.

Even if all possible measures to improve depth resolution are applied, there still exist the effects of the primary ion beam-induced damage experienced by the substrate as a result of the sputtering process. These are discussed in Section 5.3.2.4.1. Crater edge effects and crater base effects can also result in the loss of depth resolution. These are discussed in Sections 5.3.2.4.2 and 5.3.2.4.3, respectively. Dynamic range pertains to the range of concentrations of a specific element or molecule that can be examined in a particular depth profile. As can be envisaged, this depends on the detection limit and on the detector type or combinations thereof (detectors are covered in Section 4.2.3.3).

*5.3.2.4.1 Ion Beam Damage* As introduced in Section 3.2, the primary ion energy, mass, and angle of incidence control the range the primary ion will traverse into the substrate. The average range, typically referred to as the *Projected Range* (Rp), can be simulated/approximated for relatively simple projectiles using methods described in Sections 3.2.1.3 and 3.2.2.1.1. As this is almost always deeper than the sputtered front, the damage front will precede the sputtered front. This situation is illustrated pictorially in Figure 5.18.

The one exception to the above rule is noted for large cluster primary ion impact. This is realized as the increased sputter yields noted under such conditions allow the sputter front to approach or even surpass the damage front. This only occurs under highly specific conditions with an example illustrated in Figure 5.4(b).

When the damage front precedes the sputter front, the loss of depth resolution will occur irrespective of the analytical conditions used. Indeed, it is this process

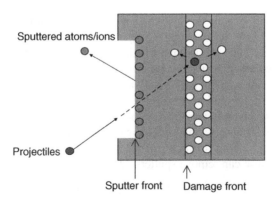

**Figure 5.18** Pictorial illustration of the subsurface damage induced by a primary ion as it travels beyond the sputtering front (this will occur unless the projectile is involved in a direct head-on collision or is scattered from the surface). The white circles represent atoms within the layer of interest.

that defines the ultimate depth resolution that can be reached when using relatively simple primary ion beams, irrespective of the analytical conditions used.

*In general*, the depth resolution scales with the primary ion energy. This results from the increased momentum transfer that occurs during the collision cascade. The term *"in general"* is used as the mass difference between the atoms within the primary ion beam and the substrate also plays a role. All of the above are discussed in greater detail in Section 3.2.2.1.1.

For a specific primary ion beam impacting at a specific angle on a specific substrate, the loss of depth resolution can be simplistically understood as arising from the fact that the damage front depth scales with the Rp of the primary ion. The Rp, in turn, scales with the impact energy, mass, and the incidence angle, as well as a greater amount of energy is dissipated within the substrate with increased energy and mass.

At this point, it is worth mentioning that higher energy primary ion analysis beams used during dual beam depth profiling in Time-of-Flight-based instruments based on pulsed primary ions *generally* have little effect on depth resolution. This is realized as the total dose of the primary ion analysis beam is minimal when compared to the dose of the primary ion sputter beam. The term *generally* is used as care must be taken when setting up depth profiles to ensure that the dose of the primary ion analysis beam remains sufficiently low relative to the sputter beam, hence the use of the term *generally*.

As mentioned in Section 5.3.2.3, depth resolution may also be affected by primary ion beam-induced modification of the substrate chemistry. This is typically noted for reactive primary ion beams. As an example, $O_2^+$ primary ion beams can oxidize the outer surface of certain materials. This can then facilitate the movement of the elements of interest within the solid, i.e. induce segregation to/from the surface. This effect is referred to as *oxidation-induced segregation*. *Radiation-induced segregation* discussed in Section 3.2.3.1 may also occur. Concomitant secondary ion yield variations resulting from these surface chemical modifications will further complicate matters.

Different secondary ions can also display different depth resolution values for the same substrate. As an example, Copper typically yields poor depth resolution because of its high diffusion coefficient, particularly when sputtered. This sputter-induced enhancement is otherwise referred to as *radiation-enhanced diffusion*. *Radiation-induced segregation* may also be initiated, with different primary ion/secondary ion combinations resulting in different trends. As a result, any emissions collected as a result of this form of sputtering will always emanate from what is termed an *altered layer*, as opposed to the initial intrinsic substrate layer. Exceptions are sometimes noted for large cluster ion impact, because, as mentioned in Section 4.1.1.3, these can remove sputter-induced damage.

*5.3.2.4.2   Crater Edge Effects*   Crater edge effects are an instrumentally introduced distortion that will result in the loss of depth resolution during depth profiling if not filtered out. These effects describe the influence of secondary ion signals emanating from the edge of the sputter crater being formed as is illustrated in Figure 5.19(a). In short, the loss of depth resolution arises from the collection of

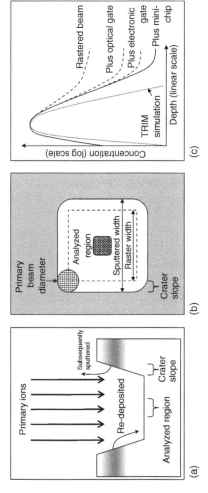

**Figure 5.19** Schematic representations of (a) the cross section of the crater formed with crater edge effects resulting from re-deposition and sputtering from shallower regions, (b) a top-down view of the crater formed revealing the relationships among the raster width, sputtered width, and analyzed region, and (c) the respective improvements in depth resolution and dynamic range that are realized using various continuous primary ion beam crater edge filtering methods (see text). Also shown in Figure 5.19(c) is a TRIM simulation of an implanted element. This displays a peak at a slightly greater depth and a more rapidly decaying signal with depth than any of the SIMS profiles. The reduced depth noted in the SIMS profiles is a distortion resulting from the sputtering process when a linear sputter rate and no sputter-induced chemical modifications are assumed.

signals from some shallower layer containing a greater concentration of the element of molecule of interest that remains exposed (that at the crater edge).

In the case of primary ion pulsed Time-of-Flight-based SIMS instruments, crater edge effects can be removed by rastering the primary ion analysis beam over a smaller region centered within the middle of the primary ion sputter beam raster pattern.

Crater edge effects noted in instruments using continuous primary ion beams can be reduced by rastering a well-focused primary ion beam over the area of interest, with the signal arising from the edge of the raster pattern filtered out. Filtering can be carried out using both/either *optical gating* and/or *electronic gating*.

*Optical gating* describes the use of secondary ion column lenses/apertures to filter out signals emanating from regions close to the crater edge. This can be carried out when using a primary ion beam of any diameter. This is specifically useful for large spot size primary ion beams noted when high primary ion currents are required.

*Electronic gating* describes the ability to filter out secondary ion signals from regions close to the crater edge through synchronization of the raster with the collection electronics. This, however, requires the use of small spot size primary ion beams, i.e. those noted when using low primary ion currents

Most typically, the analyzed region represents between 4 and 20% of the sputtered area, i.e. should be less than 33% (factor of 3 rule). Pictorial examples of the analyzed to sputtered regions accessed along with the improvements in depth resolution are demonstrated in Figure 5.19(b) and 5.19(c), respectively. Moreover, when using a Gaussian beam, the raster size must be greater than four times the beam diameter.

Also shown in Figures 5.19(c) is the further improvement in dynamic range noted through *Mini-chip analysis*. This is an alternative analysis methodology discussed further in Section 5.3.2.3.5.

Retrospective approaches can be applied. In such cases, the analyzed region (that used to construct the depth profile) is defined following data collection. There exist two methodologies that can be used. These are otherwise referred to as *Region of Interest* (ROI) analysis or *Checkerboard* analysis.

*ROI analysis* describes a methodology in which the region from which the depth profile is to be constructed (the region of interest) is defined from within the sputtered region using a more or less free hand approach. This methodology is typically found on Time-of-Flight-based SIMS instruments.

*Checkerboard analysis* describes a methodology in which the sputtered region is divided into a checkerboard-like surface ($16 \times 16$ to $128 \times 128$ are common) with the depth profile then constructed from the secondary ion signal from one or a combination of the freely selectable squares. This filtering capability can be found on some Quadrupole-based and Magnetic Sector-based SIMS instruments.

For both *ROI analysis* and *checkerboard analysis*, the spatial distribution of the secondary ions emanating from the sputtered surface must be imaged spatially such that the region from which the depth profile is to be reconstructed can be visualized.

As this is carried out using the *microprobe* mode, the image quality is defined by the primary ion beam diameter at the substrate's surface, i.e. the spot size.

*5.3.2.4.3  Crater Base Effects*  Crater base effects are designated within this text as surface topography-induced effects (crater base topography) that result in the loss of depth resolution during SIMS depth profiling. Such topography may be preexisting before sputtering or may be induced and/or modified by the sputtering process. Such effects are noted irrespective of the instrument type used.

The loss of depth resolution ensues as the sputtering front follows the initial surface topography, i.e. if the surface is not parallel to any subsurface interface or feature of interest, the apparent sharpness of the interface/feature will be a convolution of the surface topography and the abruptness of the interface/features being examined.

Preexisting topography within the surface region to be sputtered can arise from:

1. Foreign objects on the surface to be sputtered, i.e. dust, particles, fibers, and so on.
2. Scratches, pits, and/or voids
3. Steps, edges, ledges, preformed features, and so on.

For best depth resolution, the region to be examined should be free of any surface topography, whether intrinsic or extrinsic to the substrate. Foreign objects can be removed using pressurized air or Nitrogen gas before placing the sample into the SIMS instrument. Scratches, pits, voids, steps, edges, preformed features, and so on should be avoided as these can also influence the secondary ion extraction field as well as ionization yields, which, in turn, will introduce additional distortions. These can be avoided through the selective choice of analysis areas either before analysis, or if not visible, following analysis (these are typically identified through the distortions suffered).

The possibility of surface topography growth during the sputtering process alone is also apparent. This can occur on both single crystal and polycrystalline substrates.

In the case of single crystal substrates (this includes single crystal regions within polycrystalline substrates), different types of surface roughening are noted. These are dependent on the substrate type and the sputtering conditions applied.

One type of sputter-induced topography describes the growth of cones, pyramids, and so on. These tend to occur on substrates that either do not amorphize or recrystallization occurs within the sputtered region during the sputtering process. An example of this is noted following extended sputtering of Zirconium with 14.5-keV $Cs^+$ primary ions, with the resulting formation of pyramids noted within individual grains apparent in the AFM images shown in Figure 3.22. These occur progressively with increasing sputtering time.

The other type of sputter-induced topography describes the formation of ripples over the crater base region that travel in the direction of the primary ion beam when primary ions are incident at some nonnormal incidence angle. These tend to occur on substrates that readily amorphize during the sputtering process, specifically if too low a primary ion impact energy and/or too high an incident angle with respect to the sample normal are applied. An example of this is shown for Silicon when sputtered with $Cs^+$ primary ions in Figures 3.17 and 3.21. These form abruptly at some depth dependent on the analytical conditions used, with the depth at which these occur decreasing with reduced primary ion impact energies and increased primary ion incident angles with respect to the sample normal. A small but discernible increase in the intensities of various atomic secondary ions is also noted when this roughening ensues.

Note: As discussed in Section 5.3.2.4.1, depth resolution is *typically*, but not always, improved using lower energy primary ion beams impacting at greater angles with respect to the sample normal. The introduction of ripple topography, however, places a limit on the incident energies and angles that can be used. Although this form of surface roughening can be controlled using sample rotation (see Section 5.3.2.4.5), beam alignment becomes increasingly difficult at the low-impact energies and large incident angles under which ripple topography is typically noted.

Surface roughening occurring on polycrystalline substrates proceeds through an entirely different mechanism, i.e. a result of the differential sputtering rates noted for different grain orientations. An example of this is shown in the AFM images acquired following the sputtering of an initially smooth (prepolished) polycrystalline Zirconium metal substrate with 14.5-keV $Cs^+$ primary ion in Figure 5.20.

The loss of depth resolution noted from polycrystalline substrates proceeds gradually with increasing sputtering time. This depth dependency has been approximated as (Hofmann and Sanz 1984):

$$\Delta d = a.d^{0.5} \tag{5.2}$$

where $\Delta d$ is the loss in depth resolution noted, $d$ the depth of interest with both in units of nanometers, and $a$ an empirically derived parameter equal to $0.86 \pm 0.22$.

Also shown in this Figure 5.20 are the secondary ion yield variations noted. These stem from the differential uptake of Cesium within the respective grains, which, in turn, results from the differential ionization yields as well as differential sputter rates. Note: As discussed in Section 3.2.2.2.2, increased Cesium uptake occurs with decreasing sputter rates, hence the greater signal intensities noted from the shallower regions in Figure 5.20.

This form of surface roughening along with the corresponding loss of depth resolution and increased variations in ion yields can be controlled using sample rotation (Zalar 1985) as will be discussed in Section 5.3.2.4.5.

### 5.3.2.4.4 *Additional Parameters of Interest*

Even after accounting for the aspects covered in Sections 5.3.2.4.1 and 5.3.2.4.3, there still remain the questions concerning the conditions (parameters) that should be applied in depth profile analysis of a specific substrate. Examples include:

1. What defines an optimal analysis time?
2. How to define the end point of a depth profile?
3. What data point density should be used?

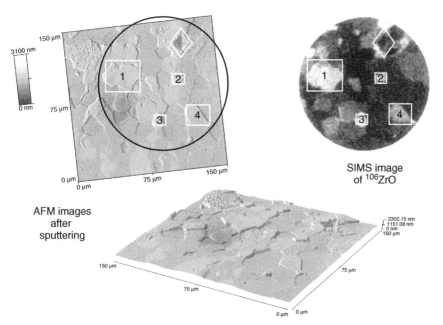

**Figure 5.20**   AFM (left and middle) and SIMS (right) images of the crater base of region of an initially smooth (polished) Yttria-Stabilized Zirconia (YSZ) substrate following prolonged 14.5 keV $Cs^+$ primary ion sputtering. Regions displaying the higher secondary ion intensities and lower sputter rates are marked numerically. The location was identified via nano-indentation marking (diamond shape marked at top right corner).

This is realized because, as covered in Section 3.2, different matrices act differently under energetic primary ion beam impact.

Analysis times should typically be less than 1 h. Optimally, these should lie in the 10–20 min range to ensure that the utmost precision as this reduces possible primary ion beam current drift issues, while allowing replicate profiles to be collected in a time efficient manner. Even so, there are situations in which much longer sputtering times are applied, i.e. when extremely deep layers are to be accessed or when high-depth resolution is required over a large depth range. Note: Depth resolution requires a high data density per unit depth, which, in turn, requires a slow sputtering rate.

An example of the time constraints in depth profile analysis is noted in high precision depth profiling of dopants as is required in the semiconductor industry. In addition to the requirement of primary ion beam stability, these typically employ the collection of at least three replicate profiles per analysis site. A minimum of three is stipulated as this represents the minimum value needed in carrying out any form of statistical analysis. Only through such methodologies can a dose precision of better than 1% be derived (this is assuming that the substrate contains sufficient dose to provide the needed counting statistics).

The primary ion stability should also be monitored during a depth profile analysis. This can be carried out in all commercially available SIMS instruments by periodically deflecting the primary ion beam into a Faraday cup situated in the primary ion column (carried out in an automated manner). In addition, measurement of some matrix signal, along with the signals of interest, during a depth profile provides further insight into the sputter rate stability while also revealing the depth scale over which steady-state sputtering prevails. Deviations from steady-state sputtering can ensue during the initial stages of sputtering and while sputtering through interfaces. Reasons for this are discussed in Section 3.2.2.2.2. Such matrix signals are also useful in quantification of the concentration scale as will be covered in Section 5.4.3.

Depth profile end points (the point at which a sputter depth profile can be terminated) can be derived through one or a combination of the following:

1. Understanding the sample structure as a function of depth
2. Application of previously defined sputter time to depth conversion factors pertinent to the matrix being examined

The latter can only be applied if the substrate of interest is of the same matrix as that of the substrate previously examined for which the sputter rate is already known. Furthermore, both substrates must be analyzed under near identical conditions. Note: The primary ion beam raster size can be altered as the size and the sputter rate are related in a quadratic manner, i.e. reducing the raster size from 500 μm × 500 μm to 250 μm × 250 μm will increase the sputter rate by a factor of four. Altering the primary ion incident energy or incident angle will affect sputter rates in a more complex manner, with an example illustrated in Figures 3.15. Such trends are further complicated by the fact that a primary ion and substrate type dependence also exists (an example is illustrated in Figure 3.15).

Data point density will depend on the depth resolution required, the depth profile end point, and what is considered an acceptable analysis time. When high-depth resolution is required, the data point density should be sufficient to easily identify the layer, marker, or interface of interest. As this will increase the analysis time, a compromise will need to be decided upon before analysis when deep layers are to be probed. Alternatively, a high sputter rate can be used to probe to a slightly shallower region. The sputter rate can then be reduced to provide the data density needed over the subsequent region of interest. Automated routines available within various commercially available SIMS instruments can prove useful in this respect.

*5.3.2.4.5 Additional Approaches of Interest* There exist a limited number of additional approaches that have been developed to control the loss of depth resolution noted under specific conditions and from specific secondary ions. These include modified and retrospective (reconstruction) approaches.

Modified approaches describe those in which the sample and/or analytical conditions have been specifically tailored. These can include, but are not limited to:

1. Sample rotation as depicted schematically in Figure 5.21(a)
2. Backside SIMS as depicted schematically in Figure 5.21(b)

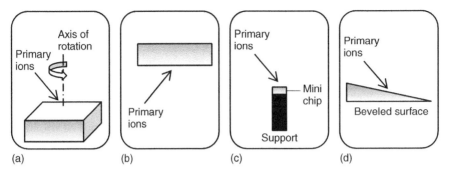

**Figure 5.21** Pictorial illustrations of (a) sample rotation, (b) Backside SIMS, (c) Mini-chip SIMS, and (d) Bevel analysis.

3. Mini-chip analysis as depicted schematically in Figure 5.21(c)
4. Bevel analysis as depicted schematically in Figure 5.21(d)

Sample rotation describes the use of azimuthal rotation of the sample such that the axis of rotation coincides with the central point at which the sputtering ion beam strikes the surface. This is sometimes referred to as Zalar™ rotation (Zalar 1985) as trade marked by Physical Electronics. This can be used to remove, or at least reduce, sputter-induced surface roughening whether in the form of cones, pyramids, ripples within single crystal regions, and differential sputtering exhibited by different crystal faces from polycrystalline substrates that are within the region being analyzed (for example, see Figure 3.22). Sample rotation is particularly effective in the analysis of polycrystalline solids as these can exhibit significant crystal orientation-specific sputter rate variations as discussed in Section 5.3.2.4.3.

Backside SIMS as the name suggests is a method in which the region of interest is profiled from the back side (Achtnich et al. 1987). This is carried out when a particular element, Copper, for example, displays a very large diffusion and/or segregation into a deeper layer during normal depth profiling conditions. In specific cases, depth profiling from the reverse direction has been shown to significantly reduce this effect, thereby resulting in acceptable results. Although front-side knock-on effects are removed, the effects of cascade mixing and radiation-enhanced diffusion remain (these are isotropic). Sample preparation for backside SIMS can, however, be a difficult and time-consuming effort as this requires careful mechanical polishing of the underlying substrate while keeping the layers of interest parallel to the polished face. As an example, the backside face should be kept parallel to the surface to within 0.06° or less. Multi-prep™ polishing tools are commonly applied along with use of red filters to ensure that polishing over the region of interest remains parallel to the front side. Owing to the stringent preparation procedures required, this method is used only in the semiconductor industry.

Mini-chip analysis is an *in situ* method (Criegern and Weitzel 1986) that describes the analysis of a chip of similar dimensions as the analysis area, i.e. in the range of $10 \, \mu m \times 10 \, \mu m$ to $100 \, \mu m \times 100 \, \mu m$. This highly specific type of analysis removes elements/molecules of interest from the surrounding regions that would otherwise form the crater wall. This can be carried out by elevating the mini-chip from its surroundings either naturally or using selective chemical etching of the surroundings and/or removing the chip from its surroundings (most commonly carried out using FIB methodologies) and depositing the mini-chip onto some support. A closely related approach is that used to form *mesa structures* (Guenther et al. 2009). This describes the formation of trenches, using a dicing saw, around the region of interest before analysis. As with Backside SIMS, these methods are only used in the semiconductor industry.

Bevel analysis, as the name suggests, describes the collection of secondary ions over a beveled region of the sample of interest (Gries 1985). The bevel angle is typically set at less than 1° with respect to the surface plane (the bevel angle, the dimensions of the analyzed region, and the spatial resolution will define the depth resolution following depth profile reconstruction). Sample preparation is carried out using Chemical etching, Mechanical polishing, or Ion beam milling.

Chemical etching to different depths can be carried out by dipping the surface vertically into an etchant solution. The bevel is then simply a function of the etch rate and the dipping time, with the bevel forming as the time over which the sample (dipped vertically) is removed from the etchant. Mechanical polishing methods would be similar to those used in Backside SIMS sample preparation. Ion beam milling would be carried out in an FIB, with lower energy ions used in subsequent steps to reduce the ion beam damage.

SIMS analysis is then carried out in spatial imaging mode or through line scan analysis along the bevel. To attain acceptable depth resolution values, the bevel angle must be sufficiently shallower and the analysis area sufficiently large. By analyzing the outer most surface, or at least that before the damage front displayed in Figure 5.18, the loss of depth resolution resulting from ion beam damage (see Section 5.3.2.4.1) is effectively circumvented. As a result, higher energy primary ions can be used. These provide the additional advantages of improved spatial resolution and a deeper damage front.

With the exception of Bevel SIMS, none of the above techniques remove the loss of depth resolution resulting from the subsurface damage introduced by the sputtering process, i.e. the two go hand in hand. This serves to explain why TRIM simulation tails never completely match SIMS profiles even when all background effects have been accounted for (see Figure 5.19(c) for a pictorial example). Only through bevel approaches or retrospective approaches can SIMS profiles be made to match TRIM simulations.

Retrospective approaches are more limited as they can only be applied in ideal cases on atomically smooth spatially homogeneous surfaces if a sufficient amount of information on the system exists. This is realized because such approaches require knowledge of either the instrument's response function (a function that describes the loss of depth resolution apparent under a predefined set of conditions) or the processes describing cascade mixing along with any ion-induced diffusion and/or segregation processes active. Smooth samples are required to ease the computational burden.

As derivation of response functions is easier than describing, in an analytical manner, the intricacies of the processes responsible for the loss of depth resolution (cascade mixing and so on) this is the general approach of choice. Even so, response functions exhibit a complex dependence on the incoming ion, the substrate, the signal recorded and the instrumental conditions used.

The derivation of a response function requires the analysis of a solid in which the original location of atoms is known, such that the migration of atoms under well-controlled conditions can be ascertained with a sufficient level of certainty. Such solids generally take the form of delta-layered structures such as that shown in Figure 5.4(a) and 5.4(b). A delta-layered structure is one in which a foreign atom (Boron in Silicon the case of Figure 5.4(a)) is present at a highly specific depth within the matrix of interest, i.e. exists as an atomically abrupt marker layer.

Knowledge of the response functions for specific analytical conditions then allows for the extrapolation back to an apparent *zero* impact energy, i.e. that in which no ion beam-induced damage is introduced. Indeed, the effectiveness of this approach has been demonstrated in various delta-layered samples, with the sharpness of the peak improving with decreasing primary ion impact energy and the peak position moving to slightly greater depths. An example of the application of this approach to an Arsenic implant profile in Silicon is shown in Figure 5.22 (McPhail and Dowsett 2009).

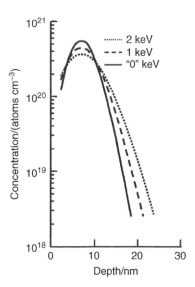

**Figure 5.22** Example of the extrapolation to zero primary ion impact energy for an As-implanted Silicon sample-based on the empirical 1 and 2 keV impact energy data along with mathematical procedures (response functions) describing the sputter-induced modifications experienced by the respective elements of interest. Reproduced with permission from McPhail and Dowsett (2009) Copyright 2009 John Wiley and Sons.

## 5.4 DATA PROCESSING

This section describes the various methodologies available for processing the as-recorded SIMS data. Data processing may be required to:

1. Understand the origin of the recorded secondary ion signal or signals
2. Reference the collected secondary ion signal to some other signal
3. Convert the sputter time to sputter depth and/or convert the secondary ion intensity to concentration

Firstly, understanding the origin of a signal recorded at some $m/q$ ratio may entail:

1. Matching the $m/q$ ratio to the isotopic mass/s of the isotopes of an element (this applies to atomic secondary ion emissions)
2. Matching the $m/q$ ratio to the sum of the isotopic masses making up a molecule (this applies to simple molecular secondary ion emissions)
3. Use of algorithms that relay molecular fragmentation pathways (this applies to heavily fragmented molecular secondary ion emissions)

The approaches used in spectral identification, whether applicable to Static or Dynamic SIMS, are discussed further in Section 5.4.1 and the subsections within.

Once the signal is identified, the signal may then be referenced to some other signal or signals. Referencing to some matrix secondary ion signal collected from the same sample during the same analytical session is a common procedure used to compensate for any sputter rate variations that may be present and in some cases topographical variations. Such variations may be intrinsic to the substrate or induced as a result of some unintended change to the analytical conditions used, i.e. primary ion current fluctuations. As relative distributions can often provide the needed information, i.e. when comparing good versus bad samples, quantification is not always necessary. Indeed, owing to the increased complexity associated with quantification over spatial or volumetric regions, such referencing methodologies alone are more common.

Isotope fractionation studies also require that the Instrument's Mass Fractionation (IMF) be accounted for. As covered in Section 3.3.2.2.3, the IMF describes the dependence of sputtering and ionization yields on the isotopic mass of the same element, with the former being the strongest. This is required in many Earth Science studies, particularly chronologically based, as the IMF can introduce deviations from actual isotope ratios (these are in the permil (ppm) range). This is rarely applied elsewhere.

Conversion of the sputter time scale to either the depth (as applies to Dynamic SIMS) or the surface coverage removed (as applies to Static SIMS) can be a relatively straightforward procedure. This is outlined in Section 5.4.2. Owing to the intricate interplay of intrinsic matrix effects with analysis-induced matrix effects, conversion of the intensity scale can be much more difficult. Indeed, effective quantification of the concentration scale requires the complete removal of all matrix effects, inclusive of those resulting from sputter rate/yield variations. This is covered in Section 5.4.3.

At this point, it is worth noting the difference between accuracy and precision. Accuracy pertains to the ability to derive the correct value. Precision pertains to the ability to repeat/reproduce a specific value as closely as possible. Note: Repeatability is the random uncertainty associated with data collection on the same equipment within the same laboratory. Reproducibility is taken as that from different laboratories.

Most analytical techniques excel in either accuracy or precision, but not both. Most manufacturing environments tend to favor precision over accuracy as this allows for fine-tuning of the various processes applied. The medical and research environments, on the other hand, are more concerned with accuracy. Indeed, any offset, shift, or bias in medical dosage or location of an incision cannot be afforded.

### 5.4.1   Spectral Identification

As outlined in Section 5.1.1.1, secondary ion emissions can be collected in the form of mass spectra. This is useful in cases where the optimal signals of interest are not known before analysis, and/or an understanding of all neighboring signals is needed. The former serves to aid in the collection of spatial or volumetric images or depth profiles, i.e. allows for the identification of the most effective signal/s to be followed, whereas the latter allows for the understanding of isobaric interferences that may be present.

In the case of atomic and simple molecular ion emissions (unfragmented ions or those displaying minimal fragmentation), identification tends to be a relatively straightforward procedure of matching their masses. This is covered in Section 5.4.1.1.

More complex molecular ion emissions, particularly those suffering extensive fragmentation can be more difficult to identify. This is further complicated by the fact that multiple analysis-specific fragmentation pathways are available for complex molecules (these tend to be primary ion momentum dependent), and multiple signal sets (those from different fragment pathways) may be generated within a single SIMS mass spectrum. As is described in Section 5.4.1.2, these can be delineated using the *Gentle SIMS* (G-SIMS) and ancillary approaches.

#### 5.4.1.1   *Atomic and Unfragmented Molecular Emissions*   As mentioned in Section 5.4, the association of a recorded signal to a specific element or molecule can be arrived at by matching the $m/q$ ratio of the signal to isotopic mass values (natural abundances of the elements are presented in Appendix A.2).

Care must, however, be taken to ensure that the mass scale (or more precisely the $m/q$ scale) of the instrument is well calibrated, and that there are no isobaric interferences under the conditions used (isobaric interferences are covered in Section 5.3.1.3). If there are isobaric interferences, these should be removed using one of the procedures outlined within subsections of Section 5.3.1.3. These include the use of HMR, KE filtering, or peak stripping.

Confidence in elemental assignment can be assured by comparison of the intensities of signals that can be ascribed to other isotopes of the element of interest, i.e. these intensity ratios should match the natural abundances isotopes listed in Appendix A.2, unless, of course, isotopic enrichment of one form or another has occurred (as is the case in nuclear reactions, extra-terrestrial samples, and so on) or has been applied (isotopic implantation). In the case of mono-isotopic elements such as Fluorine and Aluminum, molecular patterns, or even multiply charged ions, can be used to confirm assignments.

When there exist molecular secondary ion signals of the same nominal mass as the atomic ion that can be separated through HMR, the atomic signal invariably resides at the lower $m/q$ value. An example of this is shown in Figure 5.2(a). This general rule of thumb arises as a result of the mass deficit exhibited by all isotopes from $^{12}$C to $^{220}$Rn. Note: Little in the way of isobaric interferences are noted for isotopes lighter than $^{12}$C.

As for molecular secondary ions, the mass is simply the sum of the constituent elements (isotopes) making up the molecule. Assignment can then be relatively straightforward for unfragmented ions or those displaying minimal fragmentation. Greater care must, however, be taken to minimize, or at least understand, the presence of potential isobaric interferences.

### 5.4.1.2 *Heavily Fragmented Molecular Emissions*   As greater molecular fragmentation occurs as the impacting primary ion momentum increases, more complex mass spectra are generated. Adding to this is the fact that multiple fragment pathways are available, some of which can occur in parallel.

Assignment of peaks thus becomes increasingly difficult particularly when analyzing unknown organics. Until recently, molecular identification from such complex spectra was almost exclusively carried out using pattern recognition procedures. In such cases, preexisting spectral libraries prove useful.

The introduction of the G-SIMS approach, applied by itself or in combination with *Fragmentation Pathway Mapping* (FPM) and *Simplified Molecular Input Line Entry Specification* (SMILES), has, however, opened up the possibility of understanding the various fragment pathways and matching SIMS spectra collected under different analytical conditions.

### 5.4.1.2.1 *Gentle SIMS*   The G-SIMS approach is a methodology introduced in 2000 (Gilmore and Seah 2000; 2003; Seah et al. 2010) to help clarify the often complex molecular fragment patterns noted in Static SIMS mass spectra. This approach also allows differences in mass spectra derived using different primary ion energies and masses (momenta) from the same molecular species to be understood. This procedure describes the fragmentation of molecules in terms of a surface plasma temperature, i.e. as specified in the *heat spike* or *thermal evaporation*-based models (see Section 3.2.1.2).

As the degree of fragmentation is primarily a function of the primary ion momenta (all other parameters are kept constant), a methodology was developed by which mass spectra from primary ions of different momenta are acquired in an interleaved manner within the same pulsed primary ion Time-of-Flight-based SIMS instrument. Primary ion combinations that have been applied include Inert gas ions of different energy, $Ar^+$, $Xe^+$, $Ga^+$, $Cs^+$ $SF_5^+$, or $Bi_5^+$ from separate ion sources, and $Mn^+$ with $Bi_n^+$, where $n$ can equal $1-7$, from the same BiMn Liquid Metal Ion Source ($Mn^+$ and $Bi^+$ appear more useful for secondary ions below 300 $m/q$, whereas $Bi^+$ and $Bi_3^+$ appear more useful above 300 $m/q$). Other combinations can also be used with the only requirement being that the primary ions selected induce different fragment patterns.

From the mass spectra collected, a mass spectrum displaying heavily reduced fragmentation is extrapolated using some $g$ factor pertinent to the system. The $g$ factor represents a parameter relating to the surface plasma temperature, with larger $g$ values pertaining to conditions under which less damage and hence less fragmentation are expected. Through extrapolation to higher $g$ values (more gentle sputtering conditions), highly simplified spectra can be reconstructed that more closely mirror the primary fragments of the molecular surface studied, hence the name G-SIMS. An example of the application of this approach in the analysis of poly-L-lysine is shown in Figure 5.23.

Although the G-SIMS approach allows for molecular secondary ion fragments from parent molecules to be identified with greater ease, it is the combination of G-SIMS with FPM approach that has opened up the possibility of identifying the original parent molecule. Before this, identification of the original parent molecule was carried out through comparative analysis (pattern recognition) of mass spectra collected from the respective molecule under the same analytical conditions. Mass spectral libraries have proved useful in this respect.

The ability of the G-SIMS-FPM procedure to reconstruct the original molecule from the observed fragment pattern variations with $g$ can be aided using SMILES. SMILES is a computer algorithm that derives all possible fragment pathways for a particular molecule. Although this is a relatively new advent, progress is being made in the development of a web-based *G-SIMS Data Base* (G-DB1) that will allow the exploration of fragmentation process for any molecule (Green et al. 2008).

### 5.4.2  Quantification of the Depth Scale

Secondary ion emissions are collected as a function of sputter time, or more precisely, the primary ion dose. This is a result of the fact that each primary ion impact has a statistical likelihood of removing atoms and/or molecules from the surface of the substrate of interest. If more than one atomic layer is removed per analytical cycle, as is carried out in Dynamic SIMS, quantification of the depth scale may be required. For obvious reasons, this requirement does not extend to the Static SIMS regime.

Conversion of the sputter time to depth units can be carried out once an understanding of the rate at which material is removed from the substrate's surface is reached. This rate is referred to as *etch rate* or the *sputter rate*, with the latter more commonly used in SIMS (for definitions, see Section 3.2.2). Commonly used sputter rate units include Å/s, nm/s, and nm/min. In some cases, the primary ion

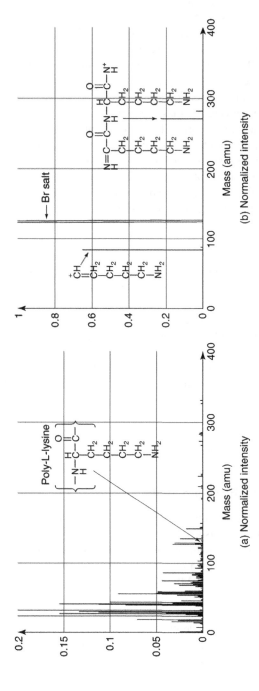

**Figure 5.23** Static SIMS mass spectra collected from poly-L-lysine resulting from (a) 10 keV Cs⁺ primary ion impact and (b) the application of the G-SIMS approach utilizing signals recorded from 10 keV Cs⁺ and Ar⁺ primary ion impact. As noted, the strong fragmentation suffered under 10 keV Cs⁺ primary ion impact alone is completely removed making visible the side chain, and the dimer repeat are clearly revealed (the Br salt is part of the substrate). Reprinted with permission from Gilmore and Seah (2003) Copyright 2003 American Chemical Society.

current is also incorporated, which then converts the units to Å/nA.s, nm/nA.s, and so on. Sputter rates can range from some negative value to positive values in excess of 5 nm/s. Negative values arise when a sizable fraction of the primary ion beam is implanted into substrates displaying very low sputter rates, i.e. when surface swelling occurs.

As discussed in Section 3.2.2, sputter rates depend on the following:

1. The dynamics (mechanism) of energy transfer which in itself depends on the masses of the colliding species, as well as the impact energy and angle, and so on.
2. The material examined, inclusive of long range lattice structure (amorphous versus crystallinity), the crystal orientation, the temperature of the sample, and so on.
3. The instrumental conditions used, i.e. the area over which the beam is scanned (stationary beams are rarely used in SIMS).
4. The rate at which the incoming ions strike the surface. Note: This dependence drops out when primary ion current units are incorporated into sputter rate units, i.e. as Å/nA.s when the analysis area remains fixed.

The sputter rate, and hence the depth accessed in a Dynamic SIMS measurement, can be deduced using one, or a combination, of procedures. These are listed in order of their application starting with the more commonly applied, as:

1. Measurements of the crater depth via some form of *surface profilometry*
2. Use of *Internal markers* at known depths within the substrate being profiled
3. *Microbalance methods*, i.e. weighing the sample before and after sputtering
4. *Theoretical methods* in the form of sputter rate/sputter yield equations

Surface profilometry along with sputter rate/yield equations can be thought of as *ex-situ* methods as these are *not routinely* applied during depth profiling. The term "*not routinely*" is used as there have been attempts to implement surface profilometry, in the form of interferometry within SIMS instruments for providing real time data. Although this has met with some success, design constraints and limitations to specific matrices have not resulted in its wide spread use. *Ex-situ* methods are discussed further in Section 5.4.1.1.

Use of internal markers can be considered a true *in situ* method as these markers can be used to relay the depth sputtered at any time during depth profile analysis. As microbalances are typically situated within the instrument, this is also considered an *in situ* method. *In situ* methods are discussed further in Section 5.4.2.2.

Note: Atomic Force microscopy (AFM), which is commonly used to derive surface topography (AFM is discussed further in Appendix A.11.4), is not typically used in evaluating sputter rates as sputtered craters typically cover areas that are at least an order of magnitude larger than the spatial range accessible to AFM (AFM areas are typically limited to less than ~10 × 10 μm).

***5.4.2.1 Ex situ Methods*** These methods allow for the translation of the sputter time scale (that recorded during a SIMS measurement) into a depth scale using prior or retrospective methods. *Ex-situ* methods most commonly applied in SIMS include the use of:

1. Surface profilometry in the form of Stylus profilometry
2. Surface profilometry in the form of Optical profilometry
3. Sputter rate or sputter yield equations/simulations

Stylus profilometry represents by far the most heavily used of all the methods used to derive sputter rates, irrespective of the application field in SIMS. This is primarily due to its simplicity, cost effectiveness, and the speed in which accurate and precise data can be attained.

As the name suggests, this method records the crater depth by scanning a stylus over the surface of the substrate analyzed, i.e. in much the same way that contact mode AFM is carried out (AFM is covered in Appendix A.11.4). Stylus profilometry is applied in contact mode after the removal of the sample from the SIMS instrument. Although stylus profilometry only provides a line scan, it is capable of a depth resolution precision approaching ∼1 nm. An example of a typical stylus profilometry scan output is portrayed along with a top-down image of the three craters measured in Figure 5.24. Note: Only one crater would typically be measured at a time.

Stylus profilometers are operated by placing the tip of the stylus over an unsputtered region adjacent to the crater of interest and then moving the sample (automated) such that the stylus scans over the crater and onto the adjacent unsputtered region. Measurement of adjacent unsputtered regions is required as this allows for subsequent leveling procedures to be applied. Leveling must be carried out before defining the depth difference between the sputtered and the unsputtered regions. If

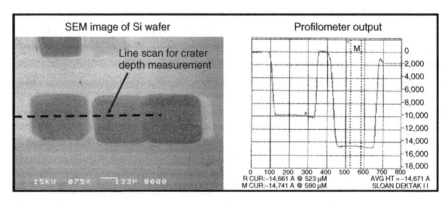

**Figure 5.24** To the left is shown a Scanning Electron Microcopy (SEM) image of multiple SIMS craters. To the right is shown the stylus profilometry scan output (line scan). The region containing the three craters the stylus is scanned over is illustrated by the dashed line within the SEM image.

this is not carried out, significant errors can be introduced into the crater depth derived. Stylus profilometry also provides the much needed information on the crater base uniformity. As covered in Section 5.3.2.4.3, the crater base should be both smooth and remain parallel to the surface throughout the depth profile as these parameters are critical to both the depth resolution and in defining the correct location of any species of interest with depth.

The sputter rate (in units of Å/nA.s or nm/nA.s) is then defined by dividing the crater depth (in units Å or nanometer), by the sum of the sputter time (in units of seconds) and the primary ion current used in forming the respective crater (in units of nanoampere). This definition assumes a uniform sputter rate over the region (depth) sputtered, hence is only applicable to a specific matrix type. Such measurements should be carried out for each crater to ensure utmost in precision (averaging procedures are often employed to reduce statistical scatter). In the case of multilayered structures, surface profilometry measurements should be carried out once each subsequent layer is sputtered, such that sputter rates pertaining to each of the different layers can be de-convoluted (recall from Section 3.2.2 that sputter rates are dependent on many parameters including those defining the matrix).

Optical profilometry, as the name suggests, uses a light source to measure the height difference between the crater base and the adjacent unsputtered regions. To reach the depth resolution values needed in SIMS, interferometric methods are applied as these are able to circumvent the diffraction limit along the depth axis (the diffraction limit is ~300 nm for visible light). Although the diffraction limit still applies in the spatial axes, this is more than sufficient considering the spatial dimensions of the regions examined, i.e. the region examined typically span more than one hundred micrometer square.

Interferometric methods operate by splitting a monochromatic coherent beam (all waves in phase) of light into two parts. One part is directed at the surface from which it then reflects and the other part directed at some reference (optically smooth mirror) from which it reflects. When the two parts are recombined, their waves will either interfere in a constructive manner (waves in phase) or destructive manner (waves out of phase). As constructive interference enhances the amplitude of the combined beam, whereas destructive interference suppresses the amplitude, an interference pattern will be produced that is dependent on the crater depth. This interference pattern can also be used to derive the surface topography of the sputtered crater in both of the spatial dimensions, with the result being an AFM-like three-dimensional topographic image.

Sputter rates (in units of Å/nA.s or nm/nA.s) are provided by dividing the crater depth (in units Å or nm), by the sum of the sputter time (in units of s) and the primary ion current (in units of nanoampere). Areas extending over regions greater than 1 × 1 mm are easily accessed within acceptable time scales. Indeed, application of interferometer-based methods offers extremely fast data acquisition to high repeatability with depth resolution values of ~2 nm. Owing to their speed, vibration canceling methods are also not required.

Although providing superior data to that of surface profilometry (three-dimensional maps are readily provided as opposed to line scans), difficulties can be experienced when examining transparent substrates (these often require two monochromatic sources of different wavelength). Likewise, extra care must be implemented in retaining sample cleanliness as the presence of dust will produce erroneous results (this is less of an issue in stylus profilometry as dust is typically swept away by the tip during data collection). Interferometer-based instruments are also more expensive relative to stylus profilometers.

Lastly, calculation methods can be useful in providing *apriori* insight into sputter rate trends, particularly if an isotropic linear cascade prevails (see Section 3.2.1.1). On the assumption that this form of kinetic sputtering prevails, the sputtered depth has been approximated as (O'Connor et al. 2003):

$$d \sim \frac{Y.J.M.t}{1000\,q.\rho.N_A.n} \tag{5.3}$$

where $Y$ is the sputter yield, $J$ the primary ion current density, $M$ the molecular weight of the substrate being sputtered, $t$ the sputtering time, $q$ the electronic charge, $\rho$ the density of the substrate, $N_A$ Avogadro's number, and $n$ the number of atoms in the molecule making up the substrate. The sputter rate $(d/t)$ can be derived through simple rearrangement of Relation 5.3.

The remaining important variable, which being the sputter yield, can be approximated through equations based on the linear cascade model (see Section 3.2.1.1) once modified to account for any ion beam modifications of the substrate (these further alter sputter rates). Such modifications are accounted for through adjustment (empirically defined) of the surface binding energy. This has been demonstrated using inert gas ion impact on a diverse range of compound substrates (Seah and Nunney 2010). Similar results have also been derived using Stopping Range of Ions in Matter (SRIM) simulations following slight modifications of the input parameters that are again empirically defined (Seah and Nunney 2010). However, owing to the great variability in the choice of analytical conditions that can be applied in SIMS, depth profile analysis, and the degree to which material properties can impact sputter rates, the use of sputter rate calculations tends not to be used during practical SIMS profiling. These tend to find greater use in fundamental studies.

And as for the other sputtering processes (kinetically assisted potential and pure potential), there exists significantly greater computational complexity.

**5.4.2.2 In situ Methods** An *in situ* method is one which provides the information of interest during the course of the analysis being carried out. *In situ* methods most commonly applied in SIMS for defining the depth scale during the course of depth profile analysis include:

1. Measurement of a substrate of a known thickness or containing some marker layer at some known depth

2. *In-situ*-based microbalance methods

3. *In-situ*-based interferometric methods

Measurement of substrates of known thicknesses or one that contains some specific and easily definable feature at some known depth is the most commonly used *in situ* method used by the SIMS community. Features that are typically used may be in the form of some interface that exists under some film of known thickness or some easily identifiable marker present at some known depth (see Figure 5.4(b), for example).

Use of interface signals for depth calibration requires that the thickness of the over-layering film is known. In many cases, these can be defined through some noninvasive and highly precise methodology such as Variable Angle Spectroscopic Ellipsometry (VASE), X-Ray Reflectivity (XRR), and Glancing Incidence X-Ray Diffraction (GID) (these are covered in Appendices A.10.1.5, A.12.2, and A.12.3). Highly accurate values can be derived through more invasive methodologies inclusive of cross-sectional SEM or Transmission Electron Microscopy (TEM) imaging (these are covered in Appendices A.11.1 and A.11.3).

The sputter rate is then derived for the matrix of interest by dividing the derived thickness (depth) by the time required to sputter down to the respective interface. The interface depth is usually taken as the position in which some secondary ion signal representative of layer either rises or drops by 50% in intensity. As matrix effects can modify secondary ion intensities over such regions, care must be taken in signal selection.

If a single substrate type is of interest, sputter rates can be derived using some marker present at some known depth. A marker in this case would represent some additional element or molecule present at some suitably low concentration such that:

1. It is easily detectable under the conditions used in the SIMS depth profile, i.e. displays a suitably high yield (see Section 3.3.2)

2. It does not influence the sputter rate or introduce any additional matrix effects (matrix effects are discussed in Section 3.3.2.2)

3. It does not suffer excessive sputter-induced diffusion or segregation characteristics (such effects are covered in Section 3.2.3.1)

A marker may be introduced through some carefully controlled atomic or molecular layer deposition process. In such cases, these form what are referred to as a *delta layer*, so named because of the theoretically defined concentration profile it should exhibit, i.e. abrupt onset and decay with a flat top.

Examples of SIMS analysis of elemental and molecular delta layers are shown in Figures 5.25 and 5.4(b), respectively. That shown in Figure 5.25 is accompanied by the cross-sectional TEM image showing the depths of the respective delta layers (TEM is covered in Appendix A.11.3). Such procedures are particularly useful in defining variable sputter rates such as those evident over the transient region (see Section 3.2.2.2.2).

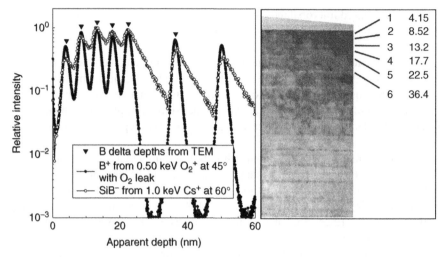

**Figure 5.25** To the left are overlaid profiles arising from $B^+$ and $SiB^-$ secondary ion emissions from a Boron delta layer Silicon structure under 0.5 keV $O_2^+$ or 1 keV $Cs^+$ impact, respectively. Both profiles, acquired on Quadrupole-based SIMS instruments in separate experiments, are normalized to the depth of the sixth delta and assuming a constant sputter rate. The intensity scale is normalized to unity to allow effective comparison of the two data sets. To the left is shown the TEM cross section of an adjacent section, with the Boron layers depth identified and listed as 1 through 6. The depths of the Boron layers defined in TEM are also signified by the inverted triangles. Note the greater discrepancies at shallower depths. Reproduced with permission from van der Heide et al. (2003) Copyright 2003 Elsevier.

Markers in the form of implanted elements of known Rp can also be used. This is realized as the Rp values can be readily derived through comparisons with:

1. Previously acquired profiles (see, for example, Figure 5.4(a))
2. Tabulated data sets (see, for example, Nishi and Doering 2007)
3. Empirical relations (see, for example, Relation 3.6a)

Sputter rates are then defined by dividing the Rp value by the sputtering time at which the implant peak is noted. Relations defining the implant peak position as a function of implant energy and angle of incidence are not only useful in providing Rp values, but they are also useful in defining the Rp value of the primary ion beam used in the SIMS measurements (see, for example, Relations 3.6(b–d)).

Another potentially useful approach is in the examination of isotopically pure layers deposited in some alternating manner. An example of this is in the multilayered Silicon material in which alternating $^{30}Si$ and $^{28}Si$ layers have been deposited (Shimazu et al. 2008). This is useful as all isotopes of the same element exhibit very similar sputter rate and ionization yields, thereby removing potential distortions resulting from matrix effects, and so on. Isotopically, pure materials are also useful in diffusion studies.

Microbalance methods, as the name suggests, defines the volume removed in a SIMS depth profile by weighing the sample before and after sputtering. The volume removed can then be converted to depth by accounting for the spatial crater size (defined from the primary ion raster pattern used). Once this is defined, the sputter rate (in units of Å/nA.s or nm/nA.s) can be defined by dividing the crater depth (in units Å or nm) by the sum of the sputter time (in units of s) and the primary ion current (in units of nanoampere).

This is most effectively carried out using a UHV compatible Quartz Crystal Microbalance (QCM). A QCM consists of a thin piezoelectric plate onto which the material of interest is deposited. Applying an AC voltage to this plate, via electrodes attached to both sides, generates an acoustic resonance of some frequency specific to the materials in contact with the Quartz crystal. Thus, any change in the amount of material will result in a change in the resonance frequency. The mass change per unit area resulting from sputtering can be defined to a sensitivity of better than $\sim 0.5$ ng/cm$^2$.

Although this method is not commonly used outside of the research environment, it does provide absolute overall sputter yield information (overall, because this includes the effects of ion implantation if apparent), a fact realized as the mass removed is measured. In addition, this can be useful in cases where neither stylus profilometry nor optical profilometry is applicable. Disadvantages associated with this method lie in the fact that this does not reveal the condition of the initial surface nor the crater base formed, both of which are important if high-depth resolution is required. Extreme care must be employed when carrying out such measurements.

Interferometric methods can be applied *in situ* to derive sputter rates using specifically designed and implemented instrumentation. This employs the use of an internally situated optical profilometer. Although this provides the advantage of understanding sputter rate variations in real time, including those noted during the initial stages of sputtering (see Section 3.2.2.2.2), and those noted on moving from one matrix to another, its continued use has been limited because of the complexities associated with its implementation. Indeed, the only commercial instrumentation in which this approach was tested was the Cameca WF and SC-ultra series (De Chambost et al. 2003).

### 5.4.3  Quantification of the Concentration Scale

Secondary ion emissions are recorded in intensity units. As a result, there can exist the need to relate secondary intensities to concentration.

The conversion of secondary ion intensities into concentration values is, however, complicated by the fact that all secondary ion emissions exhibit a strong sensitivity to the chemistry of the substrate's surface during sputtering. This is otherwise referred to as the matrix effect. Matrix effects (discussed in Section 3.3.2.2) can result in secondary ion yield variations that span orders of magnitude.

Quantification is further complicated by the fact that a vast array of analytical conditions can be used in SIMS. This is realized as these can influence the surface chemistry (primary ion implantation, damage, and segregation) and thus introduce additional matrix effects (see Figure 3.19, for example). Indeed, this dependence on surface chemistry represents the most problematic issue in SIMS. Conversion of the secondary ion intensity to concentration units requires the removal of the matrix effect.

As SIMS measures the number of secondary ions striking the detector, concentrations are most typically reported in units of atoms per some spatial or volumetric dimension, i.e. atoms/cm$^2$ (at/cm$^2$) in Static SIMS or atoms/cm$^3$ (at/cm$^3$ or at/cc) in Dynamic SIMS. These can be converted into percentage concentration values if the spatial or volume density of the substrate analyzed is known ($\sim 1 \times 10^{15}$ at/cm$^2$ and $\sim 5 \times 10^{22}$ at/cm$^3$ for Silicon). Likewise, weight percent values can be defined by accounting for the mass of the respective signals.

In the case of Dynamic SIMS analysis of elemental distributions in some constant matrix, quantification is almost exclusively carried out through the analysis of matrix-matched *reference materials* in which the concentration of the element of interest is known to a high degree of accuracy. Only through this approach, can the matrix effect, whether substrate or analysis induced, be fully accounted for. This approach, called the *Relative Sensitivity Factor* (RSF) approach, is covered in Section 5.4.3.1. The success of the RSF method is due in part to the fact that any element/molecule present in some specific matrix does not induce any additional matrix effects of their own if present at below 1 atomic %. If above 1 atomic %, concentration dependent calibration curves need to be derived.

Use of the Cesium cluster ion method (analysis of MCs$^+$ and/or MCs$_2^+$ secondary ions in which M represents the element of interest with MCs$^+$ used for the electropositive elements and MCs$_2^+$ used for electronegative elements) can prove useful in understanding the distribution of major or minor elements in unknown matrices. This stems from the significantly reduced matrix effects suffered by these populations. Downsides to this method includes the fact that a Cs$^+$ primary ion beam must be used, and both the MCs$^+$ and the MCs$_2^+$ populations provide poorer sensitivity and detection limits to the elements of interest than their elemental secondary ion counterparts.

As for molecular distributions noted in Dynamic SIMS, quantification is eased by the weaker variability of matrix effects noted, which is particularly evident for organic matrices. Indeed, the intensities of specific molecular emissions, whether fragment or otherwise, from organic substrates can be seen to follow the concentrations of the respective constituents. The useful emissions will be specific to the substrate being analyzed and exceptions will be noted when there exist variations in the concentration of chemically active species such as Oxygen, Fluorine, and Cesium.

In the case of Static SIMS, quantification can be eased by the reduced matrix variations noted when the element or molecule of interest exists in submonolayer coverage on some well-defined support, i.e. Gold, Silver, or Silicon. The reduced matrix effect variations stem from the fact that all of the secondary ions essentially see the same matrix during their departure, i.e. that of the support.

**5.4.3.1  The RSF Method**    The RSF method is a comparative method requiring some *reference material* containing the element or molecule of interest at some known concentration. The reference material, also referred to as an *internal standard*, must be of the same matrix as that of the sample of interest and must be analyzed under identical conditions to that used for the sample of interest. The term *relative* in RSF designates the fact that some matrix signal is also recorded during the same measurement such that a ratio between the matrix signal and the signal of interest can be generated.

Concentrations are related through the following equation:

$$c_i = \text{RSF} \left( \frac{I_i}{I_m} \right) \qquad (5.4)$$

where $c_i$ is the concentration of the element or molecule of interest (usually in atoms/cm$^2$ or atoms/cm$^3$ depending on whether surface or bulk concentration units are desired, which, in turn, defines the RSF units), $I_i$ the recorded intensity of the elemental or molecular signal of interest measured (in units of cps), and $I_m$ the recorded intensity of the matrix elemental or molecular signal measured (in units of cps). Note: The matrix signal should be from an element or molecule of constant concentration.

Rearrangement of Relation 5.4 allows for the derivation of a sample and analysis condition-specific RSF from a matrix-matched reference material. This is then applied to the intensities of the same secondary ion recorded from the sample of interest.

In Dynamic SIMS, the reference sample may be implanted with the element of interest. Reasons for this along with their fabrication are covered in Section 5.4.2.2. In such cases, the RSF can be defined based on the aerial dose of the implanted element via:

$$\text{RSF} = \frac{(\delta \cdot I_m \cdot C \cdot t)}{(d\Sigma I_i - d\Sigma I_{ib} \cdot C)} \cdot \frac{\text{(EM)}}{\text{(FC)}} \qquad (5.5)$$

where $\delta$ is the dose (in units of atoms/cm$^2$), $C$ the number of analytical cycles equivalent to the number of data points over the course of the depth profile, $t$ the analysis time (sputtering time), $d$ the crater depth (in units of cm), $\Sigma I_i$ the sum of the counts of the signal used over the applicable depth region, $I_{ib}$ the background signal (in units of cps), and EM/FC the ratio of the Electron Multiplier to Faraday Cup counting efficiency (this is only used when either of the signals traverses between the EM to FC counting ranges and when this ratio has not previously been accounted for). This approach integrates the area under the curve and equates this to the implanted dose.

The crater depth value is included with Relation 5.5 as aerial dose values are used, i.e. this allows for the conversion of atoms/cm$^2$ to atoms/cm$^3$. Of note is the fact that a constant sputtering rate is assumed during the course of the depth profile. As any variation in the sputter rate will result in the introduction of errors, implants should remain within a particular film, i.e. should not cross any interface and should not exist within the surface transient region (discussed within Section 3.3.2.2). If a multilayered substrate is to be examined, matrix-matched reference samples for each layer must be examined, with the associated RSF derived. In highly simplified cases, a merging of the RSF from the different layers over the interface region can be attempted.

As a specific isotope of a specific element is typically introduced in ion implantation, quantification via Relation 5.5 provides the concentration of the respective isotopic. If the substrate of interest contains all of the isotopes of the element, and the overall concentration is required, then the RSF must be corrected for the respective isotopic abundance present in the reference material.

To ensure accuracy, all samples and the reference material should be examined under identical conditions, preferably in the same analysis session by the same operator. Operator-specific variations arise from disparities in the perception of what constitutes a good data set, and where various limits should be applied, i.e. the depth range, intensity range, retrospective gating applied, and so on.

A matrix signal should be included during data collection to account for any slight variations in sputter rates in the analysis of the reference sample and/or the sample of interest. The presence of large sputter rate variations should be avoided as these variations can potentially result in the introduction of significant quantification errors. If this matrix signal is not included, then the term *Sensitivity Factor* (SF) should be applied. In such cases, Relations 5.4 and 5.5 are still used but with $I_m$ set to unity.

The signals of interest (impurity and matrix signals) within the reference material and the sample of interest should exist over the same depth range. This is stipulated as the requirement of near identical sputter rates for all samples would otherwise extend analysis times, sometimes significantly. Likewise, all signals should lie within the linear range of the respective secondary ion detection unit used (see Section 4.2.3.3). Lastly, all information to be used in the RSF approach should exist over the steady-state sputtering regime. The last requirement is stipulated for situations in which sputter rate variations noted during the onset of sputtering cannot be well matched.

The referencing procedure required can be applied in a different manner according to the type of sample examined. Those more commonly used are illustrated in Figure 5.26. Most typically, point-by-point referencing of the intensity of the signal of interest ($^{11}B^+$ in Figure 5.25) to the intensity of the matrix signal ($^{30}Si^+$) at every cycle is carried out once steady-state sputtering ensues (this is defined as method 1 in Figure 5.26). Other methods may require the use of an averaged matrix signal whether it be over the entire steady-state sputter time region (method 2 in Figure 5.26), or at some interval deeper within the substrate (Method 3 in Figure 5.26). Once the ideal method is defined, this should be used henceforth in all analysis of the respective sample type.

The onset of steady-state sputtering is defined as the sputtering time at which the matrix signal stabilizes. This limit should be placed as close to the sputtering onset as possible to ensure that the majority of the implanted signal is recorded. When using chemically active primary ions such as $O_2^+$ or $Cs^+$, this should approximate closely to $2–2.5$ times the primary ion $R_p$ value ($R_p$ values for these ions are approximated by Relations 3.6b and 3.6c). This then becomes the lower limit or boundary over which all data to be integrated should extend from (that to be used

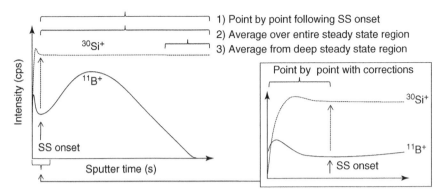

**Figure 5.26**  Pictorial illustration of procedures used in deriving RSFs for implant reference materials examined under Dynamic SIMS conditions. The profiles shown are for the $^{11}B^+$ and $^{30}Si^+$ secondary ions noted before and following steady-state sputtering conditions when using an $O_2^+$ primary ion beam (intensities are plotted on a Logarithmic scale). $^{30}Si^+$ is used as the reference as its intensities are closer to those of $^{11}B^+$. The different $^{30}Si^+$ regions that can be used to reference the $^{11}B^+$ signal against are signified as regions (1) through (3). In the inset is the transient behavior before steady-state sputtering (defined as SS).

in Relation 5.5) once the depth scale is applied. Application of this lower sputtering time (depth) boundary is required to filter out the $^{11}B^+$ signal noted from surface adsorbates. Note: Both Boron and Phosphorus also exist in the ambient environment, hence readily adsorb onto Silicon surfaces (this is not commonly observed for Arsenic). Such adsorbed populations should not be included in calculations in which the dose from an implant reference material is used in deriving the RSF (Relation 5.5). If not, this will result in an erroneous RSF. This filtering approach must also be applied in a highly consistent manner.

Data over the surface transient region noted during the onset of sputtering (inset shown in Figure 5.26) can only be quantified if the sputter and ionization yield variations for the respective secondary ions under the analytical conditions used in deriving the data are well known/well matched. The same applies if quantification over any interface region is to be carried out. In such cases, a point-by-point approach should be used. The *Point-by-point CORrection* (PCOR) method utilized by the *Evans Analytical Group* (EAG) is an attempt at this. Effective descriptions of this procedure are, however, limited. Further discussion on the PCOR method can be found in Appendix A.9.4.

The RSF method, like all quantification methods, must be employed with extreme care. Indeed, there are numerous potential sources of error that can arise in significant deviations/errors. Potential sources of error can include, but are not limited to:

1. Ineffective choice of the reference material. In short, the reference material to be used should match as closely as possible:

    a. The matrix of the sample of interest

    b. The depth scale of the sample of interest

    c. The surface topography of the sample of interest

2. Ineffective characterization of the concentration of the element/molecules present within the reference material/s before its use as a reference material (fabrication of reference materials is covered in Section 5.4.3.2).

3. Application of incorrect depth scales (as noted in Relation 5.5, this comes into defining the dose, present in whether the reference materials or the substrates of interest)

4. Ineffective use of isotope abundances. Note: Substrates from the nuclear industry and those used in cosmological studies will likely exhibit isotope abundances that deviate markedly from those listed in Appendix A.2

5. Incorporation of signals from surface adsorbates into the derivation of implanted ion dose measurements. Note: Airborne molecules readily adsorb onto both reference materials and substrates of interest. This also occurs within UHV chambers (see Section 4.2.1.1)

6. Use of ineffective analytical conditions. In short, the signals of interest should:

    a. Be free of isobaric interferences (see Sections 5.1.1.1 and 5.3.1.3)

    b. Remain within the linear range of the respective detector used (see Section 4.2.3.3)

    c. Be of sufficient intensity and density per unit depth to provide statistical significance (see Sections 4.2.3.3 and 5.3.2.4.4)

7. Primary ion beam stability. Note: Any variation in the primary ion current will result in not only the variations in the intensities of the signals of interest but also the sputter rate. Potential sources include:

    a. Primary ion source stability (this applies to the primary ion current and point of ion formation as covered in Section 4.2.2.1)

    b. Primary ion column stability (this applies to all lenses and filters)

8. Secondary ion beam stability. This represents any variation occurring even when good primary ion beam stability is noted. Potential sources include:

    a. Secondary ion column stability (this applies to all lenses and filters)

    b. Electrical charge buildup occurring on insulating substrates and those not effectively earthed to the sample holder (see Section 5.3.1.2)

    c. Inclusion of data suffering surface or interface matrix (transient) effects (see Section 3.3.1.2). Note: Unless the variations are well understood, only steady-state signals should be used.

9. Vacuum stability. Any fluctuations in the vacuum during or between the analysis of the reference materials and/or the substrate of interest can affect:

    a. The adsorption rate of gas phase molecules during analysis (see Section 4.2.1.1). Aside from their physical presence, such adsorbates can affect sputter rates and secondary ion yields

    b. The production of neutrals from the respective ion beams (see Section 5.3.2.3). If apparent in the primary ion beam, this can also affect sputter rates and depth resolution (see Section 5.3.2.3)

    c. The instrument transmission function (see Section 4.2.1.1)

10. Ion beam-induced modification of the reference materials and/or the substrate of interest. This can include sputter induced:

    a. Redistribution of elements/molecules of interest (see Section 3.2.3.1)

    b. Modification of molecular signals of interest (see Section 5.4.1.1)

    c. Surface topography variations during analysis (see Section 3.2.3.3)

Assuming all of the best practices are used, an ultimate precision to within 1% is possible. Accuracy, on the other hand, can extend to $\pm 10\%$. This poorer value stems primarily from the difficulty in quantifying the elements/molecules present in matrix-matched reference materials. This represents one part in the fabrication of effective reference materials. Reference material fabrication is covered in Section 5.4.3.2.

When different SIMS instruments as well as operators are used for the same measurement, both precision and accuracy will suffer. Indeed, it has been reported that reproducibility variations of up to $\pm 60\%$ can result (Wilson et al. 1989). This variation is, however, expected to be reduced with the:

1. Increased sophistication in the SIMS instrumentation (improved stability)
2. Improvements in best-known methods and/or practices
3. Enhanced cross-referencing methods for reference material calibration

The requirement for matrix-matched reference materials highlights the limitations of the RSF method. Without these, quantification via the RSF method is not possible. As a result, other SIMS quantification methodologies that do not require reference samples, or some form of referencing procedure, have been suggested. These are, however, not covered within this section as these have not found extensive use, a fact arising from the realization that these alternative procedures are not capable of providing the same level of precision, sensitivity, or detection limits as the RSF method. Some examples of alternative methods are covered in Appendices A.9.1–A.9.3.

### 5.4.3.2 Fabrication of Reference Materials

The effectiveness of the RSF method for SIMS quantification hinges critically on the quality of the matrix-matched reference material (internal standard) used.

Fabrication of reference materials, whether for elemental or for molecular studies, are typically carried out via:

1. Effective solid-state solution chemistry. This pertains to the fabrication of bulk reference materials as used in Dynamic SIMS as well as to the fabrication of surface reference materials as used in Static SIMS

2. Implantation of prespecified elements (implanted as ions) into stable solid substrates able to lock in the element of interest over long periods of time (thermally driven diffusion and/or segregation must be minimal)
3. Use of preexisting homogeneous materials that have been specifically prepared for SIMS and have been well characterized by one or more of the other applicable analytical techniques

Solution-based chemistry can involve the use/application of Langmuir–Blodgett films. These can be deposited on Silicon, Silver, or Gold surfaces with the latter two yielding favorable molecular sputter yields particularly when large cluster primary ion beams are used. Fragmentation patterns can prove useful in defining the presence and content of specific species. For such studies, the disappearance cross section (Relation 3.5) must be well characterized. In all cases, such reference materials are prepared in-house, with the most appropriate cross-referencing methodologies applied.

Implantation is well suited to the fabrication of the matrix-matched reference samples required for the quantification of elemental distributions during Dynamic SIMS analysis. This is realized because:

1. Any ion/substrate combination can be fabricated to suit the analytical need. Note: Many of the element–substrate combinations generated/examined may not exist in nature
2. Dose levels of implanted species can be well controlled to ensure that peak concentrations remain below 1 atomic %. As covered in Section 3.3.2, additional matrix effects can be introduced if the concentration lies above this limit particularly if chemically reactive elements are introduced
3. Energy levels can be tuned to provide implantation profiles over the appropriate depth region (Relation 3.6a reveals interrelations among $R_p$, energy, ion, and matrix mass)

In addition, implantation instrumentation available within the semiconductor industry is capable of providing ion beams of:

1. High purity (mass filtering of the specific isotope is ensured)
2. High spatial homogeneity (many are used in 300 mm wafer processing)
3. Well-controlled dose (continuously measured using Faraday Cups)

Indeed, the National Institute for Standards and Technology (NIST) presently provides three traceable implant reference materials, these being Boron, Phosphorus, and Arsenic implanted into Silicon. Closely aligned are the implant reference materials available from the *Evans Analytical Group* (EAG), and so on. Other implant reference material combinations require custom fabrication and validation.

Best results are obtained when implanting a single species into the substrate of interest and when the implanted species is well contained within the region of interest. Containment can be ensured through the correct matching of implant conditions, preamorphization of the substrate (only applies to crystalline substrates as this controls channeling effects), and control of any diffusion or segregation process that may occur under the analytical conditions to be used. Out-diffusion can be controlled by capping the respective reference material with a thin layer of the same matrix, i.e. Silicon on Silicon.

Preparation of reference materials for following elemental and/or molecular distributions closer to the outer surface, across some interface within a solid substrate, and/or at concentrations exceeding 1 atomic % is faced with greater challenges. These challenges are associated with not only their fabrication but also the validation of the elemental/molecular distributions. These must also exhibit a high degree of stability, i.e. out-diffusion etc. must be minimized.

As for their preparation, various semiconductor industry-based technologies or solution-based chemistry in which the species of interest is deposited on some applicable substrate can be applied. Those commonly applied in the semiconductor industry include, but are not limited to:

1. Atomic Layer Deposition (ALD)
2. Molecular Beam Epitaxy (MBE)
3. Chemical Vapor Deposition (CVD) of which there exist a myriad of methods

Which is to be used will depend on the substrate/film combination of interest, the thicknesses required, and the chemistry allowed. Films can be as thin as one atomic layer (ALD) to many atomic layers (CVD and MBE).

As for their validation, such reference materials can contain species introduced either at the low concentrations amenable to the detection limits afforded by SIMS or at higher concentrations where they themselves can induce significant matrix effects. In addition, the use of thin films places further limits on the array of analytical techniques to those displaying high surface specificity.

For elemental distributions, some of the analytical techniques that have found use include (listed alphabetically), but are not limited to:

1. The ion scattering methods, inclusive of
   a. Rutherford Back Scattering (RBS)
   b. Medium Energy Ion Scattering (MEIS)
   c. Low-Energy Ion Scattering (LEIS)
2. Nuclear Reaction Analysis (NRA)
3. Particle-Induced X-ray Emission (PIXE)
4. Elastic Recoil Detection (ERD) also referred to as Elastic Recoil Detection Analysis (ERDA), Elastic Recoil Scattering (ERS), Forward Recoil Scattering (FRS), Forward Recoil Spectrometry (FReS), and Hydrogen Forward Scattering (HFS)
5. Total Reflection X-ray Fluorescence (TXRF)

6. Variable Angle Spectroscopic Ellipsometry (VASE)
7. X-Ray Diffraction (XRD) in its various forms

As for molecular distributions, the array of analytical techniques that can/have been applied are further limited. Some contenders (listed alphabetically) can include:

1. Fourier Transform InfraRed spectroscopy (FTIR)
2. Raman Spectroscopy
3. UltraViolet-visible (UV-vis) spectroscopy
4. XPS

Note: Some of these techniques are used to follow specific constituents or signals reflective of concentrations of the elemental or molecular constituents of interest, as opposed to following the elemental or molecular constituents themselves. In addition, not all of the techniques listed are quantitative in their own right, i.e. these have been cross-correlated using other methodologies/analytical techniques.

As a result, the strengths and weaknesses of the respective analytical techniques used in the validation of the respective SIMS reference materials must be recognized as this will define the relative error/confidence levels of the resulting quantified profile will have when derived using the RSF method.

Lastly, in cases where the species of interest (elemental or molecular) represent a sizable portion of the matrix, calibration curves should be generated over a concentration range that extends over that being analyzed. Only from these, can the systematics of secondary ion yield variations from the specific matrix of interest be fully understood and accounted for. These tend to be sample and analysis condition specific, thereby requiring methodologies applied on a case-by-case (piecemeal) basis.

## 5.5  SUMMARY

The objective in SIMS is to record the secondary ion intensities over time (the secondary ion current) and/or the secondary ion intensities over space (the secondary ion current relative to the location of emission). This allows for insight into the relative concentration and/or concentration gradients of the respective isotopes, elements, or molecules of interest on or within the respective solid or made to be solid substrate.

There are, however, a large variety of modes under which a SIMS instrument can be operated, with each highly specific to a particular type of analysis. The analysis may require the collection of electropositive elements, electronegative elements, and/or molecules of one or the other polarity. Which analysis is required will dictate the type of primary ion beam used, the primary ion incident energy, angle of incidence and dose, the secondary ion column conditions applied, and the secondary ions collected.

The data can be collected in the form of mass spectra, depth profiles, or images in two or three dimensions depending on the information required. Each mode also has an effect on one or more of the following:

1. Sensitivity (the ability to measure small differences in concentration)
2. Detection limit (the ability to detect a signal when present at low concentrations)
3. Mass resolution (the ability to separate signals in close proximity to each other in mass)
4. Energy resolution (the ability to separate signals in close proximity of each other in energy)
5. Spatial resolution (the ability to separate signals in close proximity of each other in space)
6. Depth resolution (the ability to separate signals in close proximity of each other in depth)

Sample preparation also has an effect on the quality of the data collected. In the case of samples commonly seen in the Material Sciences, little in the way of sample preparation is required. All that is needed is that the sample be of a sufficient size to fit into the instrument and does not outgas excessively during analysis. As for the Earth Sciences, samples are most commonly encapsulated within some support (an epoxy resin puck) with the sample then exposed on polishing down to the region of interest. In the case of the Biosciences, sample preparation can be more intricate. This is realized as most all cellular structures contain water when in their natural state (this will evaporate on exposure to vacuum). Samples must thus be prepared through some form of chemical fixation that does not result in the redistribution or removal of any of the elemental or molecular information of interest. In all cases, the sample must exist, or made to exist in the solid phase, even under ultra high vacuum conditions employing in SIMS.

Sample handling must also be carried out to ensure cleanliness. This is not only to retain the integrity of the sample but also to retain the vacuum conditions required during SIMS analysis.

As for the instrument, effective calibration of the mass, energy, and intensity scales must be ensured. This is typically carried out before analysis through various alignment procedures. In the case of insulators, additional charge compensation procedures must be employed to ensure that uncontrolled electrical charge buildup does not occur during data collection. If this does, the intensities of all secondary ions will be affected. Charge compensation is typically carried out through coirradiation of the sputtered region with an electron beam.

The type of primary ion beam to be used will depend not only on the element or molecule of interest but also on the information required. Instrumental limitations may also play a part. For example, electronegative element depth profiles are best collected using $Cs^+$ primary ions. Electropositive elements, on the other hand, should be examined using $O_2^+$ or $O^-$ primary ions. Molecular depth profiles should be collected using a large cluster primary ion beam. Moreover, there are many more examples.

The three commercially available SIMS platforms are centered around the:

1. Quadrupole (Quad) secondary ion mass filter
2. Magnetic Sector (MS) secondary ion mass filter
3. Time-of-Flight secondary ion mass filter

With the exception of one variant of the last, all use a single continuous primary ion beam that can be operated over a large range of primary ion dose values. Time-of-Flight instruments utilizing pulsed primary ion beams require a low-dose analysis primary ion beam. To contend with this, a second primary ion beam is often employed. This is referred to as the sputter primary ion beam. The presence of two primary ion beams introduces further variables dependent on how these are operated, which will enhance either the mass resolution or the spatial image resolution, but at the expense of the other. Additional modes can, however, be implemented to minimize the loss.

Spatial imaging can be carried out via one of two approaches, these being the *microscope* mode or the *microprobe* mode. These are specific to the secondary ion mass filter and the secondary ion optics used. The former translates the aerial position of the secondary ion emanating from the substrate's surface to a position-sensitive detector, whereas the latter scans a finely focused beam across the substrate's surface with the beams position relayed to the detector. Three-dimensional imaging can be made possible by stacking spatial images collected as a function of sputtering time.

Depth profiling is the preferred option when depth distributions are of interest. For this, the sample must be spatially homogeneous over the area analyzed. Good depth resolution requires that the substrate be free of topography (intrinsic to the substrate or induced during analysis) and that crater edge effects be filtered out. As the name suggests, crater edge effects describe inclusion of signals from the crater edge, i.e. those from shallower regions. These are removed using various gating methods. Data point density is also of importance.

Secondary ion isobaric interferences must also be considered, as these can contribute to the signal/s of interest unless effectively removed. Isobaric interferences describe the presence of other signals at the same nominal mass as the signal of interest (an example is $^{12}C^{16}O$ and $^{28}Si$). Although three procedures exist (peak stripping, HMR, and KE filtering), the use of HMR is the most commonly applied irrespective of the instrument type. This is sometimes combined with KE filtering to further enhance the removal of isobaric interferences.

Additional data processing may be required dependent on the purpose of the analysis. Data processing may be required to:

1. Understand the origin of the signal
2. Derive isotope fractions

3. Construct images, whether in the spatial or volumetric sense
4. Derive concentration profiles

Understanding the origin of the signal/s measured in SIMS can be as simple as comparing masses of the signals measured to those listed in natural abundance tables such as that listed in Appendix A.2. This applies to atomic and simple molecular ions.

For more complex molecular spectra, particularly those collected under Static SIMS conditions, the situation becomes more complicated. In such cases, comparative analysis of spectra or even the use of the *Gentle SIMS* approach may be required. The original parent molecule responsible for the respective signals can also be defined when the *Gentle SIMS* approach is combined with the FPM along with the use of the *Simplified Molecular Input Line Entry Specification* algorithm.

Processing methodologies used in deriving isotope fractions require that the *Instruments Mass Fractionation* be corrected for.

Processing methodologies used in image generation tend to concern the ratioing of signals representative of the element or molecules of interest.

Quantification can take the form of converting the sputter time to depth, and/or converting the recorded secondary ion intensities to concentrations. Conversion of the sputter time to depth (only required in Dynamic SIMS) typically entails the post measurement of the sputtered crater via stylus profilometry or optical profilometry (interferometry). Other methodologies include the measurement of substrates bearing interfaces or markers at known depths, the use of a Quartz microbalance, or the use of sputter rate calculations. Note: The latter two are only used in highly specific cases. The depth is then converted to a sputter rate assuming a constant sputter rate throughout the layer measured. Point-by-point correction schemes can also be applied in cases where variable sputter rates are noted.

Conversion of the recorded secondary ion signal into concentration is presently best carried out through the coanalysis of matrix-matched reference materials. Matrix-matched reference materials are required as, if analyzed during the same session, these allow for the effective removal of all matrix effects when applying the RSF approach. This approach, in essence, directly relates the intensity ratios (the signal of interest and some matrix signal) from the sample of interest to those recorded from some reference sample in which the element or molecule of interest exists at some known concentration. Note: As the signal of interest can be compared to the matrix signal via various different methods, care must be taken to ensure consistency in the approach used. If a matrix signal is not recorded, the RSF simply reverts to an SF.

The RSF approach can, however, only be applied if matrix-matched reference materials exist. As this is rarely the case in nature, these have to be fabricated. Although a number of approaches exist, the most common relies on the implantation of the element of interest into some matrix. As for molecular signals, alternative solution chemistry approaches are required. In all cases, the reference material must be well characterized using other analytical techniques. The requirement of well-characterized matrix-matched reference materials represents one of the primary limitations of the RSF approach, and SIMS in general. Other approaches have been suggested but not have to date been able to provide the precision levels supplied by the RSF approach.

Many of the quantification difficulties noted in Dynamic SIMS are not noted in Static SIMS. This is particularly true of situations in which there exists sub-monolayer coverage of some element or molecule of interest on some well-defined support, i.e. Gold, Silver, or Silicon (a result of the reduced matrix variations often noted). In cases where the species of interest (elemental or molecular) represents a sizable portion of the matrix, calibration curves should also be generated over a concentration range that extends over that being analyzed. Only from these, can the systematics of secondary ion yield variations from the specific matrix of interest be fully accounted for. As a result, many of the methodologies used in SIMS tend to be applied in a case by case or in a case-specific manner.

It should be noted, however, that full quantification of the SIMS data is not always required. Indeed, when comparing good versus bad samples, or when an understanding of concentration variations is required, intensity variations alone can generally suffice. In such cases, it is, however, still advisable to reference the signal of interest to some stable internal signal (a matrix signal).

# ▰▰▰ APPENDIX A

## A.1  PERIODIC TABLE OF THE ELEMENTS

The periodic table of the elements is based on atomic number and reactivity. The elements are represented by their chemical symbol, which is listed in large font below their respective full name. *Atomic number* and *standard atomic weight* (definitions are provided in Sections 2.1.1 and 2.1.1.1) are listed above and below the respective chemical symbol. Atomic weights of elements given within square brackets do not occur naturally.

| Hydrogen 1 **H** 1.0079 | | | | | | | | | | | | | | | | | | Helium 2 **He** 4.0026 |
|---|---|---|---|---|---|---|---|---|---|---|---|---|---|---|---|---|---|---|
| Lithium 3 **Li** 6.941 | Beryllium 4 **Be** 9.0122 | | | | | | | | | | | Boron 5 **B** 10.811 | Carbon 6 **C** 12.011 | Nitrogen 7 **N** 14.007 | Oxygen 8 **O** 15.999 | Fluorine 9 **F** 18.998 | Neon 10 **Ne** 20.180 |
| Sodium 11 **Na** 22.990 | Magnesium 12 **Mg** 24.305 | | | | | | | | | | | Aluminium 13 **Al** 26.982 | Silicon 14 **Si** 28.086 | Phosphorus 15 **P** 30.974 | Sulfur 16 **S** 32.065 | Chlorine 17 **Cl** 35.453 | Argon 18 **Ar** 39.948 |
| Potassium 19 **K** 39.098 | Calcium 20 **Ca** 40.078 | Scandium 21 **Sc** 44.956 | Titanium 22 **Ti** 47.867 | Vanadium 23 **V** 50.942 | Chromium 24 **Cr** 51.996 | Manganese 25 **Mn** 54.938 | Iron 26 **Fe** 55.845 | Cobalt 27 **Co** 58.933 | Nickel 28 **Ni** 58.693 | Copper 29 **Cu** 63.546 | Zinc 30 **Zn** 65.39 | Gallium 31 **Ga** 69.723 | Germanium 32 **Ge** 72.61 | Arsenic 33 **As** 74.922 | Selenium 34 **Se** 78.96 | Bromine 35 **Br** 79.904 | Krypton 36 **Kr** 83.80 |
| Rubidium 37 **Rb** 85.468 | Strontium 38 **Sr** 87.62 | Yttrium 39 **Y** 88.906 | Zirconium 40 **Zr** 91.224 | Niobium 41 **Nb** 92.906 | Molybdenum 42 **Mo** 95.94 | Technetium 43 **Tc** [98] | Ruthenium 44 **Ru** 101.07 | Rhodium 45 **Rh** 102.91 | Palladium 46 **Pd** 106.42 | Silver 47 **Ag** 107.87 | Cadmium 48 **Cd** 112.41 | Indium 49 **In** 114.82 | Tin 50 **Sn** 118.71 | Antimony 51 **Sb** 121.76 | Tellurium 52 **Te** 127.60 | Iodine 53 **I** 126.90 | Xenon 54 **Xe** 131.29 |
| Caesium 55 **Cs** 132.91 | Barium 56 **Ba** 137.33 | 57-70 * | Lutetium 71 **Lu** 174.97 | Hafnium 72 **Hf** 178.49 | Tantalum 73 **Ta** 180.95 | Tungsten 74 **W** 183.84 | Rhenium 75 **Re** 186.21 | Osmium 76 **Os** 190.23 | Iridium 77 **Ir** 192.22 | Platinum 78 **Pt** 195.08 | Gold 79 **Au** 196.97 | Mercury 80 **Hg** 200.59 | Thallium 81 **Tl** 204.38 | Lead 82 **Pb** 207.2 | Bismuth 83 **Bi** 208.98 | Polonium 84 **Po** [209] | Astatine 85 **At** [210] | Radon 86 **Rn** [222] |
| Francium 87 **Fr** [223] | Radium 88 **Ra** [226] | 89-102 ** | Lawrencium 103 **Lr** [262] | Rutherfordium 104 **Rf** [261] | Dubnium 105 **Db** [262] | Seaborgium 106 **Sg** [266] | Bohrium 107 **Bh** [264] | Hassium 108 **Hs** [269] | Meitnerium 109 **Mt** [268] | Ununnilium 110 **Uun** [271] | Unununium 111 **Uuu** [272] | Ununbium 112 **Uub** [277] | Ununquadium 114 **Uuq** [289] | | | | | |

| * Lanthanide series | Lanthanum 57 **La** 138.91 | Cerium 58 **Ce** 140.12 | Praseodymium 59 **Pr** 140.91 | Neodymium 60 **Nd** 144.24 | Promethium 61 **Pm** [145] | Samarium 62 **Sm** 150.36 | Europium 63 **Eu** 151.96 | Gadolinium 64 **Gd** 157.25 | Terbium 65 **Tb** 158.93 | Dysprosium 66 **Dy** 162.50 | Holmium 67 **Ho** 164.93 | Erbium 68 **Er** 167.26 | Thulium 69 **Tm** 168.93 | Ytterbium 70 **Yb** 173.04 |
|---|---|---|---|---|---|---|---|---|---|---|---|---|---|---|
| ** Actinide series | Actinium 89 **Ac** [227] | Thorium 90 **Th** 232.04 | Protactinium 91 **Pa** 231.04 | Uranium 92 **U** 238.03 | Neptunium 93 **Np** [237] | Plutonium 94 **Pu** [244] | Americium 95 **Am** [243] | Curium 96 **Cm** [247] | Berkelium 97 **Bk** [247] | Californium 98 **Cf** [251] | Einsteinium 99 **Es** [252] | Fermium 100 **Fm** [257] | Mendelevium 101 **Md** [258] | Nobelium 102 **No** [259] |

## A.2  ISOTOPIC MASSES, NATURAL ISOTOPE ABUNDANCES, ATOMIC WEIGHTS, AND MASS DENSITIES OF THE ELEMENTS

List of naturally occurring isotopes terrestrially, relative isotopic mass (in Da), natural distributions (in fractions relative to unity), atomic weights (in g/mol) as

*Secondary Ion Mass Spectrometry: An Introduction to Principles and Practices*, First Edition.
Paul van der Heide.
© 2014 John Wiley & Sons, Inc. Published 2014 by John Wiley & Sons, Inc.

available from http://www.nist.gov/pml/data/comp.cfm. Atomic weights within square brackets are for elements not occurring naturally. For full definitions, see Sections 2.1.1 and 2.1.1.1. Also included are mass densities in g/cm$^3$ from the CRC Handbook of Chemistry and Physics (1985).

| Z | Isotope | | Relative Isotopic Mass | Natural Distribution | Atomic Weight | Mass Density |
|---|---|---|---|---|---|---|
| 1 | H | 1 | 1.00782503207(10) | 0.999885(70) | 1.00794(7)* | 0.089 |
| | D | 2 | 2.0141017778(4) | 0.000115(70) | | |
| | T | 3 | 3.0160492777(25) | | | |
| 2 | He | 3 | 3.0160293191(26) | 0.00000134(3) | 4.002602(2) | 0.179 |
| | | 4 | 4.00260325415(6) | 0.99999866(3) | | |
| 3 | Li | 6 | 6.015122795(16) | 0.0759(4) | 6.941(2)* | 0.53 |
| | | 7 | 7.01600455(8) | 0.9241(4) | | |
| 4 | Be | 9 | 9.0121822(4) | 1.0000 | 9.012182(3) | 1.85 |
| 5 | B | 10 | 10.0129370(4) | 0.199(7) | 10.811(7)* | 2.34 |
| | | 11 | 11.0093054(4) | 0.801(7) | | |
| 6 | C | 12 | 12.0000000(0) | 0.9893(8) | 12.0107(8)* | 2.26 |
| | | 13 | 13.0033548378(10) | 0.0107(8) | | |
| | | 14 | 14.003241989(4) | | | |
| 7 | N | 14 | 14.0030740048(6) | 0.99636(20) | 14.0067(2)* | 1.03 |
| | | 15 | 15.0001088982(7) | 0.00364(20) | | |
| 8 | O | 16 | 15.99491461956(16) | 0.99757(16) | 15.9994(3)* | 1.43 |
| | | 17 | 16.99913170(12) | 0.00038(1) | | |
| | | 18 | 17.9991610(7) | 0.00205(14) | | |
| 9 | F | 19 | 18.99840322(7) | 1.0000 | 18.9984032(5) | 1.97 |
| 10 | Ne | 20 | 19.9924401754(19) | 0.9048(3) | 20.1797(6) | 1.56 |
| | | 21 | 20.99384668(4) | 0.0027(1) | | |
| | | 22 | 21.991385114(19) | 0.0925(3) | | |
| 11 | Na | 23 | 22.9897692809(29) | 1.0000 | 22.98976928(2) | 0.97 |
| 12 | Mg | 24 | 23.985041700(14) | 0.7899(4) | 24.3050(6) | 1.74 |
| | | 25 | 24.98583692(3) | 0.1000(1) | | |
| | | 26 | 25.982592929(30) | 0.1101(3) | | |
| 13 | Al | 27 | 26.98153863(12) | 1.0000 | 26.9815386(8) | 2.70 |
| 14 | Si | 28 | 27.9769265325(19) | 0.92223(19) | 28.0855(3)* | 2.33 |
| | | 29 | 28.976494700(22) | 0.04685(8) | | |
| | | 30 | 29.97377017(3) | 0.03092(11) | | |
| 15 | P | 31 | 30.97376163(20) | 1.0000 | 30.973762(2) | 1.82 |
| 16 | S | 32 | 31.97207100(15) | 0.9499(26) | 32.065(5)* | 2.07 |
| | | 33 | 32.97145876(15) | 0.0075(2) | | |
| | | 34 | 33.96786690(12) | 0.0425(24) | | |
| | | 36 | 35.96708076(20) | 0.0001(1) | | |
| 17 | Cl | 35 | 34.96885268(4) | 0.7576(10) | 35.453(2)* | 2.09 |
| | | 37 | 36.96590259(5) | 0.2424(10) | | |
| 18 | Ar | 36 | 35.967545106(29) | 0.003365(30) | 39.948(1) | 1.78 |
| | | 38 | 37.9627324(4) | 0.000632(5) | | |
| | | 40 | 39.9623831225(29) | 0.996003(30) | | |
| 19 | K | 39 | 38.96370668(20) | 0.932581(44) | 39.0983(1) | 0.86 |
| | | 40 | 39.96399848(21) | 0.000117(1) | | |
| | | 41 | 40.96182576(21) | 0.067302(44) | | |

| Z | Isotope | | Relative Isotopic Mass | Natural Distribution | Atomic Weight | Mass Density |
|---|---|---|---|---|---|---|
| 20 | Ca | 40 | 39.96259098(22) | 0.96941(156) | 40.078(4) | 1.54 |
| | | 42 | 41.95861801(27) | 0.00647(23) | | |
| | | 43 | 42.9587666(3) | 0.00135(10) | | |
| | | 44 | 43.9554818(4) | 0.02086(110) | | |
| | | 46 | 45.9536926(24) | 0.00004(3) | | |
| | | 48 | 47.952534(4) | 0.00187(21) | | |
| 21 | Sc | 45 | 44.9559119(9) | 1.0000 | 44.955912(6) | 2.99 |
| 22 | Ti | 46 | 45.9526316(9) | 0.0825(3) | 47.867(1) | 4.51 |
| | | 47 | 46.9517631(9) | 0.0744(2) | | |
| | | 48 | 47.9479463(9) | 0.7372(3) | | |
| | | 49 | 48.9478700(9) | 0.0541(2) | | |
| | | 50 | 49.9447912(9) | 0.0518(2) | | |
| 23 | V | 50 | 49.9471585(11) | 0.00250(4) | 50.9415(1) | 6.1 |
| | | 51 | 50.9439595(11) | 0.99750(4) | | |
| 24 | Cr | 50 | 49.9460442(11) | 0.04345(13) | 51.9961(6) | 7.19 |
| | | 52 | 51.9405075(8) | 0.83789(18) | | |
| | | 53 | 52.9406494(8) | 0.09501(17) | | |
| | | 54 | 53.9388804(8) | 0.02365(7) | | |
| 25 | Mn | 55 | 54.9380451(7) | 1.0000 | 54.938045(5) | 7.43 |
| 26 | Fe | 54 | 53.9396105(7) | 0.05845(35) | 55.845(2) | 7.86 |
| | | 56 | 55.9349375(7) | 0.91754(36) | | |
| | | 57 | 56.9353940(7) | 0.02119(10) | | |
| | | 58 | 57.9332756(8) | 0.00282(4) | | |
| 27 | Co | 59 | 58.9331950(7) | 1.0000 | 58.933195(5) | 8.9 |
| 28 | Ni | 58 | 57.9353429(7) | 0.680769(89) | 58.6934(4) | 8.9 |
| | | 60 | 59.9307864(7) | 0.262231(77) | | |
| | | 61 | 60.9310560(7) | 0.011399(6) | | |
| | | 62 | 61.9283451(6) | 0.036345(17) | | |
| | | 64 | 63.9279660(7) | 0.009256(9) | | |
| 29 | Cu | 63 | 62.9295975(6) | 0.6915(15) | 63.546(3) | 8.96 |
| | | 65 | 64.9277895(7) | 0.3085(15) | | |
| 30 | Zn | 64 | 63.9291422(7) | 0.48268(321) | 65.38(2) | 7.14 |
| | | 66 | 65.9260334(10) | 0.27975(77) | | |
| | | 67 | 66.9271273(10) | 0.04102(21) | | |
| | | 68 | 67.9248442(10) | 0.19024(123) | | |
| | | 70 | 69.9253193(21) | 0.00631(9) | | |
| 31 | Ga | 69 | 68.9255736(13) | 0.60108(9) | 69.723(1) | 5.91 |
| | | 71 | 70.9247013(11) | 0.39892(9) | | |
| 32 | Ge | 70 | 69.9242474(11) | 0.2038(18) | 72.64(1) | 5.32 |
| | | 72 | 71.9220758(18) | 0.2731(26) | | |
| | | 73 | 72.9234589(18) | 0.0776(8) | | |
| | | 74 | 73.9211778(18) | 0.3672(15) | | |
| | | 76 | 75.9214026(18) | 0.0783(7) | | |
| 33 | As | 75 | 74.9215965(20) | 1.0000 | 74.92160(2) | 5.72 |
| 34 | Se | 74 | 73.9224764(18) | 0.0089(4) | 78.96(3) | 4.79 |
| | | 76 | 75.9192136(18) | 0.0937(29) | | |
| | | 77 | 76.9199140(18) | 0.0763(16) | | |
| | | 78 | 77.9173091(18) | 0.2377(28) | | |
| | | 80 | 79.9165213(21) | 0.4961(41) | | |
| | | 82 | 81.9166994(22) | 0.0873(22) | | |
| 35 | Br | 79 | 78.9183371(22) | 0.5069(7) | 79.904(1) | 4.10 |
| | | 81 | 80.9162906(21) | 0.4931(7) | | |

| Z | Isotope | | Relative Isotopic Mass | Natural Distribution | Atomic Weight | Mass Density |
|---|---|---|---|---|---|---|
| 36 | Kr | 78 | 77.9203648(12) | 0.00355(3) | 83.798(2) | 3.07 |
| | | 80 | 79.9163790(16) | 0.02286(10) | | |
| | | 82 | 81.9134836(19) | 0.11593(31) | | |
| | | 83 | 82.914136(3) | 0.11500(19) | | |
| | | 84 | 83.911507(3) | 0.56987(15) | | |
| | | 86 | 85.91061073(11) | 0.17279(41) | | |
| 37 | Rb | 85 | 84.911789738(12) | 0.7217(2) | 85.4678(3) | 1.53 |
| | | 87 | 86.909180527(13) | 0.2783(2) | | |
| 38 | Sr | 84 | 83.913425(3) | 0.0056(1) | 87.62(1) | 2.60 |
| | | 86 | 85.9092602(12) | 0.0986(1) | | |
| | | 87 | 86.9088771(12) | 0.0700(1) | | |
| | | 88 | 87.9056121(12) | 0.8258(1) | | |
| 39 | Y | 89 | 88.9058483(27) | 1.0000 | 88.90585(2) | 4.46 |
| 40 | Zr | 90 | 89.9047044(25) | 0.5145(40) | 91.224(2) | 6.46 |
| | | 91 | 90.9056458(25) | 0.1122(5) | | |
| | | 92 | 91.9050408(25) | 0.1715(8) | | |
| | | 94 | 93.9063152(26) | 0.1738(28) | | |
| | | 96 | 95.9082734(30) | 0.0280(9) | | |
| 41 | Nb | 93 | 92.9063781(26) | 1.0000 | 92.90638(2) | 8.4 |
| 42 | Mo | 92 | 91.906811(4) | 0.1477(31) | 95.96(2) | 10.2 |
| | | 94 | 93.9050883(21) | 0.0923(10) | | |
| | | 95 | 94.9058421(21) | 0.1590(9) | | |
| | | 96 | 95.9046795(21) | 0.1668(1) | | |
| | | 97 | 96.9060215(21) | 0.0956(5) | | |
| | | 98 | 97.9054082(21) | 0.2419(26) | | |
| | | 100 | 99.907477(6) | 0.0967(20) | | |
| 43 | Tc | 97 | 96.906365(5) | | [98] | 11.5 |
| | | 98 | 97.907216(4) | | | |
| | | 99 | 98.9062547(21) | | | |
| 44 | Ru | 96 | 95.907598(8) | 0.0554(14) | 101.07(2) | 12.2 |
| | | 98 | 97.905287(7) | 0.0187(3) | | |
| | | 99 | 98.9059393(22) | 0.1276(14) | | |
| | | 100 | 99.9042195(22) | 0.1260(7) | | |
| | | 101 | 100.9055821(22) | 0.1706(2) | | |
| | | 102 | 101.9043493(22) | 0.3155(14) | | |
| | | 104 | 103.905433(3) | 0.1862(27) | | |
| 45 | Rh | 103 | 102.905504(3) | 1.0000 | 102.90550(2) | 12.4 |
| 46 | Pd | 102 | 101.905609(3) | 0.0102(1) | 106.42(1) | 12.0 |
| | | 104 | 103.904036(4) | 0.1114(8) | | |
| | | 105 | 104.905085(4) | 0.2233(8) | | |
| | | 106 | 105.903486(4) | 0.2733(3) | | |
| | | 108 | 107.903892(4) | 0.2646(9) | | |
| | | 110 | 109.905153(12) | 0.1172(9) | | |
| 47 | Ag | 107 | 106.905097(5) | 0.51839(8) | 107.8682(2) | 10.5 |
| | | 109 | 108.904752(3) | 0.48161(8) | | |
| 48 | Cd | 106 | 105.906459(6) | 0.0125(6) | 112.411(8) | 8.65 |
| | | 108 | 107.904184(6) | 0.0089(3) | | |
| | | 110 | 109.9030021(29) | 0.1249(18) | | |
| | | 111 | 110.9041781(29) | 0.1280(12) | | |
| | | 112 | 111.9027578(29) | 0.2413(21) | | |
| | | 113 | 112.9044017(29) | 0.1222(12) | | |
| | | 114 | 113.9033585(29) | 0.2873(42) | | |
| | | 116 | 115.904756(3) | 0.0749(18) | | |

| Z | Isotope | | Relative Isotopic Mass | Natural Distribution | Atomic Weight | Mass Density |
|---|---|---|---|---|---|---|
| 49 | In | 113 | 112.904058(3) | 0.0429(5) | 114.818(3) | 7.31 |
| | | 115 | 114.903878(5) | 0.9571(5) | | |
| 50 | Sn | 112 | 111.904818(5) | 0.0097(1) | 118.710(7) | 7.30 |
| | | 114 | 113.902779(3) | 0.0066(1) | | |
| | | 115 | 114.903342(3) | 0.0034(1) | | |
| | | 116 | 115.901741(3) | 0.1454(9) | | |
| | | 117 | 116.902952(3) | 0.0768(7) | | |
| | | 118 | 117.901603(3) | 0.2422(9) | | |
| | | 119 | 118.903308(3) | 0.0859(4) | | |
| | | 120 | 119.9021947(27) | 0.3258(9) | | |
| | | 122 | 121.9034390(29) | 0.0463(3) | | |
| | | 124 | 123.9052739(15) | 0.0579(5) | | |
| 51 | Sb | 121 | 120.9038157(24) | 0.5721(5) | 121.760(1) | 6.62 |
| | | 123 | 122.9042140(22) | 0.4279(5) | | |
| 52 | Te | 120 | 119.904020(10) | 0.0009(1) | 127.60(3) | 6.24 |
| | | 122 | 121.9030439(16) | 0.0255(12) | | |
| | | 123 | 122.9042700(16) | 0.0089(3) | | |
| | | 124 | 123.9028179(16) | 0.0474(14) | | |
| | | 125 | 124.9044307(16) | 0.0707(15) | | |
| | | 126 | 125.9033117(16) | 0.1884(25) | | |
| | | 128 | 127.9044631(19) | 0.3174(8) | | |
| | | 130 | 129.9062244(21) | 0.3408(62) | | |
| 53 | I | 127 | 126.904473(4) | 1.0000 | 126.90447(3) | 4.92 |
| 54 | Xe | 124 | 123.9058930(20) | 0.000952(3) | 131.293(6) | 3.77 |
| | | 126 | 125.904274(7) | 0.000890(2) | | |
| | | 128 | 127.9035313(15) | 0.019102(8) | | |
| | | 129 | 128.9047794(8) | 0.264006(82) | | |
| | | 130 | 129.9035080(8) | 0.040710(13) | | |
| | | 131 | 130.9050824(10) | 0.212324(30) | | |
| | | 132 | 131.9041535(10) | 0.269086(33) | | |
| | | 134 | 133.9053945(9) | 0.104357(21) | | |
| | | 136 | 135.907219(8) | 0.088573(44) | | |
| 55 | Cs | 133 | 132.905451933(24) | 1.0000 | 132.9054519(2) | 1.90 |
| 56 | Ba | 130 | 129.9063208(30) | 0.00106(1) | 137.327(7) | 3.5 |
| | | 132 | 131.9050613(11) | 0.00101(1) | | |
| | | 134 | 133.9045084(4) | 0.02417(18) | | |
| | | 135 | 134.9056886(4) | 0.06592(12) | | |
| | | 136 | 135.9045759(4) | 0.07854(24) | | |
| | | 137 | 136.9058274(5) | 0.11232(24) | | |
| | | 138 | 137.9052472(5) | 0.71698(42) | | |
| 57 | La | 138 | 137.907112(4) | 0.00090(1) | 138.90547(7) | 6.17 |
| | | 139 | 138.9063533(26) | 0.99910(1) | | |
| 58 | Ce | 136 | 135.907172(14) | 0.00185(2) | 140.116(1) | 6.77 |
| | | 138 | 137.905991(11) | 0.00251(2) | | |
| | | 140 | 139.9054387(26) | 0.88450(51) | | |
| | | 142 | 141.909244(3) | 0.11114(51) | | |
| 59 | Pr | 141 | 140.9076528(26) | 1.0000 | 140.90765(2) | 6.77 |
| 60 | Nd | 142 | 141.9077233(25) | 0.272(5) | 144.242(3) | 7.00 |
| | | 143 | 142.9098143(25) | 0.122(2) | | |
| | | 144 | 143.9100873(25) | 0.238(3) | | |
| | | 145 | 144.9125736(25) | 0.083(1) | | |
| | | 146 | 145.9131169(25) | 0.172(3) | | |
| | | 148 | 147.916893(3) | 0.057(1) | | |
| | | 150 | 149.920891(3) | 0.056(2) | | |

| Z | Isotope | | Relative Isotopic Mass | Natural Distribution | Atomic Weight | Mass Density |
|---|---|---|---|---|---|---|
| 61 | Pm | 145 | 144.912749(3) | | [145] | -- |
| | | 147 | 146.9151385(26) | | | |
| 62 | Sm | 144 | 143.911999(3) | 0.0307(7) | 150.36(2) | 7.54 |
| | | 147 | 146.9148979(26) | 0.1499(18) | | |
| | | 148 | 147.9148227(26) | 0.1124(10) | | |
| | | 149 | 148.9171847(26) | 0.1382(7) | | |
| | | 150 | 149.9172755(26) | 0.0738(1) | | |
| | | 152 | 151.9197324(27) | 0.2675(16) | | |
| | | 154 | 153.9222093(27) | 0.2275(29) | | |
| 63 | Eu | 151 | 150.9198502(26) | 0.4781(6) | 151.964(1) | 4.61 |
| | | 153 | 152.9212303(26) | 0.5219(6) | | |
| 64 | Gd | 152 | 151.9197910(27) | 0.0020(1) | 157.25(3) | 8.23 |
| | | 154 | 153.9208656(27) | 0.0218(3) | | |
| | | 155 | 154.9226220(27) | 0.1480(12) | | |
| | | 156 | 155.9221227(27) | 0.2047(9) | | |
| | | 157 | 156.9239601(27) | 0.1565(2) | | |
| | | 158 | 157.9241039(27) | 0.2484(7) | | |
| | | 160 | 159.9270541(27) | 0.2186(19) | | |
| 65 | Tb | 159 | 158.9253468(27) | 1.0000 | 158.92535(2) | 8.54 |
| 66 | Dy | 156 | 155.924283(7) | 0.00056(3) | 162.500(1) | 8.78 |
| | | 158 | 157.924409(4) | 0.00095(3) | | |
| | | 160 | 159.9251975(27) | 0.02329(18) | | |
| | | 161 | 160.9269334(27) | 0.18889(42) | | |
| | | 162 | 161.9267984(27) | 0.25475(36) | | |
| | | 163 | 162.9287312(27) | 0.24896(42) | | |
| | | 164 | 163.9291748(27) | 0.28260(54) | | |
| 67 | Ho | 165 | 164.9303221(27) | 1.0000 | 164.93032(2) | 9.05 |
| 68 | Er | 162 | 161.928778(4) | 0.00139(5) | 167.259(3) | 9.37 |
| | | 164 | 163.929200(3) | 0.01601(3) | | |
| | | 166 | 165.9302931(27) | 0.33503(36) | | |
| | | 167 | 166.9320482(27) | 0.22869(9) | | |
| | | 168 | 167.9323702(27) | 0.26978(18) | | |
| | | 170 | 169.9354643(30) | 0.14910(36) | | |
| 69 | Tm | 169 | 168.9342133(27) | 1.0000 | 168.93421(2) | 9.31 |
| 70 | Yb | 168 | 167.933897(5) | 0.0013(1) | 173.054(5) | 6.97 |
| | | 170 | 169.9347618(26) | 0.0304(15) | | |
| | | 171 | 170.9363258(26) | 0.1428(57) | | |
| | | 172 | 171.9363815(26) | 0.2183(67) | | |
| | | 173 | 172.9382108(26) | 0.1613(27) | | |
| | | 174 | 173.9388621(26) | 0.3183(92) | | |
| | | 176 | 175.9425717(28) | 0.1276(41) | | |
| 71 | Lu | 175 | 174.9407718(23) | 0.9741(2) | 174.9668(1) | 9.84 |
| | | 176 | 175.9426863(23) | 0.0259(2) | | |
| 72 | Hf | 174 | 173.940046(3) | 0.0016(1) | 178.49(2) | 13.1 |
| | | 176 | 175.9414086(24) | 0.0526(7) | | |
| | | 177 | 176.9432207(23) | 0.1860(9) | | |
| | | 178 | 177.9436988(23) | 0.2728(7) | | |
| | | 179 | 178.9458161(23) | 0.1362(2) | | |
| | | 180 | 179.9465500(23) | 0.3508(16) | | |
| 73 | Ta | 180 | 179.9474648(24) | 0.00012(2) | 180.94788(2) | 16.6 |
| | | 181 | 180.9479958(19) | 0.99988(2) | | |

| Z | Isotope | | Relative Isotopic Mass | Natural Distribution | Atomic Weight | Mass Density |
|---|---------|---|------------------------|----------------------|---------------|--------------|
| 74 | W | 180 | 179.946704(4) | 0.0012(1) | 183.84(1) | 19.3 |
| | | 182 | 181.9482042(9) | 0.2650(16) | | |
| | | 183 | 182.9502230(9) | 0.1431(4) | | |
| | | 184 | 183.9509312(9) | 0.3064(2) | | |
| | | 186 | 185.9543641(19) | 0.2843(19) | | |
| 75 | Re | 185 | 184.9529550(13) | 0.3740(2) | 186.207(1) | 21.0 |
| | | 187 | 186.9557531(15) | 0.6260(2) | | |
| 76 | Os | 184 | 183.9524891(14) | 0.0002(1) | 190.23(3) | 22.6 |
| | | 186 | 185.9538382(15) | 0.0159(3) | | |
| | | 187 | 186.9557505(15) | 0.0196(2) | | |
| | | 188 | 187.9558382(15) | 0.1324(8) | | |
| | | 189 | 188.9581475(16) | 0.1615(5) | | |
| | | 190 | 189.9584470(16) | 0.2626(2) | | |
| | | 192 | 191.9614807(27) | 0.4078(19) | | |
| 77 | Ir | 191 | 190.9605940(18) | 0.373(2) | 192.217(3) | 22.5 |
| | | 193 | 192.9629264(18) | 0.627(2) | | |
| 78 | Pt | 190 | 189.959932(6) | 0.00014(1) | 195.084(9) | 21.4 |
| | | 192 | 191.9610380(27) | 0.00782(7) | | |
| | | 194 | 193.9626803(9) | 0.32967(99) | | |
| | | 195 | 194.9647911(9) | 0.33832(10) | | |
| | | 196 | 195.9649515(9) | 0.25242(41) | | |
| | | 198 | 197.967893(3) | 0.07163(55) | | |
| 79 | Au | 197 | 196.9665687(6) | 1.0000 | 196.966569(4) | 19.3 |
| 80 | Hg | 196 | 195.965833(3) | 0.0015(1) | 200.59(2) | 13.6 |
| | | 198 | 197.9667690(4) | 0.0997(20) | | |
| | | 199 | 198.9682799(4) | 0.1687(22) | | |
| | | 200 | 199.9683260(4) | 0.2310(19) | | |
| | | 201 | 200.9703023(6) | 0.1318(9) | | |
| | | 202 | 201.9706430(6) | 0.2986(26) | | |
| | | 204 | 203.9734939(4) | 0.0687(15) | | |
| 81 | Tl | 203 | 202.9723442(14) | 0.2952(1) | 204.3833(2)* | 11.85 |
| | | 205 | 204.9744275(14) | 0.7048(1) | | |
| 82 | Pb | 204 | 203.9730436(13) | 0.014(1) | 207.2(1) | 11.4 |
| | | 206 | 205.9744653(13) | 0.241(1) | | |
| | | 207 | 206.9758969(13) | 0.221(1) | | |
| | | 208 | 207.9766521(13) | 0.524(1) | | |
| 83 | Bi | 209 | 208.9803987(16) | 1.0000 | 208.98040(1) | 9.8 |
| 84 | Po | 209 | 208.9824304(20) | | [209] | 9.4 |
| | | 210 | 209.9828737(13) | | | |
| 85 | At | 210 | 209.987148(8) | | [210] | -- |
| | | 211 | 210.9874963(30) | | | |
| 86 | Rn | 211 | 210.990601(7) | | [222] | ~4.4 |
| | | 220 | 220.0113940(24) | | | |
| | | 222 | 222.0175777(25) | | | |
| 87 | Fr | 223 | 223.0197359(26) | | [223] | -- |
| 88 | Ra | 223 | 223.0185022(27) | | [226] | ~5.0 |
| | | 224 | 224.0202118(24) | | | |
| | | 226 | 226.0254098(25) | | | |
| | | 228 | 228.0310703(26) | | | |
| 89 | Ac | 227 | 227.0277521(26) | | [227] | 10.1 |
| 90 | Th | 230 | 230.0331338(19) | | 232.03806(2) | 11.7 |
| | | 232 | 232.0380553(21) | 1.0000 | | |

| Z | Isotope | | Relative Isotopic Mass | Natural Distribution | Atomic Weight | Mass Density |
|---|---|---|---|---|---|---|
| 91 | Pa | 231 | 231.0358840(24) | 1.0000 | 231.03588(2) | 15.4 |
| 92 | U | 233 | 233.0396352(29) | | 238.02891(3) | 19.07 |
| | | 234 | 234.0409521(20) | 0.000054(5) | | |
| | | 235 | 235.0439299(20) | 0.007204(6) | | |
| | | 236 | 236.0455680(20) | | | |
| | | 238 | 238.0507882(20) | 0.992742(10) | | |
| 93 | Np | 236 | 236.046570(50) | | [237] | 20.3 |
| | | 237 | 237.0481734(20) | | | |
| 94 | Pu | 238 | 238.0495599(20) | | [244] | 19.8 |
| | | 239 | 239.0521634(20) | | | |
| | | 240 | 240.0538135(20) | | | |
| | | 241 | 241.0568515(20) | | | |
| | | 242 | 242.0587426(20) | | | |
| | | 244 | 244.064204(5) | | | |
| 95 | Am | 241 | 241.0568291(20) | | [243] | 11.8 |
| | | 243 | 243.0613811(25) | | | |
| 96 | Cm | 243 | 243.0613891(22) | | [247] | -- |
| | | 244 | 244.0627526(20) | | | |
| | | 245 | 245.0654912(22) | | | |
| | | 246 | 246.0672237(22) | | | |
| | | 247 | 247.070354(5) | | | |
| | | 248 | 248.072349(5) | | | |
| 97 | Bk | 247 | 247.070307(6) | | [247] | -- |
| | | 249 | 249.0749867(28) | | | |
| 98 | Cf | 249 | 249.0748535(24) | | [251] | -- |
| | | 250 | 250.0764061(22) | | | |
| | | 251 | 251.079587(5) | | | |
| | | 252 | 252.081626(5) | | | |
| 99 | Es | 252 | 252.082980(50) | | [252] | -- |
| 100 | Fm | 257 | 257.095105(7) | | [257] | -- |
| 101 | Md | 258 | 258.098431(5) | | [258] | -- |
| | | 260 | 260.10365(34) | | | |
| 102 | No | 259 | 259.10103(11) | | [259] | -- |

Note: Due to terrestrial variations, listed isotope distributions represent average terrestrial values (Wieser 2011)

## A.3   1ST AND 2ND IONIZATION POTENTIALS AND ELECTRON AFFINITIES OF THE ELEMENTS

1st and 2nd ionization potentials along with electron affinities for the naturally occurring elements in units of eV. For definitions, see Section 2.2.1.3. Data from the CRC Handbook of Chemistry and Physics (1985).

| Z | Element | 1st Ionization Potential | 2nd Ionization Potential | Electron Affinity |
|---|---------|--------------------------|--------------------------|-------------------|
| 1 | H | 13.598 | -- | 0.754 |
| 2 | He | 24.587 | 54.416 | 0 |
| 3 | Li | 5.392 | 75.638 | 0.62 |
| 4 | Be | 9.322 | 18.211 | 0 |
| 5 | B | 8.298 | 25.154 | 0.28 |
| 6 | C | 11.260 | 24.383 | 1.27 |
| 7 | N | 14.534 | 29.601 | 0 |
| 8 | O | 13.618 | 35.116 | 1.46 |
| 9 | F | 17.422 | 34.970 | 1.4 |
| 10 | Ne | 21.564 | 40.962 | 0 |
| 11 | Na | 5.139 | 47.286 | 0.546 |
| 12 | Mg | 7.646 | 15.035 | 0 |
| 13 | Al | 5.986 | 18.828 | 0.46 |
| 14 | Si | 8.151 | 16.345 | 1.39 |
| 15 | P | 10.486 | 19.725 | 0.743 |
| 16 | S | 10.360 | 23.33 | 2.08 |
| 17 | Cl | 12.967 | 23.81 | 3.62 |
| 18 | Ar | 15.759 | 27.629 | 0 |
| 19 | K | 4.341 | 31.625 | 0.501 |
| 20 | Ca | 6.113 | 11.871 | 0 |
| 21 | Sc | 6.54 | 12.80 | 0 |
| 22 | Ti | 6.82 | 13.58 | 0.2 |
| 23 | V | 6.74 | 14.65 | 0.5 |
| 24 | Cr | 6.766 | 16.50 | 0.66 |
| 25 | Mn | 7.435 | 16.640 | 0 |
| 26 | Fe | 7.870 | 16.18 | 0.25 |
| 27 | Co | 7.86 | 17.06 | 0.7 |
| 28 | Ni | 7.635 | 18.168 | 1.15 |
| 29 | Cu | 7.726 | 20.292 | 1.23 |
| 30 | Zn | 9.394 | 17.964 | 0 |
| 31 | Ga | 5.999 | 20.51 | 0.3 |
| 32 | Ge | 7.899 | 15.934 | 1.2 |
| 33 | As | 9.81 | 18.633 | 0.8 |
| 34 | Se | 9.752 | 21.19 | 2.02 |
| 35 | Br | 11.814 | 21.8 | 3.36 |
| 36 | Kr | 13.999 | 24.359 | 0 |
| 37 | Rb | 4.177 | 27.28 | 0.486 |
| 38 | Sr | 5.695 | 11.030 | 0 |
| 39 | Y | 6.38 | 12.24 | 0 |
| 40 | Zr | 6.84 | 13.13 | 0.5 |

| Z | Element | 1st Ionization Potential | 2nd Ionization Potential | Electron Affinity |
|----|---------|---------|---------|---------|
| 41 | Nb | 6.88 | 14.32 | 1 |
| 42 | Mo | 7.099 | 16.15 | 1 |
| 44 | Ru | 7.37 | 16.76 | 1.1 |
| 45 | Rh | 7.46 | 18.08 | 1.2 |
| 46 | Pd | 8.34 | 19.43 | 0.6 |
| 47 | Ag | 7.576 | 21.49 | 1.3 |
| 48 | Cd | 8.993 | 16.908 | 0 |
| 49 | In | 5.786 | 18.869 | 0.3 |
| 50 | Sn | 7.344 | 14.632 | 1.25 |
| 51 | Sb | 8.641 | 16.53 | 1.05 |
| 52 | Te | 9.009 | 18.6 | 1.05 |
| 53 | I | 10.451 | 19.131 | 3.06 |
| 54 | Xe | 12.130 | 21.21 | 0 |
| 55 | Cs | 3.894 | 25.1 | 0.472 |
| 56 | Ba | 5.212 | 10.004 | 0 |
| 57 | La | 5.577 | 11.06 | 0.5 |
| 58 | Ce | 5.47 | 10.85 | 0.5 |
| 59 | Pr | 5.42 | 10.55 | 0.5 |
| 60 | Nd | 5.49 | 10.72 | 0.5 |
| 62 | Sm | 5.63 | 11.07 | 0.5 |
| 63 | Eu | 5.67 | 11.5 | 0.5 |
| 64 | Gd | 6.14 | 12.1 | 0.5 |
| 65 | Tb | 5.85 | 11.52 | 0.5 |
| 66 | Dy | 5.93 | 11.67 | 0.5 |
| 67 | Ho | 6.02 | 11.80 | 0.5 |
| 68 | Er | 6.10 | 11.93 | 0.5 |
| 69 | Tm | 6.18 | 12.05 | 0.5 |
| 70 | Yb | 6.254 | 12.17 | 0.5 |
| 71 | Lu | 5.426 | 13.9 | 0.5 |
| 72 | Hf | 7.0 | 14.9 | 0 |
| 73 | Ta | 7.89 | -- | 0.6 |
| 74 | W | 7.98 | -- | 0.6 |
| 75 | Re | 7.88 | -- | 0.15 |
| 76 | Os | 8.7 | -- | 1.1 |
| 77 | Ir | 9.1 | -- | 1.6 |
| 78 | Pt | 9.0 | 18.563 | 2.13 |
| 79 | Au | 9.225 | 20.5 | 2.31 |
| 80 | Hg | 10.437 | 18.756 | 0 |
| 81 | Tl | 6.108 | 20.428 | 0.3 |

| Z | Element | 1st Ionization Potential | 2nd Ionization Potential | Electron Affinity |
|---|---------|--------------------------|--------------------------|-------------------|
| 82 | Pb | 7.416 | 15.032 | 1.1 |
| 83 | Bi | 7.289 | 16.69 | 1.1 |
| 90 | Th | -- | 11.5 | -- |
| 92 | U | -- | -- | -- |

Energy conversions: 1 eV = 8065.73 cm$^{-1}$, = 96.485 kJ/mol, =23.06 kcal/mol, =11604.9 K, =1.6022 × 10$^{-19}$ J, = 0.036749 Hatrees (au).

## A.4   WORK–FUNCTION VALUES OF ELEMENTAL SOLIDS

Electron work–functions for the listed elemental solids. Values for crystal-specific faces are also provided where available along with those for different phases. All data is in units of electron volts. For work–function definition, see Section 2.2.2.2. Data compiled from the CRC Handbook of Chemistry and Physics (1985) with corroboration, adjustments, and/or additions from Hüfner (2003)*. Data in italics is not well substantiated.

| Z | Element | Work–function | Crystal face |
|---|---------|---------------|--------------|
| 3 | Li | 2.93* | Poly |
| 4 | Be | 4.98 | Poly |
| 5 | B | 4.45 | |
| 6 | C | 5.0 | |
| | | 4.6 HOPG | |
| 11 | Na | 2.36* | |
| 12 | Mg | 3.66* | Poly |
| 13 | Al | 4.28* | |
| | | 4.41 | 100 |
| | | 4.28* | 110 |
| | | 4.23* | 111 |
| 14 | Si | 4.85n | |
| | | 4.91p | 100 |
| | | 4.60p | 111 |
| 19 | K | 2.28* | Poly |
| 20 | Ca | 2.87* | Poly |
| 21 | Sc | 3.5* | |
| 22 | Ti | 4.33* | Poly |
| 23 | V | 4.3* | |

| Z | Element | Work-function | Crystal face |
|---|---------|---------------|--------------|
| 24 | Cr | 4.5* | Poly |
| 25 | Mn | 4.1* | Poly |
| 26 | Fe | 4.5* | |
| | | 4.67* | 100 |
| | | 4.81$\alpha$ | 111 |
| | | 4.70$\alpha$ | |
| | | 4.62$\beta$ | |
| | | 4.68$\chi$ | |
| 27 | Co | 5.0 | Poly |
| 28 | Ni | 5.15* | Poly |
| | | 4.89* | 100 |
| | | 4.64* | 110 |
| | | 5.22* | 111 |
| 29 | Cu | 4.65* | Poly |
| | | 4.59* | 100 |
| | | 4.48* | 110 |
| | | 4.94* | 111 |
| | | 4.53 | 112 |
| 30 | Zn | 3.63* | Poly |
| | | 4.9 | 0001 |
| 31 | Ga | 4.32* at 273 K | Poly |
| 32 | Ge | 5.0 | |
| | | *4.80* | 111 |
| 33 | As | *3.75* | |
| 34 | Se | 5.9 | |
| 37 | Rb | 2.16* | Poly |
| 38 | Sr | *2.59* | |
| 39 | Y | 3.1* | |
| 40 | Zr | 4.05* | |
| 41 | Nb | 4.3* | Poly |
| | | 4.02* | 001 |
| | | 4.53* | 100 |
| | | 4.87* | 110 |
| | | 4.36* | 111 |
| | | 4.63 | 112 |
| | | 4.29 | 113 |
| | | 3.95 | 116 |
| | | 4.18 | 310 |
| 42 | Mo | 4.6* | Poly |

| Z | Element | Work–function | Crystal face |
|---|---|---|---|
| 42 | Mo | 4.53* | 100 |
| | | 4.95* | 110 |
| | | 4.55* | 111 |
| | | 4.36 | 112 |
| | | 4.50 | 114 |
| | | 4.55 | 332 |
| 43 | Tc | 4.88 | Poly |
| 44 | Ru | 4.71* | Poly |
| 45 | Rh | 4.98 | |
| 46 | Pd | 5.22* | Poly |
| | | 5.16 | 111 |
| 47 | Ag | 4.26 | Poly |
| | | 4.64 | |
| 100 | | | |
| | | 4.52 | 110 |
| | | 4.74 | 111 |
| 48 | Cd | 4.08* | Poly |
| 49 | In | 4.09* | |
| 50 | Sn | 4.42* | |
| 51 | Sb | 4.55* | amorphous |
| | | 4.7* | 100 |
| 52 | Te | 4.95 | |
| 55 | Cs | 1.95* | Poly |
| 56 | Ba | 2.52* | Poly |
| 57 | La | 3.5* | Poly |
| 58 | Ce | 2.9 | Poly |
| 60 | Nd | 3.2* | Poly |
| 62 | Sm | 2.7* | |
| 63 | Eu | 2.5* | |
| 64 | Gd | 3.1* | |
| 65 | Td | 3.0 | |
| 68 | Er | 2.97* | Poly |
| 70 | Yb | 2.9 | Poly |
| 71 | Lu | 3.3 | |
| 72 | Hf | 3.9* | |
| 73 | Ta | 4.25 | |
| | | 4.15* | 100 |
| | | 4.8* | 110 |
| | | 4.0* | 111 |

| Z | Element | Work–function | Crystal face |
|---|---------|---------------|--------------|
| 74 | W | 4.6* | Poly |
| | | 4.63* | 100 |
| | | 5.25* | 110 |
| | | 4.47* | 111 |
| | | 4.18 | 113 |
| | | 4.30 | 116 |
| 75 | Re | 4.72* | Poly |
| | | 5.75 | 1011 |
| 76 | Os | 5.93* | |
| 77 | Ir | 5.27 | |
| | | 5.67 | 100 |
| | | 5.42* | 110 |
| | | 5.76* | 111 |
| | | 5.00 | 210 |
| 78 | Pt | 5.5* | Poly |
| | | 5.84* | 100 |
| | | 5.93* | 111 |
| 79 | Au | 5.1 | Poly |
| | | 5.47 | 100 |
| | | 5.37 | 110 |
| | | 5.31 | 111 |
| 80 | Hg | 4.43 at 173 K | Poly |
| 81 | Tl | 3.84 | |
| 82 | Pb | 4.25* | |
| 83 | Bi | 4.34* | Poly |
| 90 | Th | 3.4 | |
| 92 | U | 3.63 | |
| | | 3.73 | 100 |
| | U | 3.90 | 110 |
| | | 3.67 | 113 |

## A.5   SIMS DETECTION LIMITS OF SELECTED ELEMENTS

Dynamic SIMS Detection Limits (DLs) when using the listed primary ions/secondary ions for Hydrogen through Krypton when present within Silicon or $SiO_2$. These were defined in Magnetic Sector or Quadrupole mass filter-based instrumentation and are listed as a significant digit in units of atoms/cm$^3$. An asterisk signifies the requirement for High Mass Resolution (HMR) with minimum values applied.

| Element | Z | Primary Ion Used | Isotope Measured | DL (at/cm$^3$) | Source |
|---------|---|------------------|------------------|----------------|--------|
| H | 1 | $Cs^+$ | $^1H^-$ | 1e17 | EAG AN339 |
| He | 2 | $O_2^+$ | $^4He^+$ | 5e17 | Wilson et al. 1989 |
| Li | 3 | $O_2^+$ | $^7Li^+$ | 5e12 | EAG AN339 |
| Be | 4 | $O_2^+$ | $^9Be^+$ | 5e13 | Wilson et al. 1989 |
| B | 5 | $O_2^+$ | $^{11}B^+$ | 2e13 | EAG AN339 |
| C | 6 | $Cs^+$ | $^{12}C^-$ | 1e16 | EAG AN339 |
| N | 7 | $Cs^+$ | $^{28}Si^{14}N^-$ | 1e15 | EAG AN339 |
| O | 8 | $Cs^+$ | $^{16}O^-$ | 5e16 | EAG AN339 |
| F | 9 | $Cs^+$ | $^{19}F^-$ | 1e14 | Wilson et al. 1989 |
| Ne | 10 | $O_2^+$ | $^{20}Ne^+$ | 2e18 | Wilson et al. 1989 |
| Na | 11 | $O_2^+$ | $^{23}Na^+$ | 5e12 | EAG AN339 |
| Mg | 12 | $O_2^+$ | $^{24}Mg^+$ | 4e12 | Wilson et al. 1989 |
| Al | 13 | $O_2^+$ | $^{27}Al^+$ | 2e13 | EAG AN339 |
| P | 15 | $Cs^+$ | $^{31}P^-$ | 1e14* | EAG AN339 |
| S | 16 | $Cs^+$ | $^{32}S^-$ | 2e14 | Wilson et al. 1989 |
| Cl | 17 | $Cs^+$ | $^{35}Cl^-$ | 2e14 | Wilson et al. 1989 |
| Ar | 18 | $Cs^+$ | $^{40}Ar^{133}Cs^+$ | 1e17 | Wilson et al. 1989 |
| K | 19 | $O_2^+$ | $^{39}K^+$ | 5e12 | EAG AN339 |
| Ca | 20 | $O_2^+$ | $^{40}Ca^+$ | 1e13 | EAG AN339 |
| Sc | 21 | $O_2^+$ | $^{45}Ca^+$ | 6e14 | Wilson et al. 1989 |
| Ti | 22 | $O_2^+$ | $^{48}Ti^+$ | 1e13 | EAG AN336 |
| V | 23 | $O_2^+$ | $^{51}V^+$ | 1e14 | Wilson et al. 1989 |
| Cr | 24 | $O_2^+$ | $^{52}Cr^+$ | 2e13 | EAG AN336 |
| Mn | 25 | $O_2^+$ | $^{55}Mn^+$ | 2e13 | EAG AN336 |
| Fe | 26 | $O_2^+$ | $^{56}Fe^+$ | 5e13$^{no\ Ca}$ | EAG AN336 |
| Co | 27 | $O_2^+$ | $^{59}Co^+$ | 2e15 | Wilson et al. 1989 |
| Ni | 28 | $O_2^+$ | $^{58}Ni^+$ | 5e14 | EAG AN336 |
| Cu | 29 | $O_2^+$ | $^{63}Cu^+$ | 2e14 | EAG AN336 |
| Zn | 30 | $O_2^+$ | $^{64}Zn^+$ | 5e15 | EAG AN336 |
| Ga | 31 | $O_2^+$ | $^{69,\,71}Ga^+$ | 1e14 | Wilson et al. 1989 |
| Ge | 32 | $Cs^+$ | $^{74}Ge^-$ | 2e14 | EAG AN336 |
| As | 33 | $Cs^+$ | $^{75}As^{28}Si^-$ | 5e13 | EAG AN336 |
| Se | 34 | $Cs^+$ | $^{80}Se^{28}Si^-$ | 6e15 | Wilson et al. 1989 |
| Br | 35 | $Cs^+$ | $^{81}Br^-$ | 7e13 | Wilson et al. 1989 |
| Kr | 36 | $Cs^+$ | $^{84}Kr^{133}Cs^+$ | 1e18 | Wilson et al. 1989 |

Note: Impurity levels in Czochralski Silicon are <1e16 atoms/cm$^3$ for Hydrogen, ~1e16 atoms/cm$^3$ for Carbon, ~5e16 atoms/cm$^3$ for Nitrogen, and ~1e18 atoms/cm$^3$ for Oxygen. Reduction in Oxygen levels is noted within the surface Denuded Zone (DZ). The Oxygen level in float-zoned Silicon is ~1e16 atoms/cm$^3$.

## A.6 CHARGED PARTICLE BEAM TRANSPORT

SIMS relies on the formation, collection, and transport of charged particle beams (most specifically ions, although electrons are also used in dealing with insulating samples), whether within the primary or secondary ion columns of the instrument.

A charged particle beam as opposed to a plasma can be thought of as a system in which the internal forces of individual charged particles are negligible with respect to the externally applied forces. Thus, an ion or an electron beam represents a collection of charged particles moving in the same direction with approximately parallel trajectories. This is otherwise referred to as a *collimated beam*. Internal forces are collectively referred to as *space charge effects*. These stem from Columbic interactions between neighboring charged particles. The fact that this is proportional to the particle density thus explains why a more finely focused beam can be arrived at when the beam is accelerated to higher kinetic energies, at least until the diffraction limit kicks in.

External forces take the form of applied electrostatic and magnetic fields. A field results in a force gradient over some spatial region. Electrostatic fields induce a force proportional to the velocity and charge of the ion within an ion beam. As a result, these are considered energy dispersive. Magnetic fields induce a force proportional to the ion's mass, velocity and charge. These are thus considered mass and energy dispersive. Magnetic fields are used either by themselves, or in combination with a crossed electrostatic fields (plane perpendicular to the magnetic field), as found in Wein filters for ion beam filtering (mass separation). Electrostatic fields can also be used to filter out unwanted species from the ion species of interest, i.e. neutral atoms/molecules. Note: Owing to their lack of charge, neutral atoms/molecules are unaffected by electrostatic or magnetic fields, thus can be filtered out by placing a small bend in the respective column.

External forces can thus be used to control (focus, defocus, or deflect) a charged particle beam or some portion thereof. As a result, electrostatic or magnetic field gradients are often described as charged particle lenses or deflectors.

### A.6.1 Ion Beam Trajectories

The trajectories of a collimated charged particle beam (ions or electrons) can be described using classical field arguments. Approaches that have found maximum use in the design of SIMS instrumentation are those of *Phase Space Dynamics* and/or *Ray Tracing Methods*. These model the internal and external forces acting on and within an ion beam as described by *Liouville's Theorem*. All are covered within Sections A.6.1.1, A.6.1.2 and A.6.1.3. Charged particle beams exhibit optical properties analogous to the optical properties of a light beam, i.e. all can be described by the focal properties of the various lenses through which the respective particles pass through. Indeed, an electrostatic or magnetic field gradient through which a charged particle beam passes can be thought of as analogous to an optical lens through which light (a stream of photons) passes through. The optical properties and subsequent aberrations that can be experienced are covered within Sections A.6.2 and A.6.2.1.

***A.6.1.1*** ***Liouville's Theorem***   Liouville's Theorem states that *provided the forces acting on the ion beam are external and conservative, the local density of points in phase space will remain constant* (Lawson 1978). The phase space in this case refers to the particle's position in space and its momentum. Because any variation in the phase space does not affect the local density of charged particles, the number of ions within the ion beam remains constant. Thus, when in a region where an ion beam does not experience any acceleration or deceleration, the area bounded by the phase space does not alter, even though the shape may change. An effective analogy to this is a balloon filled with a fixed amount of water. If you squeeze it in one direction, it will expand in the other with the density of water within the balloon remaining constant.

***A.6.1.2*** ***Phase Space Dynamics***   Phase space dynamics describes a methodology for defining the optimal set of external forces for transferring a charged particle beam between two points in space.

The phase space of interest refers to the particle's position in space and its momentum. As momentum can be described by classical statistical mechanics, equations of motion can be expressed in terms of the particle's position in space and its momentum, hence the term *Phase Space Dynamics*. Considering that an ion beam is composed of a large number of charged particles, it then follows that the optical properties of the beam can be described as a collection of such parameters.

The parameters include the particle's position ($x$, $y$, and $z$) with respect to the axis of motion and the particle's momentum ($p_x$, $p_y$, and $p_z$) in each of the three spatial dimensions. The boundaries of the phase space, which can otherwise be considered the walls of a balloon containing a fixed amount of water, are defined by the hypervolume bounded by the maximum displacement and momentum of the ions within the charged particle beam at any point in space.

With the boundaries known, the beam properties of the charged particles can be defined at any point throughout the entire length of their trajectories using *matrix formalisms* of vectors representing their boundaries. The first term of the vector describes the maximum beam displacement, whereas the second term describes the maximum beam divergence at that point. As the transformations experienced by the charged particle beam as it passes from one point to another are described by *transformation matrices*, the properties of the beam along its entire trajectory can be represented as a linear transformation of these transformation matrices.

Optimizing the transmission of such a charged particle beam is then simply a case of optimizing the phase space shape at that point in space (this is referred to as the beam's *emittance*) with the various obstacles the beam may experience (this is referred to as the beam's *acceptance*). This matching procedure is otherwise referred to as *emittance–acceptance matching*.

***A.6.1.3*** ***Ray Tracing Methods***   Ray tracing methods employ the use of Laplace equations to describe the individual trajectories of all ions across some region in space. As this is more computationally complex, such simulations are carried out using computers. Laplace equations describe the second-order partial differential equations. In this case, they are used to solve Maxwell's equations of charged particles of specific density in any electrostatic and/or electromagnetic field. These are particularly useful when the charged particle beam experiences nonlinear acceleration/deceleration.

Although ray tracing methods are very powerful, these describe only the trajectories of charged particles through specific predefined regions in space. Defining the optimal external forces (those applied by specific lens geometries) thus becomes an iterative trial-and-error procedure. Optimal beam transmission is thus best defined using ray tracing methods following phase space emittance–acceptance matching. One of the ray tracing programs used in SIMS-charged particle system design is SIMION. This was developed from its predecessor Mac-SIMION, which was first introduced in the late 1970s (McGilvery 1978).

## A.6.2  Optical Properties

Ion, electron, and photon beams all share the same optical properties, i.e. all can be formed into collimated beams (a stream of particles following more or less parallel trajectories), which can be focused at some position in space. Likewise, all can suffer from the same set of deformations, some more so than others. This is inclusive of:

1. Aberrations, the most important of which come in the form of:
   a. Chromatic aberration
   b. Spherical aberration
   c. Astigmatism
   d. Distortion
   e. Coma
2. Diffraction

Note: Diffraction is generally restricted to photon and electron beams with the effects suffered being inversely proportional to the particles mass.

*A.6.2.1  Aberrations*    Aberrations refer to the departure of the particle's trajectory from that expected. This can occur for a number of reasons, with each specific to the type of aberration. These lead to a blurring of the image, or probe spot, as described henceforth.

Chromatic aberrations arise when the energy of the respective particles (ions, electrons, or photons) spans some finite range. This is realized because the focal point, and hence the focusing characteristics of any optical element, is a direct function of the probe particle's energy as is illustrated in Figure A.1 (Recall: energy, wavelength, and frequency are all functions of each other). Note: There remains some intrinsic spread in energy even in the most heavily monochromatized beam. This fundamental limit exists because of the Heisenberg uncertainty principle, a particle's position and momentum (and hence energy) cannot all be absolutely fixed at the same time.

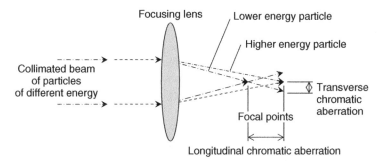

**Figure A.1** Chromatic aberration as suffered by particles of different energy exiting a lens at the same location relative to the beam axis. Reprinted with permission from van der Heide (2012). Copyright 2012 John Wiley and Sons.

For a beam of fixed current, these effects can be minimized by reducing the energy spread or accelerating the probe beam to higher energies as this minimizes the effective energy as a fraction of the beam's energy. The latter approach also has the advantage of reducing space charge effects by effectively reducing their proximity to each other. Reducing the beam current density has the same effect for the same reason.

Spherical aberrations describe the increased focusing/defocusing action experienced by the particles that make up the beam that are further from the optical axis as is illustrated in Figure A.2.

For a beam of fixed current, these effects can be minimized by confining the beam cross section to be as close to the optical axis as possible, increasing the acceptance areas of the respective lenses, and/or altering the lens shape along the beam axis. Additional lenses can also be inserted with the aim of correcting such aberrations.

Coma refers to the deviation in the paths of some portion of the respective particles from that of the remainder of the particles with the result appearing as a coma

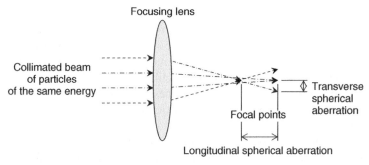

**Figure A.2** Spherical aberration as suffered by particles of the same energy entering a lens at different locations relative to the beam axis. Reprinted with permission from van der Heide 2012. Copyright 2012 John Wiley and Sons.

or a tail on one side of the image. This aberration, similar to spherical aberrations, is most evident when generating an image of some point source. This differs from spherical aberrations in that it arises when a particle beam enters a lens at some off-normal angle as under such conditions, focal points of particles from different points relative to the optical axis deviate. Coma can be minimized by either ensuring that the particles do not enter the lens region at some off-normal angle, or by modifying the lens curvature.

Distortion refers to the deviation in the paths of the respective particles from that necessary to project a distortion-free image, the result being an image that is distorted from that of the object that is being portrayed. These arise from ineffective use of the ion–optical properties of the respective lens. The two primary types are *Barrel distortion* and *Pincushion distortion*. Barrel distortion refers to the decrease in magnification suffered by the image on moving away from the optical axis, whereas pincushion distortion refers to the increase in magnification suffered by the image on moving away from the optical axis.

Astigmatism refers to the deviation in the focal points of particles traveling along one plane (say the $x$–$z$ plane with $z$ the direction of travel) from those traveling along the other plane (the $y$–$z$ plane). Like distortion, these deviations arise from ineffective use of the ion–optical properties of the respective lens. Astigmatism falls into two groups. One is a third-order aberration that is experienced by particles away from the optical axis. This is sometimes referred to as *monochromatic aberration* as this can occur for particles of the same energy. The other is *ophthalmic astigmatism*. This occurs for particles on the optical axis as a result of some unintentional asymmetry in the lens.

### A.6.2.2 *Diffraction and the Diffraction Limit*   Diffraction describes a process experienced by any stream of particles displaying wave-like properties that pass through an aperture of dimensions comparable to the wavelength of particles. Note: Apertures are typically used to confine the spatial spread normal to the axis of travel of any particle beam. This would be required in minimizing spherical aberrations and also to control the current density of a charged particle beam passing some point in space. Diffraction describes the apparent bending of a particle's trajectory when passing close to an obstacle, and hence, the spreading out of the trajectories of a particle beam on passing through a pinhole. This phenomenon is associated with the interaction of the particle's waves with the obstacle, as well as the waves of other particles. The latter is referred to as *interference*. Diffraction thus occurs for all forms of matter displaying wavelike properties. However, as its effects scale with the inverse of the particles mass, these effects are only observed for photons, electrons, and any combination of particles displaying long range wavelike properties.

The diffraction limit describes the spatial limit (spot size) below which a particle beam cannot be focused. This limit relates back to the interaction of the respective particle's wave, i.e. the interference experienced. Although this limit is not of any significance for ion beams used in SIMS (the spatial dimensions of these beams are significantly larger), the diffraction limit is covered for the sake of completeness.

This can be understood as the spatial distance over which particles interact scales with the associated particle's wavelength and the fact that wavelength is inversely proportional to particle mass. Another way of expressing this is that more massive bodies experience a lesser uncertainty in their position at a given energy (Heisenberg uncertainty principle). This explains why:

1. An optical microscope operating in the visible range cannot provide images to a spatial resolution much below ~0.5 μm (photons have zero rest mass and a wavelength in the visible region that spans 400–700 nm). Optical microscopes operating in the UV range push this limit to ~300 nm.
2. A Field Emission Scanning Electron Microscope (FE-SEM) operated at 15 keV produces images with a spatial resolution of ~1 nm (electrons have a mass of $9.109 \times 10^{-28}$ g).
3. A He Ion Microscope (HIM) operated at an accelerating voltage of 45 keV with a point source can relay images to ~0.25 nm (He has a mass ~8000 times that of an electron).

That being said, the spatial dimensions over which a charged particle beam can be focused (the minimum spot size) is also a function of the space charge effects and the aberrations (primarily spherical and chromatic) suffered. This explains why:

1. As the current of an ion beam is increased, so too is the minimum spot size (this explains why imaging in the microprobe mode, whether in SIMS, SEM, or HIM, must be carried out at very low ion currents).
2. Spot sizes increase as the energy is decreased.

## A.7  SOME STATISTICAL DISTRIBUTIONS OF INTEREST

There are several types of distributions used to model the likelihood (statistics) of particular types of events displaying high frequency and nonzero spread. These exist because the distribution of specific types of events can vary, i.e. may be symmetric or nonsymmetric. Those of specific interest are:

1. Gausian distributions
2. Poisson distributions
3. Lorentzian distributions

Gaussian and Poisson distributions are related in that they are extreme forms of the *Binomial distribution*. The binomial distribution describes the probability distribution for any number of discrete trials. A Gaussian distribution is therefore used when the probability of an event is large (this results in more symmetric bell-shaped curves), whereas a Poisson distribution is used when the probability is small (this results in asymmetric curves). The Lorentzian distribution represents the exact solution to the differential equation describing forced resonance, thus it is the most symmetric of the three.

The likelihood of a particular event is *generally* described by the mean, median, and standard deviation ($\sigma$). The word *generally* is used as Lorentzian distributions do not have a mean, and hence $\sigma$ values, whereas Gaussian and Poisson distributions do. The median represents the middle of the range of values modeled. Note: A mean is needed to derive $\sigma$. The mean, if it exists, represents the average value exhibited by the distribution and $\sigma$ describes the likelihood from the mean value over which a particular event will occur, i.e. 68.3% occur within $\pm 1\sigma$ of the mean value, 95.5% occur within $\pm 2\sigma$, and 99.7% occur within $\pm 3\sigma$ for Gaussian distributions.

### A.7.1   Gaussian Distribution

This frequency-type distribution, also called a *Normal distribution*, is applied when describing events that exhibit a highly symmetric array of possible outcomes (mean and median can differ) with populations falling into the standard deviations listed earlier. An example of this is the distribution of the number of ions noted across an ion beam's cross section. The FWHM of a Gaussian distribution is commonly used to describe the spread (width). This equates to $2\sqrt{(2 \ln 2)}\sigma$.

### A.7.2   Poisson Distribution

This frequency-type distribution is applied when describing events that have a non-symmetric probability of occurring around some mean value. Typical examples are in particle decay and particle detection. The event/nonevent is then expressed in units over which the signal decays, i.e. $e^{-n}$, where $n$ is some integer number. It was through an extension of this that dead time effects noted in primary ion-pulsed Time-of-Flight instruments can be partially corrected for (see Section 4.2.3.3.2.3).

### A.7.3   Lorentzian Distributions

This frequency-type distribution, also called a *Lorentz distribution* or a *Cauchy distribution* (named after Hendrick Lorentz and Augustin Cauchy), is applied when describing events that exhibit a truly symmetric array of possible outcomes as in forced resonance, i.e. that due to homogeneous broadening. As a result, these distributions have a narrower FWHM but wider tails than comparative Gaussian distributions and do not have a distinct mean value nor $\sigma$ (the latter is defined from the former).

## A.8   SIMS INSTRUMENT DESIGNS

SIMS has experienced extensive growth and sophistication over the past few decades with its commercialization resulting in the availability of numerous stand-alone instruments (price tags range from several hundred thousand to several million dollars US), and a market sufficient to support multiple different vendors. Presently available instruments are all based around Quadrupole mass analyzers, Magnetic Sector mass analyzers, or Time-of-Flight mass analyzers. As covered in Chapter 4, there also exist alternative experimental platforms ranging from combinations of the above (hybrid instruments), to those using Ion Trap and Ion cyclotron Resonance mass analyzers.

Owing to their relative simplicity and effectiveness, Quadrupole SIMS instruments found greatest commercial popularity throughout the 1970–1990s. Their market share has, however, dwindled as a result of the development in Time-of-Flight-based SIMS instrumentation. That being said, Magnetic Sector-based SIMS instruments have remained the instrument of choice when sensitivity and detection limits are of utmost importance in the area of Dynamic SIMS. It should also be noted that there exists many niche areas where specific types of SIMS instrumentation excel.

In order to provide a flavor of the various SIMS instrument geometries available (complete stand-alone instruments), schematics are presented in Sections A.8.1–A.8.10. These represent case examples of the diverse geometries available. Note: The omission of specific instruments from this list is in no way reflective of their capabilities or otherwise. The geometries are listed in the order of commercially available Quadrupole, Magnetic Sector, and Time-of-Flight-based SIMS instruments. These are then followed by two examples of developmental SIMS instruments (the J105 and Q-star). The Q-Star was derived through the conversion of a commercially available Mass Spectrometer into a SIMS instrument through the incorporation of a $C_{60}^+$ primary ion source. Note: This is not the only example of a SIMS instrument developed from a standalone Mass Spectrometer. Another example was realized on adding an LMIG to a VG ZAB-2f Mass Spectrometer (Schuetzle et al. 1989). This instrument, developed by the Ford Motor company, Dearborn, Michigan, was moved at the hand of the Author to the University of Western Ontario, in London Ontario, Canada, in 1994.

A brief instrument comparison table is then presented in Section A.8.11 along with a vendor contact information list in Section A.8.12. The vendor list also includes suppliers of individual bolt-on components, i.e. primary ion guns and/or secondary ion mass analyzers. Owing to size constraints, the mass analyzers are limited to Quadrupole and Time-of-Flight systems. Although not sharing the same capabilities of their stand-alone counterparts, bolt-on components do allow for the development of specifically tailored multi-technique instrumentation with many possibilities still to be examined. Indeed, one can envisage the prospect of generating/collecting x-ray-induced photo-electrons within the secondary ion flight time of a pulsed primary ion Time-of-Flight-based SIMS instrument as the flight time ($\sim$100 µs) per primary ion pulse ($\sim$1 ns) affords this.

## A.8.1 Physical Electronics 6600

The 6600 instrument (Figure A.3), developed by Physical Electronics (previously Perkin Elmer but now part of the ULVAC group) born out of the 6100 and since

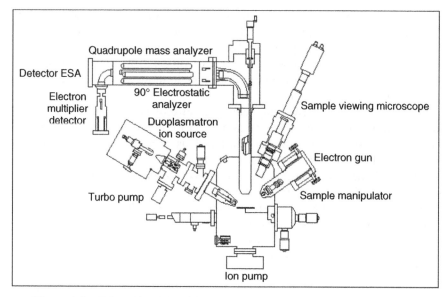

**Figure A.3**  Schematic of the Phi 6600 instrument care of Physical Electronics.

morphed into the Adept 1010, is a microprobe-based Dynamic SIMS instrument designed for elemental analysis. This utilizes a Quadrupole mass filter situated after a 90° energy filter with a secondary ion flight path of ~1 m. The primary competitor in this market was the Atomica (bought out by FEI in 2002 and then Cameca in 2005) series of Quadrupole SIMS instruments.

Primary advantages of these instrument types revolve around:

1. Fast peak switching times compared to Magnetic Sector instruments
2. The low vacuums attainable ($10^{-11}$ Torr range)
3. The relatively low cost of ownership

The primary disadvantages associated with this instrument-type concern:

1. their lack of HMR capabilities
2. the sequential mode of data collection (only selected signals are recorded)
3. the poorer detection limits due to the lower secondary ion transmission.

### A.8.2  ASI SHRIMP I, II, and IV

The SHRIMP IV (commercialized version of the SHRIMP SI) was born out of the SHRIMP I (1989), SHRIMP II (1990), and SHRIMP IIe (mid 2000s). SHRIMP stands for Sensitive High-Resolution Ion Micro Probe. These instruments, designed to provide isotopic ratios specifically for the Earth Sciences, utilizes a Magnetic Sector mass filter situated after the energy filter (forward geometry) to

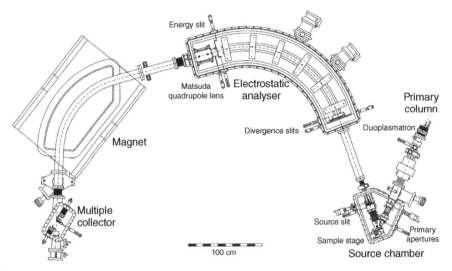

**Figure A.4**    Schematic of the SHRIMP SI instrument care of ASI.

provide extremely high sensitivity at HMR. These are the largest of the SIMS instruments with secondary ion flight paths around 7 meters and a weight of ~12 tons (Figure A.4).

Primary advantages of this instrument type revolve around:

1. The HMR capabilities
2. The high secondary ion transmission even at HMR
3. The use of Kohler illumination of the sample by the primary ion gun

The primary disadvantages associated with this instrument type concern:

1. The sequential mode of data collection (only selected signals are recorded)
2. The high cost of ownership

### A.8.3  SHRIMP RG

The SHRIMP RG (Reverse Geometry), also referred to as the SHRIMP III, is a Dynamic SIMS instrument designed by ASI in the mid-1990s to provide isotopic ratios specifically for the Earth Sciences (Figure A.5). The unique Reverse Geometry (Magnetic Sector mass filter is situated before the energy filter) provides greater magnet dispersion at the same magnification (~4×) translates to improved mass resolution at high transmission. The disadvantage associated with this geometry lies in the fact that only one mass-to-charge ion can be studied at a time. The size and weight are similar to SHRIMP I, II, and IV.

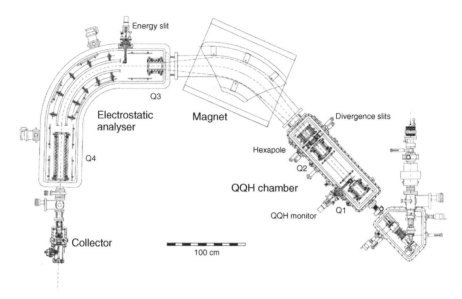

**Figure A.5** Schematic of the SHRIMP RG instrument care of ASI.

Primary advantages of this instrument type revolve around:

1. The HMR capabilities
2. The high secondary ion transmission even at HMR
3. The use of Kohler illumination of the sample by the primary ion gun

The primary disadvantages associated with this instrument type concern:

1. The sequential mode of data collection (only a single signal can be recorded)
2. The high cost of ownership

### A.8.4 Cameca IMS-1280

The IMS-1280, like the ims-1270, is a Dynamic SIMS instrument developed by Cameca designed to provide isotopic ratios primarily for the Earth Sciences. These use a forward geometry Magnetic Sector/energy filter combination to provide high detection limits at HMR. The secondary ion flight path in these instruments is around 5 m (Figure A.6).

Primary advantages of these instrument types revolve around:

1. The HMR capabilities
2. The high secondary ion transmission even at HMR

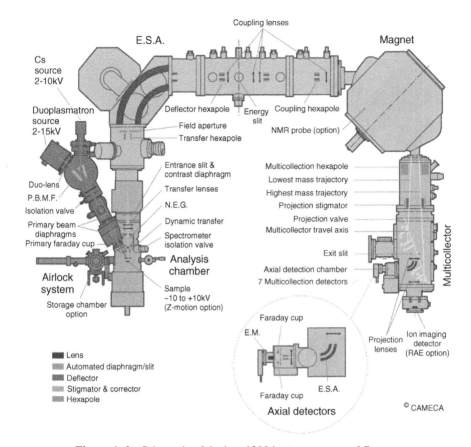

**Figure A.6** Schematic of the ims-1280 instrument care of Cameca.

The primary disadvantages associated with these instrument types concern:

1. The sequential mode of data collection (only selected signals are recorded)
2. The high cost of ownership

## A.8.5 Cameca IMS 7f

The IMS-7f developed by Cameca is the latest in the 3f, 4f, 5f, and 6f series of instruments. This Dynamic SIMS instrument is designed to provide elemental distributions at HMR using a forward geometry Magnetic Sector/energy filter combination. The secondary ion flight path in these instruments is around 2 m (Figure A.7).

Primary advantages of these instrument types revolve around:

1. The HMR capabilities
2. The high secondary ion transmission (but not at HMR)

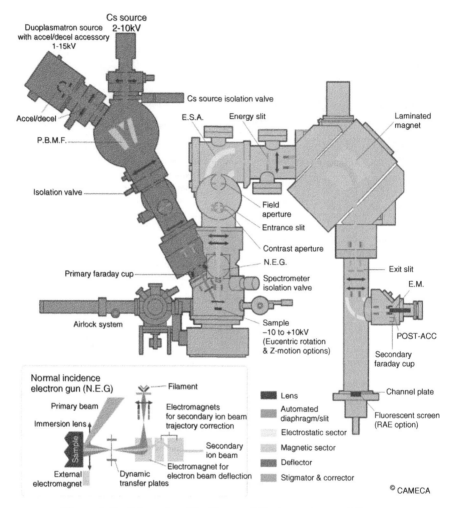

**Figure A.7**    Schematic of the Cameca 7f instrument care of Cameca.

The primary disadvantages associated with these instrument types concern:

1. The sequential mode of data collection (only selected signals are recorded)
2. The high cost of ownership

### A.8.6    Cameca nanoSIMS 50

The nanoSIMS50 (NS50) developed by Cameca is a Dynamic SIMS instrument designed to provide elemental/isotopic distributions at HMR, sensitivity, and spatial resolution of multiple signals. This does so using a forward geometry Magnetic Sector/energy filter combination along with normal incidence primary ion beams. The secondary ion flight path in these instruments is around 2 m (Figure A.8).

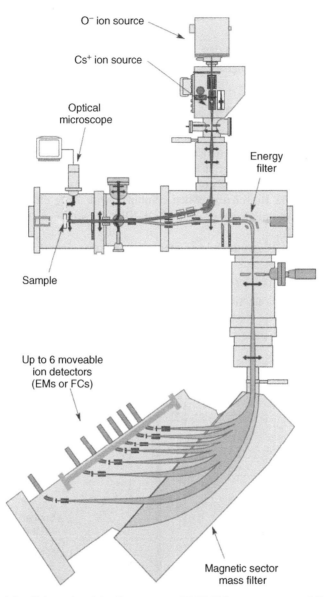

O⁻ ion source

Cs⁺ ion source

Optical
microscope

Energy
filter

Sample

Up to 6 moveable
ion detectors
(EMs or FCs)

Magnetic sector
mass filter

**Figure A.8**    Schematic of the Cameca nanoSIMS 50 instrument care of Cameca.

Primary advantages of this instrument type revolve around:

1. The HMR capabilities
2. The high secondary ion transmission (but not at HMR)
3. The smallest $O^-$, $O_2^+$, and $Cs^+$ primary ion spot sizes (50 nm or less for $Cs^+$)

The primary disadvantages associated with this instrument type concern:

1. The sequential mode of data collection (only selected signals are recorded)
2. The high cost of ownership

### A.8.7   Ion-Tof TOF-SIMS 5

The TOF-SIMS 5 (Figure A.9) is a reflectron-style Time-of-Flight SIMS instrument from Ion-TOF designed to provide HMR Dynamic and Static SIMS capabilities of elemental and/or molecular distributions. A 300 model is also available for full wafer analysis. Although highly effective in all of the sciences, these have spawned new interest in the Biosciences when combined with large cluster primary ion sources. This is the predecessor of the TOF-SIMS 4 with a secondary ion flight path of almost 2 m.

Primary advantages of these instrument types revolve around:

1. The HMR with high secondary ion transmission
2. Simultaneous collection of all secondary ion signals

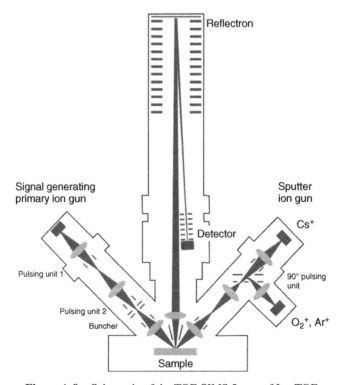

**Figure A.9**   Schematic of the TOF-SIMS 5 care of Ion TOF.

The primary disadvantages associated with these instrument types concern:

1. The poorer detection limits when operated in dynamic mode than Magnetic Sector-based instruments (this loss is not evident in Static SIMS)
2. The high cost of ownership

### A.8.8   Physical Electronics nano-*SIMS*

The Phi nanoSIMS (Figure A.10) is a TRIple Focusing Time-of-Flight (TRIFT)-style Time-of-Flight SIMS instrument developed by Physical Electronics designed to provide HMR Dynamic and Static SIMS capabilities of elemental and/or molecular distributions. Although highly effective in all of the sciences, these have spawned new interest in the Biosciences when combined with a large cluster primary ion source. This follows the Phi TRIFT series of instruments with a secondary ion flight path of ~2 m.

Primary advantages of these instrument types revolve around:

1. The HMR with high secondary ion transmission
2. Simultaneous collection of all secondary ion signals

The primary disadvantages associated with these instrument types concern:

1. The poorer detection limits when operated in dynamic mode than Magnetic Sector-based instruments (this loss is not evident in Static SIMS)
2. The high cost of ownership

### A.8.9   Ionoptika J105-3D Chemical Imager

The J105 chemical imager (Figure A.11) marketed by Ionoptika is a Dynamic SIMS instrument aimed at providing molecular distributions in all three dimensions from organic materials to high spatial resolution (a spatial resolution of 1 μm has been displayed (Fletcher et al. 2011)). This is carried out using a continuous primary on beam (40 keV $C_{60}{}^+$ source), with secondary ions pulsed within a buncher region, situated before a Reflectron Time-of-Flight mass filter. This research-based instrument was developed to maximize weak molecular secondary ion signals.

Primary attributes associated with this instrument geometry include:

1. The ability to map organic molecular distributions in all three dimensions
2. The ability to map at high spatial resolution while under HMR
3. The ability to carry out MS-MS experiments

Disadvantages associated with this instrument geometry are:

1. The highly specific fields to which this instrument is applied
2. The high cost of ownership

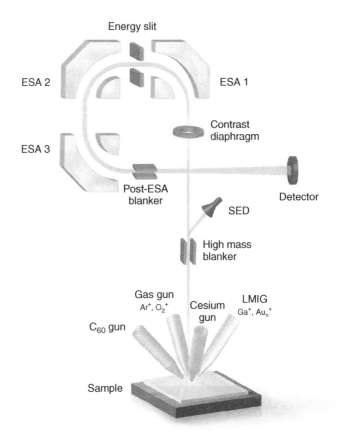

**Figure A.10** Schematic of the nano*SIMS* care of Physical Electronics.

## A.8.10 Q-Star Chemical Imager

The Q-STAR (Figure A.12) is another Dynamic SIMS instrument for providing molecular information from organic materials in all three dimensions using a continuous 20 keV $C_{60}^+$ primary ion source. Secondary ions are pulsed before being sampled by an orthogonal Time-of-Flight secondary ion mass filter. Pulsing along with MS/MS experiments are carried out using three Quadrupole mass filters situated before the Time-of-Flight mass filter. This research-based SIMS instrument was developed from a commercially available MS/MALDI system marketed by Applied Biosystems.

Primary attributes of this instrument geometry include:

1. The ability to map organic molecular distributions in all three dimensions
2. The ability to map at high spatial resolution while under HMR
3. The ability to carry out MS/MS experiments

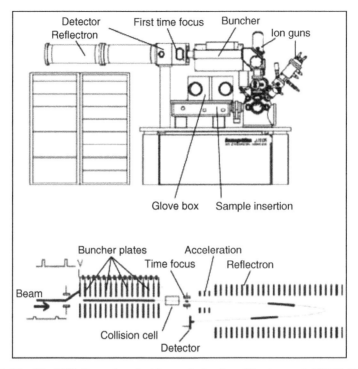

**Figure A.11**     The J105. Reproduced with permission from Fletcher et al. (2011). Copyright 2011 Elsevier.

Disadvantages associated with this instrument geometry include:

1. The highly specific field this instrument is applied in
2. The high cost of ownership

### A.8.11     SIMS Instrument Capability Table

The following table provides a cursory list of selected parameters approximating various standalone SIMS instrument capabilities (available commercially or otherwise), along with areas these are most commonly applied, and any additional comments. The omission of specific instruments is in no way reflective of their capabilities or otherwise. These are listed in the order of Quadrupole-based SIMS instruments, Magnetic Sector-based SIMS instruments, Time-of-Flight-based SIMS instruments, and a Hybrid instrument. These instruments are listed alphabetically within each sub-group. Additional information can be acquired from the respective vendors (see Section A.8.12).

| Instrument | Primary Ion Source/s | Secondary Ion Mass Analyzer | Mass range/for the Analysis of | Mass Resolution ($m/\Delta m$) | Primary Application Area (Application in Others Possible) | Additional Comments |
|---|---|---|---|---|---|---|
| Cameca (*Atomica) 4500 and 4550 | $O_2^+$, $Cs^+$ | Quadrupole | 1 – < 500 u/elements | Unit | Materials Science | *Atomica design, LMIG optional |
| Millbrook miniSIMS alpha | $O_2^+$, $Cs^+$ | Quadrupole | 3 – 300 u/elements | Unit | Materials Science | Bench-top design |
| Physical Electronics Adept 1010 | $O_2^+$, $Cs^+$ | Quadrupole | 1 – < 500 u/elements | Unit | Materials Science | Optimized from 6600 design LMIG optional |
| ASI SHRIMP IV | $O^-$, $O_2^-$, $O_2^+$, $Cs^+$ | Magnetic Sector | 1 – 350 u/elements | ~10,000 | Earth Sciences | Optimized from SHRIMP I, II design |
| ASI SHRIMP RG | $O^-$, $O_2^-$, $O_2^+$, $Cs^+$ | Magnetic Sector | 1-350 u/Elements | ~40,000 Theoretical | Earth Sciences | Also called SHIMP III |
| Cameca 1280 | $O^-$, $O_2^-$, $O_2^+$, $Cs^+$ | Magnetic Sector | 1 – < 500 u/elements | ~34,000 | Earth Sciences | Optimized from ims1270 design |
| Cameca Ims-7f Geo | $O_2^+$, $Cs^+$ | Magnetic Sector | 1 – < 500 u/Elements | ~20,000 | Materials Science | Optimized from ims 3, 4, 5, and 6 designs |
| Cameca WF, and SC ultra | $O_2^+$, $Cs^+$ | Magnetic Sector | 1 – 560 u/elements | ~25,000 | Materials Science in the semiconductor industry | WF is for full wafers and SC is for samples |

| Instrument | Primary ion source | Mass analyzer | Mass range | $m/\Delta m$ | Application | Notes |
|---|---|---|---|---|---|---|
| Cameca nanoSIMS 50 (NS50) | $O_2^+$, $Cs^+$ | Magnetic Sector | 1– < 500 u/elements | ~15,000 | Materials Science, Biosciences | Specifically designed for elemental nanoscale imaging |
| Ion-tof TOF-SIMS-5 | $O_2^+$, $Cs^+$, LMIG, $C_{60}^+$ | Time of Flight | 1–10,000 u/elements, molecules | ~10,000 | Materials Science, Biosciences | Mass range is pulse sequence dependent, $Ar_n^+$ optional |
| Kore Technologies Surface Seer | $Cs^+$, $Ar^+$ | Time of Flight | 1–1000 u/elements, molecules | ~2000 | Materials Science | Mass range is pulse sequence dependent, $Ar_n^+$ optional |
| Millbrook miniSIMS TOF | LMIG | Time of Flight | 1–1200 u/elements, molecules | ~650 | Materials Science | Mass range is pulse sequence dependent, Bench top design |
| Physical Electronics nanoSIMS | $O_2^+$, $Cs^+$, LMIG, $C_{60}^+$ | Time of Flight | 1–10,000 u/elements, molecules | ~10,000 | Materials Science, Biosciences | Mass range is pulse sequence dependent, $Ar_n^+$ optional |
| Ionoptika J105 | $C_{60}^+$ | Time of Flight | 1– > 1,000 u/molecules | ~6,000 | Biosciences | Mass range is pulse sequence dependent |
| Q-Star SIMS/MALDI | $C_{60}^+$, $N_2$ laser | Hybrid | 1– > 1,500 u/molecules | ~14,000 | Biosciences | Mass range is pulse sequence dependent |

LMIG → These primary ion sources can provide $Ga^+$, $In^+$, $Au_n^+$, $Bi_n^+$, and so on.

Hybrid → Multiple types of mass filters (see respective instrument schematic).

$m/\Delta m$ values pertain to those using 10% peak widths. Note: These vary with $m/q$ for all but magnetic sector instruments.

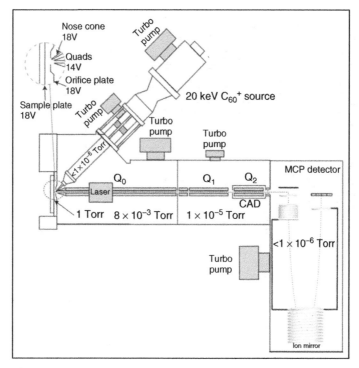

**Figure A.12** Q-Star. Reproduced with permission from Passarelli et al. (2011). Copyright 2011 Elsevier.

### A.8.12 SIMS Instrument/Component Vendor List

Atom Sciences
114 Ridgeway Center
Oak Ridge, TN 37830
U.S.A.
+1 865-483-1113
www.atom-sci.com

Atomica Instruments
Note: Atomica was bought out by FEI in 2002 and then by Cameca Instruments
in 2005
Bruckmannring 40
D-85764 Oberschleissheim
Munich,
Germany
49 89 315891-0
www.cameca.fr

Australian Scientific Instruments (ASI)
111-113 Gladstone St.
Fyshwick, Canberra
ACT 2609,
Australia
612 6280 7570
www.asi-pl.com

Cameca Instruments
103 Boulevard Saint-Denis, BP 6
92403 Courbevoie Cedex
France
203-459-0623 (U.S.)
www.cameca.fr

Hiden Analytical
75 Hancock Rd., Ste. H
Peterborough, NH 03458
U.S.A.
+1 888-964-4336
www.hidenanalytical.com

Ionoptika
Unit 7 Warrior Park,
Eagle Close,
Chandlers Ford, Hampshire,
U.K.
44 23 8027 0735
www.ionoptika.com

Kore Technology
e-Space North
Wisbech Rd.
Littleport, Ely
Cambridgeshire CB6 1RA
U.K.
44 01353 865330
www.kore.co.com

ION-TOF GmbH
Mendelstrasse 11
D-48149, Muenster
Germany
49 251 980 1636
www.ion-tof.com

Ionwerks
2472 Bolsover, Ste. 255
Houston, TX 77005
U.S.A.
+1 713-522-9880
www.ionwerks.com

Kimball Physics
311 Kimball Hill Road
New Hampshire, 03086-9742
U.S.A
www.kimballphysics.com

Millbrook Instruments, Ltd.
Note: The MiniSIMS subsidiary was acquired by Scientific Analysis
    Instruments (SAI) in 2009
Blackburn Technology Centre
Challenge Way,
Blackburn, BB1 5QB,
U.K.
44 1254 699606
www.minisims.com

Scientific Analysis Instruments (SAI)
Hadfield House
9 Hadfield Road
Old Trafford
Manchester, M16 9FE
U.K
www.saiman.co.uk

SPECS Surface Nano Analysis Gmbh
Voltastrasse 5
13355 Berlin
Germany
+49 304678240
www.specs.de

Peabody Scientific
P.O. Box 2009
Peabody, MA 01960,
U.S.A.
+1 978-535-0444
www.peabody-scientific.com/

Physical Electronics, Inc.
18725 Lke Drive East,
Chanhassen, MN 55317
U.S.A.
+1 952-828-6100
www.phi.com

Note: This is not an exhaustive list (there are other manufacturers from other areas that should also be considered where applicable).

## A.9  ADDITIONAL SIMS METHODS OF INTEREST

This addendum covers some additional methods that have been developed, or at the very least, proposed, for quantifying or aiding the quantification of the signals (conversion of intensities to concentration values) arising from atomic secondary ion emissions noted under Dynamic SIMS conditions, i.e. depth profiling.

With the exception of one of these methods, these are all based on the use/modification of Relative Sensitivity Factors (RSFs) such that regions typically outside the scope of the RSF method can be quantified, even if it is only at a semi-quantitative level.

Indeed, it has long been recognized that one of the primary limitations of SIMS revolves around the difficultly in quantifying the recorded data. As indicated in Section 5.4.3.1, the RSF method is by far the most common and heavily utilized method for quantification of the secondary ion signals recorded in SIMS, whether in the form of elemental or molecular emissions collected under Static or Dynamic conditions. This methodology, however, imposes some severe limitations. The most notable limitations are that:

1. Quantification is only possible within a specific layer of interest, when matrix-matched reference materials exist that contain the element of interest (Note: These must be examined in the same analytical session to avoid instrument-induced variations)

2. Quantification cannot be carried out over surface (the transient region) or interfacial regions, unless the respective RSFs are modified in a manner pertinent to the region of interest (this is to account for the secondary ion yield variations noted).

In an attempt to broaden the applicability of the RSF method to substrates for which reference materials were not available, the *matrix-transferable RSF* approach was suggested. This is covered in Section A.9.1. Shortly thereafter, another approach that also did not require matrix-matched reference materials was suggested. This method, termed the *Infinite Velocity method*, is discussed in Section A.9.2. These, however, suffered in that they could not provide the precision levels required for high sensitivity/detection limits commonly expected of SIMS.

As a result, two additional methodologies describing site (depth)-specific RSF modifications were introduced allowing full utilization of the sensitivity/detection limits afforded by SIMS. These are described within the *Lattice Valency model* and the *Point-by-point CORrection* of SIMS profiles (PCOR-SIMS™) approach. These are covered in Sections A.9.3 and A.9.4, respectively. Although the former has found limited mention in the literature, the latter has become a commonly used/advertised protocol by the Evans Analytical Group (EAG).

### A.9.1    Matrix Transferable RSFs

As noted in Figure 3.25, rough correlations exist between the positively charged atomic secondary ion yields and the ionization potential ($I$) of the respective element when present at some concentration below one atomic percent when present within some specific substrate (Wilson et al. 1989). The same can be said for the negatively charged atomic secondary ion emissions from the same substrate and their electron affinities ($EA$).

On changing the substrate, the systematic secondary ion yield trends with $EA$ and $I$ tend to remain, albeit shifted along the yield axis (Wilson et al. 1989). As RSFs reflect the secondary ion yields from specific substrates examined under specific analytical conditions, these too relay systematic trends when plotted against the respective elements $I$ or $EA$ values.

The observation that RSF versus $I$ or $EA$ plots for the atomic secondary ion emissions from a specific substrate follow trends that shift up or down on the Yield or RSF scale with the substrate (matrix) type (Wilson et al. 1989) lead to the proposal that matrix-transferable RSFs could be predicted for multicomponent substrates based on some scaling factor (Grasserbauer et al. 1989; Friedbacher 1988). This scaling factor, however, required that the concentration of the major elements present within the respective substrate be known.

Although the assumption within this matrix-transferable RSF proposal appeared sound at the time, the extreme variability in the atomic secondary ion yields with the local chemistry within the region over which secondary ion emission took place, i.e. the matrix effect (this is discussed in further detail in Section 3.3.2.2), along with the extreme sensitivity to the analytical conditions applied, could not be fully accounted in the highly precise manner typically expected of SIMS. As a result, little in the way of additional work has been communicated within the literature in this area.

### A.9.2    The Infinite Velocity Method

The Infinite Velocity (IV) method represents an alternative approach in that RSFs are not derived/required in defining the concentration/s of the element/s of interest. As the name suggests, the IV method describes the extrapolation of data to infinite emission velocity. This is made possible by collecting the intensities of secondary ions of the elements of interest at several different emission energies and then extrapolating to infinite emission velocity. This method was first introduced in 1994 (van der Heide et al. 1994) and refined thereafter (van der Heide et al. 1998).

Extrapolation to infinite emission velocity is made possible by re-plotting the data in inverse velocity form ($1/v$) as infinite energy equates to zero $1/v$. This extrapolation is carried out as matrix effects are assumed to be removed at infinite emission velocity. Note: This assertion was first recognized when examining relations describing the yields of secondary ions on emission velocity, such as those relayed in Relations 3.10a and 3.10b. In addition, evidence for the reduction in matrix effects has been reported in various emission velocity studies (Yu et al. 1978; Yu 1991; van der Heide 1994a).

As secondary ion yield values as a function on emission velocity, or at least proportional to emission velocity are required, the instrument transmission function and the sputtered atomic neutral to sputtered atomic ion dependence on emission energy had to be accounted for. This had been derived for a Cameca ims-3f used in the initial studies (van der Heide 1994b). The transmission function needs to be accounted for as all SIMS instruments display an emission energy-dependent transmission function. The sputtered neutral to sputtered ion dependence on emission energy, on the other hand, is required as this provides values proportional to the secondary ion yield. Within the original version of the IV method, the sputtered neutral to sputtered ion dependence on emission energy was defined using the Sigmund–Thompson relation.

Also recognized at the time was the fact that errors can be introduced if optimized conditions are not implemented, i.e. if data is not collected over optimal energy ranges (van der Heide et al. 1998). An optimal energy range was specified to account for the appearance of additional secondary ion populations at lower emission energies (nonresonance charge transfer processes appear to dominate at emission energies $<\sim 50$ eV), and to remove the effect to the roll-off noted at high emission energies. The roll-off results from the fact that only a finite amount of energy can be imparted during primary ion impact, a fact not accounted for within the Sigmund–Thompson relation.

In addition, adjustment factors were reported to be required for specific elements whose concentrations were being over estimated, i.e. those of H, C, and O (van der Heide et al. 1998). Subsequent studies carried out under improved vacuum conditions within a Cameca ims-6f revealed that these factors were most likely to account for the adsorption of $H_2$, $O_2$, and various gas phase-based organics onto the respective sputtered surfaces during acquisition of the energy-dependent depth profiles (van der Heide, unpublished observation). Note: As covered in Section 4.2.1.1, molecular adsorption occurs even under high and ultrahigh vacuum conditions, i.e. one monolayer can form in as short a time as 10 s at $10^{-7}$ Torr. It is only the rate that is affected (see Relations 4.3 and 4.4).

This method gained initial favorable attention as this presented the possibility of quantification without the requirement for reference materials. Indeed, early results from independent labs appeared to corroborate the initial assertions (Losing et al. 1998). Likewise, for a derivative thereof referred to as the *High Energy* method (Higashi et al. 1998). Further studies, however, identified significant limitations, the primary limitation being the loss of sensitivity and detection limits associated with the requirement of collecting multiple signals at higher emission energies (velocities). Note: Best detection limits are only noted when data is collected from the low-emission energy populations as is routinely carried out within the RSF method (this quantification method is described within Section 5.4.3.1). The possibility of a hybrid quantification approach may, however, circumvent some of these limitations.

### A.9.3 Lattice Valency Model

The *Lattice Valency model* describes a methodology based on the modification of bulk material-based RSFs (previously defined from associated reference materials), to allow for high sensitivity/detection limit quantification over interfacial region. This was put forward to account for the variability in atomic secondary ion intensities from minor and/or trace elements with the valence state of atoms (cations) making up the substrate (Iltgen 1997). Although trends consistent with expectations are noted within the few publications available, this method does not account for sputter rate variations, nor the impact of any segregation processes, if active.

Within this model, the variability in the negative atomic secondary ion intensities of minor and/or trace elemental emissions is assumed to arise from variations in the concentration of highly electronegative elements (anions) within the near surface region of the substrate being sputtered. An example of this would be the effect the presence of Oxygen (the anion) has on the emissions of the Arsenic secondary ions from a Silicon (the cation) lattice (Gehre et al. 2001).

The *Lattice Valency model* attempts to correct for these variations by, firstly, deriving the valency of the cations making up the lattice, and then, secondly, applying valency-specific RSFs to the respective secondary ion emissions. The cation valency is derived within the same measurement, by recording the atomic Silicon secondary ions (measured as $Si^-$), and its various oxide secondary ions (measured as $SiO^-$, $SiO_2^-$, $SiO_3^-$, $SiO_4^-$, etc.). These intensities are then plotted against the valency (for Silicon, SiO, and $SiO_2$, this would equate to 0, 2, and 4, respectively, although fractions of are noted in the sputtered population) with the resultant plot exhibiting a more or less Gaussian shape. The valency of the near surface region during sputtering is then assumed to equate to the value at which the intensity versus valency plot maxima is noted. As this can be derived over every sputter cycle, a valency versus sputtering time is constructed. If the sensitivity of the RSF of the secondary ion emission of interest with the lattice valency is known, it is then asserted that the secondary ion intensity dependence on valency can be corrected, i.e. by applying the valence-specific RSFs.

### A.9.4  PCOR-SIMS™ Method

The *Point-by-point CORrection* of SIMS profiles (PCOR-SIMS™) approach describes another methodology (unrelated to the Lattice Valency model) based on the modification of bulk material-based RSFs (previously defined from associated reference materials), to allow for high sensitivity/detection limit quantification over interfacial region. This approach was first put forward to account for the variability in the secondary ion signals of $Si^-$ and $AsSi^-$ across the $SiO_2$/Si interface and during the initial stages of sputtering with a $Cs^+$ primary ion beam (Buyuklimanli et al. 2004). Note: The $AsSi^-$ signal resulting from $Cs^+$ primary ion impact is used as this provides the best sensitivity and detection limits to that of Arsenic in Silicon.

Within this method, the variability in recorded secondary ion intensities resulting from chemically reactive primary ions over the transient region is assumed to arise from the variations in the concentration of the respective primary ions within the substrate. This effect extends over a depth equivalent of $2R_p$ to $2.5R_p$, where $R_p$ is the projected range of the respective primary ion within the substrate of interest. The modification of the substrate also has the effect of adjusting sputter rates.

The PCOR-SIMS™ approach attempts to correct for both the secondary ion intensity and sputter rate variations noted over the associated transient and/or interfacial region by, first, deriving the actual sputter rates and RSFs at each point of the respective depth profile for each specific matrix as analyzed under a specific set of conditions and then, secondly, applying the derived sputter rates and RSFs on a point-by-point basis to the substrate being examined such that the appropriate corrections are applied. The first step is often aided using specifically fabricated reference materials that have been fully characterized, often by various other cross-referencing techniques.

Owing to its initial success, the PCOR-SIMS™ method has since been extended to multiple different secondary ions from differing substrate types whether sputtered by $Cs^+$ or $O_2^+$ primary ions over depths ranging from tens to thousands of nanometers in depth. In all cases, the derived results appear consistent to those relayed by MEIS, NRA, ERDA, and other cross-referencing techniques inclusive of Ultra Low Energy (ULE) SIMS. The details concerning the application of this method to specific substrates, however, appear to remain somewhat limited (this is believed to arise from the fact that this method is considered an intellectual property).

## A.10   ADDITIONAL SPECTROSCOPIC/SPECTROMETRIC TECHNIQUES

There is no *one end all be all* technique, i.e. one technique that provides all possible combinations of the information of interest. This stems from the fact that each source/signal combination provides different forms of information, with the technique in question specialized toward maximizing the respective information content contained within. The techniques covered in this section are subdivided according to whether photons, electrons, or ion emissions are recorded. These include:

1. The photon spectroscopies (those that analyze photon emissions) of:
   a. IR-based techniques inclusive of RAIRS, ATR, and DRIFTS
   b. Raman including SERS and TERS
   c. EDX, WDX, and LEXES
   d. XRF and TXRF
   e. VASE (this can also be considered a reflection technique)
2. The electron spectroscopies (those that analyze electron emissions) of:
   a. XPS and UPS
   b. AES and SAM
   c. EELS, including REELS and HREELS
3. The ion spectroscopies/spectrometries (those that analyze ion emissions) of:
   a. GD-MS, GD-OES, and ICP-MS
   b. MALDI and ESI-MS
   c. SNMS and RIMS
   d. APT (this can also be considered a microscopy)
   e. The ion scattering methods inclusive of LEIS, MEIS, and RBS
   f. ERD, NRA, NAA, and PIXE

The techniques chosen in this section are by no means a complete selection, rather representative in the eyes of the Author, of some of the more popular/more interesting spectroscopies/spectrometries that can or have been applied to provide chemical information over localized volumes. The original difference between a spectroscopy and spectrometry lies in the fact that in a spectrometry, only one parameter in the collected data set varies (this is generally the intensity of the signal), whereas in a spectroscopy, more than one parameter can vary (the other could be the energy of the signal).

Within this definition, XPS is considered a spectroscopy. This is realized as both the intensity and the binding energy within the recorded data can vary with the intensity revealing the concentration and the peak binding energy revealing the element and its speciation. SIMS, within this context, is a spectrometry. This is realized as only the intensity of the signal of interest can vary (this relates to concentration once matrix effects are accounted for). Mass, whether pertaining to an element or molecule, remains constant within the energy range accessed by SIMS. Ion scattering methods, on the other hand, can be considered a spectroscopy or a spectrometry depending on how it is used. For example, if the energy of the signal is used to elucidate the element and its depth of origin (intensity relates to concentration), it is a spectroscopy. If, however, the energy is used to elucidate the element alone, then the respective ion scattering method could be considered a spectrometry. This is realized as this energy pertains to the energy loss on scattering, which is mass dependent.

APT is included in this Appendix as this has recently experienced extensive commercial development that, in essence, has moved it from a purely academic methodology (this originated from FIM) to a routinely utilizable technology. Of the techniques listed, only a few are capable of providing compositional information at the atomic scale. Indeed, APT can map these out all three dimensions for every isotope of all the elements. This, in essence, makes APT both a spectrometry and a microscopy. Traditional microscopies along with diffraction- and reflection-based techniques used for imaging are discussed in Sections A.11 and A.12, respectively. Note: Some of the techniques discussed can fall into more than one of the above-designated groups, i.e. VASE is a spectroscopy that derives the elemental composition from the reflected light. Likewise, APT is a spectrometry that maps out the elements in space as a microscopy would.

### A.10.1    Photon Spectroscopies

Photon spectroscopies are techniques that provide chemical information by sampling photon emissions.

***A.10.1.1    IR, RAIRS, ATR, and DRIFTS***    InfraRed (IR)-based techniques are those that derive information on the chemical bonding that occurs between bound atoms, whether in the liquid/solid or gas phase. This information is derived by irradiating the sample with photons over the IR range ($<1$ eV) and sampling the adsorption/transmission that occurs. Adsorption occurs as a result of excitation of valence electrons into higher vibrational states with the energy specific to the masses of the bound atoms and the bonding present. The extent of adsorption also relates to the amount of the elements/molecules present with detection limits of the order of 1%. The IR beam can also be focused to allow imaging (diffraction-limited spatial resolution approaching 1 μm). Fourier transform (FT) analysis is typically used as this enhances the data intensity/quality with the acronym *FTIR* used. FTIR is, however, not intrinsically surface sensitive, as the IR beam passes many micrometers through many solids. IR beams do, however, reflect off of metal layers, thereby depositing an IR-transparent film on a metal layer and recording the reflected signal, some degree of surface specificity can be realized. Although specific sample preparation is required, analysis can be carried out under ambient conditions. Greater surface specificity can be realized using IR in the form of RAIRS, ATR, or even DRIFTS.

Reflection-Adsorption Infra-Red Spectroscopy (RAIRS), also called IRAS (IR Adsorption Spectroscopy), is a form of FT-IR that can provide bonding information and concentration if the surface coverage is $> \sim 1 \times 10^{13}$ atoms/cm$^2$ of any atom/molecule adsorbed on a highly reflective low surface area solid, e.g. CO on Pt. This does so by recording the adsorption that occurs upon specular reflection of photons directed at low incidence angles. No prior sample preparation is required. Although not requiring UHV conditions, this helps in controlling the environment.

Attenuated Total Reflection (ATR) is a form of FT-IR that can provide bonding information and concentration if the surface coverage is $> \sim 1 \times 10^{13}$ atoms/cm$^2$ of atoms/molecules adsorbed on low-surface area solids. In this form, a photon beam is passed through a thin IR-transmitting high density crystal with a high refractive index (ZnSe or Ge) that is in direct contact with the sample of interest. Upon reflection from the crystal face, photon absorption occurs with molecules present at the crystal/sample interface. This occurs as some fraction of the radiation penetrates a small distance into the sample (evanescent wave). The Multiple Internal Reflection (MIR) technique enhances the sensitivity of ATR by bouncing the photons off the surface numerous times. Minimal sample preparation is required, and analysis can be carried out under ambient conditions. Research has also been carried out in the area of Surface Enhanced IR Spectroscopy (SEIRS) using an ATR-type setup after the development of SERS, the Raman analog.

Diffuse Reflectance Infra-red Fourier Transform Spectroscopy (DRIFTS) is a form of FT-IR that can provide bonding information and concentration if the surface coverage is $> 1 \times 10^{13}$ atoms/cm$^2$ of atoms/molecules adsorbed on high surface area solids, i.e. finely dispersed catalysts. This technique does so by sampling photons experiencing diffuse reflectance. In this mode, some fraction of the photons impinging on a surface are transmitted into the solid. These photons may then be absorbed, further transmitted, or reflected out of the solid. The surface-reflected and bulk re-emitted components, which when summed, represent the recorded signal. No prior sample preparation is required, and analysis can be carried out under ambient conditions.

***A.10.1.2  Raman, SERS, and TERS***    Raman-based techniques are those that derive information on the chemical bonding that occurs between bound atoms, whether in the liquid/solid or gas phase. This information is derived by irradiating the sample with photons over the visible/UV range ($> 1$ eV) and sampling the adsorption/emissions induced. These occur as a result of excitation of valence electrons into higher electronic levels that then relax back into their original or related vibrational/rotational state. The difference in energy between the original and related states is then measured (this is specific to the masses of the bound atoms and bonding present). The signal intensity also relates to the amount of the elements/molecules present with detection limits of the order of 1%. The laser used can also be focused to allow imaging with a spatial resolution that is diffraction limited to ~0.3 μm. Raman is closely related to FTIR spectroscopy, with the data being highly complementary. Like FTIR, Raman is not intrinsically surface specific. Raman also suffers from a lack of sensitivity. Recent developments have, however, resulted in the resurgence of Raman in the form of SERS and TERS, both of which are highly region specific. No prior sample preparation is required, and analysis can be carried out under ambient conditions.

Surface Enhanced Raman Spectroscopy (SERS) is a technique capable of providing information from individual molecules. This does so by introducing Silver or Gold nanoparticles as this significantly enhances the Raman signal over highly localized regions. The fact that enhancement factors greater than $10^{10}$ are realized opens up the possibility of noninvasively examining individual molecules. The significant enhancement is believed to be due to the introduction of a surface plasmon resonance resulting in an electromagnetic enhancement that decays exponentially with distance from the nanoparticle. These can thus be thought of as nanoprobes. Although circumventing the diffraction limit, placement of the particles is random by nature. Minimal sample preparation is required, and analysis can be carried out under ambient conditions. Detection limits are poorer than the original form of Raman.

Tip Enhanced Raman Spectroscopy (TERS) is a technique capable of providing images to a resolution of ~20 nm over prespecified areas on a solid substrate. Although closely related to SERS (TERS relies on the same enhancement provided by the presence of Silver), imaging over predefined regions is made possible by supporting a Silver or a Gold nanoparticle on the end of an AFM-like probe tip, and scanning this over the surface of interest that is also irradiated by photons in the visible/UV region (that needed to generate the Raman signal). Localized Raman signal enhancement is then noted from highly localized regions, i.e. those regions within ~10 nm from the tip. This represents another technique that has been able to circumvent the diffraction limit. Minimal sample preparation is required, and analysis can be carried out under ambient conditions. Detection limits are poorer than the original form of Raman.

**A.10.1.3  EDX, WDS, and LEXES**   Energy Dispersive X-ray (EDX) spectroscopy (also called Energy Dispersive x-ray Spectroscopy (EDS)) and Wavelength Dispersive x-ray Spectroscopy (WDS) are techniques capable of providing the elemental composition of the outer ~1–5 μm from any solid. Detection limits are around 0.1% (these are element dependent). All elements from B–U are detectable. These techniques do so by directing a medium (5–30 keV if carried out in SEM or EPMA) to high-energy (up to 300 keV if carried out in TEM) electron beam at the solid of interest. This induces core electron emission, along with Auger electrons and/or fluorescence (the latter two stem from the core hole generated by core electron emission). Both EDX and WDS sample the fluorescence (the energy is element specific), but in different manners. EDX uses either Si(Li) or Silicon Drift Detectors with the latter gaining in popularity. WDS uses a crystal with the angle of reflection based on the impinging photon's energy, i.e. by satisfying the Bragg relation. The difference being that EDX generates a full spectrum of all elements, whereas WDS collects data from specific elements. No prior sample preparation is needed (TEM is the exception), but a minimum of HV is required. That being said, there do exist EDX in SEM instruments, referred to as Environmental SEM (E-SEM) or Variable Pressure SEM (VP-SEM), applicable over the MV range.

Low-Energy X-ray Emission Spectroscopy (LEXES) is a low-energy form of WDS optimized to provide the maximum in sensitivity and detection limits for specific elements within wafers, i.e. it is a technique specifically developed for the semiconductor industry. The term, applied by Cameca, was used to differentiate this technique from that of WDS. Concentration is provided in units of areal dose, i.e. atoms/cm$^2$ as derived over the volume accessed by the respective electron beam (greater depths are accessed using greater energy electrons). Volumes extending from 1 nm to 1 μm in depth and 10 μm or greater in diameter are typical. Detection limits can approach $5 \times 10^{13}$ atoms/cm$^2$. Like WDS, no prior sample preparation is needed, but a minimum of HV is required.

**A.10.1.4  XRF and TXRF**  X-Ray Fluorescence (XRF) is a technique capable of providing the elemental composition throughout a specifically derived location with all elements from Na to U detectable with some down to 0.01 and 0.001% (detection limits are element dependent). It is not a surface-specific technique (the volume probed can extend to 10 μm below the surface) but is covered here as it also identifies the elements based off of the Binding Energies of the respective electrons. The binding energy is derived in much the same way as in EDX/WDS, the difference being that the initial core hole is generated by an x-ray as opposed to an electron. This induces core electron emission along with Auger electrons and/or fluorescence (the latter two stem from the core hole generated by core electron emission). XRF samples the fluorescence. No prior sample preparation or vacuum conditions are required.

Total reflectance X-Ray Fluorescence (TXRF) is a variant of XRF capable of providing the elemental composition of the outer ∼1–5 nm from any highly reflective solid to detection limits of between 0.01 and 0.001% (detection limits are element dependent). All elements from S to U are typically detectable (recent developments have extended this range to Na–U). This technique does so by directing a high-energy photon beam (x-rays) at the solid of interest at some small glancing angle. This induces photoelectron emission, along with Auger electrons and fluorescence (the latter two stem from the core hole generated) from the surface region alone. TXRF samples the fluorescence produced (the energies are element specific). No prior sample preparation is needed nor are vacuum conditions required.

**A.10.1.5  VASE**  Variable Angle Spectroscopic Ellipsometry (VASE) is an optical technique belonging to the ellipsometry family capable of providing the thickness and elemental concentration of the highly regular film (typically that present on a wafer). Most elements are detectable if at or above 1 atomic % (film dependent) from films of thicknesses ranging from 5 nm to >5 μm. The area analyzed is typically in the mm range. VASE measures the Fresnel reflection coefficients (changes in s and p polarized light recorded as $\Psi$ and $\Delta$) of a polarized optical beam reflected off of a sample's surface. This technique does so by scanning the linearly polarized light over a relatively large wavelength and angle of incidence range. The reflected light relays the refractive index and extinction coefficient of the film. The extraction process relies on the effective simulation (this must be physically meaningful) of the recorded data, i.e. by obtaining a "best fit." No prior sample preparation or vacuum conditions are required in data collection. Note: As this technique measures the properties of the reflected light, this can also be considered a reflection-based technique.

### A.10.2 Electron Spectroscopies

Electron spectroscopies are techniques that provide chemical information by sampling electron emissions.

***A.10.2.1 XPS and UPS*** X-ray Photoelectron Spectroscopy (XPS) is a technique capable of providing the elemental composition of the outer $\sim 1-5$ nm from any solid, although insulators are difficult, to detection limits down to $\sim 0.1\%$ (detection limits are element dependent) with some speciation information also available. All elements from Li to U are detectable. The spatial resolution can be down to $\sim 2$ μm. This technique does so by directing an x-ray beam (Al-k$\alpha$ is most often used) at the solid of interest. This induces core electron emission (valence electrons are also produced). Elemental identification is made possible as the core electron energies are element/level specific. No prior sample preparation is needed, but analysis must be carried out under UHV conditions.

Ultraviolet Photoelectron Spectroscopy (UPS) is a technique that is capable of providing the chemical bonding information, and in very limited cases elemental composition of the outer $\sim 1-5$ nm from any solid. The spatial resolution can be down to $\sim 100$ nm. This technique does so by directing a low-energy photon beam (UV energy most typically from an He-I or He-II lines) at the solid of interest and recording the photoelectron emissions. No prior sample preparation is needed, but analysis must be carried out under UHV conditions.

The primary difference between UPS and XPS lies in the energy of the photons used to induce the emissions, i.e. UPS can only sample valence electrons, whereas XPS samples both valence and core electrons. As the photon energy used in UPS is more closely matched to the valence electron binding energies, this technique is far more sensitive to the valence region than XPS. No prior sample preparation is needed, but UHV is required.

***A.10.2.2 AES and SAM*** Auger Electron Spectroscopy (AES) is a technique capable of providing the elemental composition of the outer $\sim 1-5$ nm from any solid, although insulators are difficult, to detection limits down to $\sim 0.5\%$ (detection limits element dependent) with some speciation information also available. All elements from Li to U are detectable. This does so by directing a $5-30$-keV electron beam at the solid of interest. This induces valence and core electron emission with the resulting holes producing Auger electrons or fluorescence (only the Auger electrons produced are recorded in AES). Elemental identification arises as Auger energies are element/level specific. No prior sample preparation is needed, but analysis must be carried out under UHV conditions.

Scanning Auger Microscopy (SAM) is the scanning form of AES, i.e. the highly focused 5–30-keV electron beam is scanned across the surface as this allows for the lateral distribution of elements present on the surface to be mapped. Spatial resolution can extend to ~10 nm or better. AES is similar to XPS in that both provide similar information using similar instrumentation. The primary difference lies in the fact that AES provides a superior spatial resolution, but at the cost of sensitivity (the sensitivity of XPS is slightly better).

***A.10.2.3 EELS, REELS, and HREELS*** Electron Energy Loss spectroscopy (EELS), Reflected Electron Energy Loss Spectroscopy (REELS), and High-Resolution Electron Energy Loss Spectroscopy (HREELS) are all techniques that directly record the energy loss experienced by an electron beam incident on a solid. The energy loss arises from the excitation induced. This may be in the form of vibrational/vibronic modes as measured by HREELS to core-level excitations as induced by REELS and EELS. As a result, HREELS is also referred to as VELS (Vibrational Energy Loss Spectroscopy) and can be considered an electron analog of Raman spectroscopy. EELS, on other hand, is sometimes referred to as CEELS (Core EELS) as core levels are accessed. As core-level binding energies are element specific, elemental identification of all the elements is in principle possible. Detection limits are typically in the 1% range but quantification can be difficult because of the high background levels present. These techniques differ primarily in the energy of the incoming electron beam and the instruments these are carried out in. In HREELS, a 1–10-eV beam is used in highly specialized HREELS instrumentation. In REELS, a 1–25-keV beam is typically used in AES-type instruments. In EELS, higher energy electrons, i.e., up to 300 keV are typically used in TEM-type instruments. As a result, REELS and HREELS are surface specific to the outer ~2 nm, and EELS is volume specific when performed in a TEM. EELS, on the other hand, displays the best spatial resolution, i.e. this can extend below 1 nm when performed in a high-resolution TEM instrument. That of REELS can, in principle, reach that of the probe size. HREELS is not typically used for imaging surfaces. No prior sample preparation is needed (with the exception of EELS if carried out in a TEM instrument), but UHV is required in all cases.

### A.10.3 Ion Spectroscopies/Spectrometries

Ion spectroscopies and spectrometries are techniques that provide chemical information by sampling ion emissions.

***A.10.3.1 GD-MS, GD-OES, and ICP-MS*** Glow Discharge Mass Spectrometry (GD-MS) is a destructive mass spectrometric technique that provides ultra-trace elemental analysis for (Li–U) down to parts per $10^{11}$ (well into the parts per trillion range). This does so by placing the sample in a noble gas atmosphere of ~1 Torr and generating a glow discharge plasma between the solid sample and an extraction plate. Sputtering ensues as plasma ions strike the sample. Positive ions form as a result of Penning Ionization of emitted atoms. These are then directed into a mass spectrometer for elemental identification. Although depth profiling can be carried out, spatial imaging is not an option. Owing to the improved detection limits, this technique has supplanted Spark Source Mass Spectrometry (SSMS) in the analysis of impurities in conducting and semiconducting samples. No prior sample preparation is needed, but analysis must be carried out under HV or better conditions.

Glow Discharge Optical Emission Spectroscopy (GD-OES) is an optical technique that provides trace elemental analysis for all elements down to parts per $10^9$ (into the parts per billion range). This does so in much the same way as in GD-MS, i.e. a glow discharge between the solid sample (the cathode) and the extraction plate (the anode) is generated. The plasma then induces the sputtering of surface bound constituents, many of which exist in the excited state. The energies of the photon emissions resulting from core level de-excitation are then measured, as these energies are element specific. No prior sample preparation is needed, but analysis must be carried out under HV or better conditions.

Inductively Coupled Plasma Mass Spectrometry (ICP-MS) is a destructive mass spectrometric technique used for measuring ultra-trace elemental impurity levels (Li–U) down to parts per $10^{12}$ (well into the low parts per trillion range) in gaseous samples. This does so by inductively heating the gas within concentric quartz tubes via an RF electric current. The ions produced are then extracted through a series of cones before entering a mass spectrometer for elemental identification. ICP-MS is applied to the analysis of solid films within the semiconducting industry by interfacing Vapor Phase Decomposition (VPD) with ICP-MS. This is otherwise referred to as VPD-ICP-MS. Solids can also be examined following complete dissolution or Laser Ablation (LA) of the region of interest. Owing to the improved sensitivity and detection limits afforded by mass spectrometry, ICP-MS has all but supplanted Inductively Coupled Optical Emission Spectroscopy (ICP-OES). Sample preparation is needed, and analysis must be carried out under HV or better conditions.

Note: Although GD-MS, GD-OES, and ICP-MS display superior detection limits to SIMS, whether in its Static or Dynamic forms, none provide the spatial or depth resolution values offered by SIMS. As an example, GD-MS and GD-OES when operated in their depth profiling modes can provide depth resolution values of 10–100 nm, but require millimeter spatial regions. VPD-ICP-MS, on the other hand, provides only values summed over the entire volume consumed.

### A.10.3.2 MALDI and ESI-MS

Matrix-Assisted Laser Desorption/Ionization (MALDI) is a destructive mass spectrometric technique for examining large molecular species and fragments thereof within solid samples. This can include Proteins, Peptides, Sugars, and other large molecules that, in many cases, are beyond the mass range of large cluster primary ion SIMS. MALDI does so by directing a UV laser at the solid of interest, which, in turn, induces the ablation that can extend over micron scale volumes. Ionization of the desorbed population, otherwise referred to as a *hot plume*, is then thought to occur through protonation/deprotonation of the molecules and/or molecular fragments. These are then directed into a mass spectrometer for molecular identification. No prior sample preparation is needed, and analysis can be carried out in close to atmospheric conditions.

Electro-Spray Ionization Mass Spectrometry (ESI-MS), also referred to as ES-MS, is a mass spectrometric technique that provides a similar information content to that of MALDI, i.e. is useful in the analysis of macro-molecules. ESI-MS can, however, access a greater mass range because a greater number of multiply charged ions are produced relative to MALDI. These ions are produced by dispersing, via electro-spray, a liquid containing the analyte of interest. Ions formed as a result of solvent evaporation are then directed into a mass spectrometer for molecular identification. No prior sample preparation is needed, and analysis can be carried out in close to atmospheric conditions.

### A.10.3.3 SNMS and RIMS

Secondary Neutral Mass Spectrometry (SNMS), also referred to as "Sputtered Neutral Mass Spectrometry," is a destructive technique primarily used for examining elemental constituents within solid samples. This technique is closely related to Dynamic SIMS in that an ion beam is used to sputter the solid of interest. The difference lies in the fact that the sputtered neutral population, once ionized, is passed through a mass spectrometer. Ionization is induced via the action of a laser, an electron beam, or plasma (ionization yields vary from ~10% for lasers to ~1% for plasmas). As the greatest fraction of the sputtered population departs in the neutral state, this methodology provides the advantage of improved detection limits and reduced matrix effects relative to SIMS. Depth resolution can extend to 1 nm. Spatial imaging is generally not carried out. No prior sample preparation is needed, but HV or better conditions are required.

Resonance Ionization Mass Spectrometry (RIMS) can be considered a specific and highly efficient form of SNMS/MALDI. This is realized in that a laser of one or more discrete wavelengths is used to induce the ionization of neutral species emanating from the solid of interest. This is then directed into a mass spectrometer for elemental and/or molecular identification. The neutral population is generally produced via the sputtering action induced through ion beam impact in the solid of interest. Element-specific Resonantly Enhanced Multi Photon Ionizations (REMPI) schemes are employed to maximize the neutral to ion conversion of the species of interest, which can extend in the parts per trillion concentration range. As REMPI is element specific, isobaric interferences from other elements/molecules are effectively removed. RIMS has experienced significant technological development over the past two decades making it a technique of interest in particulate analysis, whether air-born or interstellar. No prior sample preparation is needed, but HV or better conditions are required.

### A.10.3.4 APT

Atom Probe Tomography (APT) also called Tomographic Atom Probe (TAP) is a destructive technique capable of providing elemental/isotopic information over any region of any solid with at least 50% of the atoms being detected. Note: As areas much smaller than most other techniques are examined, the concept of detection limits scales with the volume examined (this is not so with most other techniques as substantially larger volumes are examined). Volume maps can be derived in which the depth and spatial resolution can be <0.3 nm. All elements and their isotopes from H to U are detectable. This technique attains this information by directing a laser at a specifically fabricated tip containing the region of interest (constructed using FIB instrumentation). This form of APT, also referred to as LA-TAP where LA stands for Laser-Assisted, induces the removal of atoms that become ionized by the high potential field surrounding the tip. This field also guides these ions through a Time-of-Flight mass filter onto a position-sensitive detector. The former allows for elemental identification (isotopic), whereas the latter allows for the tip volume to be reconstructed at the atomic level. Significant sample preparation (construction of the tip), as well as UHV, is required. This heavily limits the type of samples that can be examined. Note: As this technique provides volume renditions of the region of interest, this can also be considered a microscopy.

### A.10.3.5  *Ion Scattering Methods*    These comprise of several techniques referred to as Low-Energy Ion Scattering (LEIS) Spectroscopy, Medium-Energy Ion Scattering (MEIS) Spectroscopy, and Rutherford Back Scattering (RBS) Spectroscopy. These are all capable of providing the elemental composition as a function of depth, and in some cases, the lattice structure over the surface region of the solid of interest. The depth accessed is a direct function of the probe used, with this depth extending from <0.3 to 100 nm. Although generally considered nondestructive, damage along with sputtering can occur when using increased doses. The spatial resolution is typically of the order of 1 mm (recent developments have, however, pushed this into the $10–50$ μm range). The elemental detection limit, which is $Z^2$ dependent, approaches 0.1 % for light elements and 10 ppm for heavier elements. All elements from H to U are in principle detectable with the low Z elements strongly dependent on the geometry and ions used. This information is attained by analyzing the energy lost by ions ($He^{2+}$, $He^+$, $Ne^+$ $Ar^+$, or $Xe^+$ depending on the technique) scattered off the surface of interest. This is possible as the energy loss in mono-energetic ion scattering is specific to the mass of the collision partners and the depth below the surface at which the collision took place. RBS is the most quantitative of the ion scattering techniques as the high energy ions used (typically $0.5–3$ MeV $He^{2+}$ ions) only allows interaction between the nuclei of the atoms within the sample of interest and the incoming ions. MEIS uses lower energy ions (typically $10–50$ keV $He^+$, $Ne^+$, $Ar^+$, or $Xe^+$ ions), whereas LEIS uses even lower energy ions (typically <10 keV $He^+$, $Ne^+$, $Ar^+$, or $Xe^+$ ions). These provide improved depth resolution. In the case of LEIS, and MEIS to a lesser extent, the possibility of deriving surface crystal structure is also realized. As an example, surface atomic positions can be measured to better than 0.01 nm using specialized variants of LEIS, otherwise referred to as TOF-SARS, SARIS, and ICISS. Quantification, however, becomes increasingly difficult as the scattered ion energy is decreased because of the greater

electronic interaction occurring between these ions and the surface probed (this increases ion neutralization probabilities dependent on the surface chemistry). In all cases, little sample preparation is needed, but UHV is required (most important for LEIS).

**A.10.3.6   ERD, NRA, NAA, and PIXE**   Elastic Recoil Detection (ERD) also called ERDA (Elastic Recoil Detection Analysis), ERS (Elastic Recoil Spectroscopy), HFS (Hydrogen Forward Scattering), FRS (Forward Recoil Scattering), or FReS (Forward Recoil Spectrometry) is a highly specialized nondestructive technique used for quantitative isotopic depth profiles (primarily Hydrogen) of a solid's surface to a depth of $\sim 1$ µm. Detection limits are in the 0.01 atomic % range. ERS shares many similarities with SIMS in that an energetic ion beam (MeV $^4$He$^+$) is directed at the substrates surface. When carried out at close to glancing angles, this induces the forward scattering the Hydrogen isotopes present within the substrate. The energy of the departing H$^+$ ions is then recorded. Advantages relative to SIMS are that it is nondestructive, fast, and matrix effect free. Disadvantages include the fact that it requires an accelerator and it is limited to specific elements (the very light elements). Indeed, it is often thought of as an extension of RBS as it uses the same instrumentation as in RBS and RBS generally only supplies information for elements heavier than the probe ion. Although detection limits for H are not as good as NRA, data interpretation is easier and data acquisition is faster. In all cases, little sample preparation is needed, but HV or better is required.

Nuclear Reaction Analysis (NRA) is a highly specialized nondestructive technique that provides essentially the same information as ERD but to improved detection limits (in favorable cases, Hydrogen can be detected to low parts per million levels). This, however, comes at the expense of data processing complexity and increased data collection times. Analysis is carried out by directing an energetic beam of particles (100 keV to MeV) at the sample of interest. The isotope used (the probe) is dependent on the isotope within the substrate that is of interest. Energetic ion impact that results in specific nuclear reactions, with the emission recorded (gamma rays). Isotopic identification and quantification are derived by measuring the energy and counts of the emissions. Depth resolution, made possible through resonance NRA, is of the order of a few nanometers. NRA complements RBS in the former is more sensitive to the lighter elements, whereas the later is more sensitive to the heavier elements. In all cases, little sample preparation is needed, but HV or better is required.

Nuclear Activation Analysis (NAA) is another highly specialized nondestructive technique closely related to NRA. This is sensitive to at least 60% of the elements within the periodic table. Analysis is carried out by directing a neutron flux at the solid of interest. This induces trace elements to become radioactive, with the activity (high-energy photon emissions) recorded via gamma-ray spectroscopy. Elemental identification and quantification are derived through the photon energy and counts recorded. As neutrons travel significant distances (>many tens of µm) in solids, this can be considered a bulk analytical technique. In all cases, little sample preparation is needed, but HV or better is required.

Particle Induced X-ray Emission (PIXE) is another highly specialized nondestructive technique used to elucidate the chemical (elemental) composition of a solids surface (all elements from Li to U are in principle detectable). This does so by directing energetic particles (generally 0.3–10 MeV protons or helium ions) at the surface of interest. This induces the emission of photons. Like XRF (which uses x-rays as the probe) and EDX (which uses electrons as the probe), measurement of the energy of these photons allows for the elucidation of the atoms they were emitted from. Emissions in PIXE are from depths between EDX and XRF. PIXE, however, requires an accelerator to produce the energetic probe particles, but provides orders of magnitude better detection limits. Both PIXE and EDX require HV, or better and minimal sample preparation is needed.

## A.11  ADDITIONAL MICROSCOPIES

The techniques chosen in this section are representative in the eyes of the Author of some of the more popular surface- or probe-based microscopies for examining solid materials. Note: Microscopies are those used to provide images of localized volumes (little or no chemical information is relayed). Spectroscopies/spectrometries are discussed in Section A.6, whereas diffraction techniques are discussed in Section A.8. The techniques covered in this section include:

1. SEM
2. HIM
3. TEM
4. SPM in the form of:
    a. AFM
    b. SNOM
    c. EFM, MFM, KFM, CAFM, and TUNA
    d. SCM and SSRM
    e. STM

The developments in this area were spurred, in essence, by the fact that the diffraction limit did not allow for optical microscopes to view regions to a spatial resolution much below 1 μm (see Section A.5). Indeed, the spatial resolution in state-of-the-art optical microscopes is diffraction limited to around 300 nm, or half the wavelength of visible light. To improve upon this requires circumventing the diffraction limit (accomplished by the SPM-based techniques) or by reducing the diffraction limit using more energetic probes (TEM uses up to 300 keV electrons to reduce this limit to values well below atomic dimensions) or using more massive probe particles (HIM uses He ions rather than electrons or photons).

TEM is included in this list as, although not a surface technique, it is capable of providing atomic scale images of any volume of interest. The many different variants of SPM are included as this is a field that has experienced extensive commercial development over the past two decades.

### A.11.1    SEM

Scanning Electron Microscopy (SEM) represents the most heavily used of the microscopies for imaging down to the nanometer scale, a region that is inaccessible to the optical microscopies. Indeed, state-of-the-art FE-SEM instruments, where FE stands for the Field Emission source used, can display a spatial resolution approaching ~1 nm. SEM operates by directing a finely focused electron beam (source) of ~15 keV at a solid surface of interest. The intensity of the low-energy secondary electrons generated are then recorded (secondary electrons are recorded as they provide much greater intensities than the Auger or scattered electrons also produced). Distribution maps are then generated by scanning the probe beam over the surface and synchronizing this with the detector. The images generated can then be useful in revealing surface structures, surface topography, and even the dispersion of different matrices. This arises from the fact that different faces, features, and even materials have different secondary electron yields. The spatial resolution is then limited by the probe beam diameter and the interaction volume within the surface. Secondary electrons are generated from depths extending 10 nm or more below the surface (see Figure 4.6). As electrons are used, all analyses must be carried out under HV conditions or better. Analysis at higher pressures can only be carried out in specialized instruments in which the pressure within a localized region around the sample can be increased. No special sample preparation is required.

### A.11.2    HIM

Helium Ion Microscopy (HIM) represents a new technology that uses many of the same principles of SEM. The primary difference is a finely focused He ion beam is used. The end result is a microscopy that can image to a spatial resolution of ~0.25 nm. HIM operates by directing a finely focused He ion beam of energy around 45 keV at the solid surface of interest. The intensity of the low-energy secondary electrons generated is then recorded. Distribution maps are generated by scanning the probe beam over the surface and synchronizing this with the detector. As with SEM, the images generated can be useful in revealing surface structures, surface topography, and even the dispersion of different matrices. The spatial resolution is limited by the probe beam diameter and the interaction volume within the surface. Both are smaller for He ions with respect to energetic electrons. As ions and electrons are used, all analysis must be carried out under HV conditions or better. The use of He ions allows for the analysis of insulating samples (can be difficult in SEM). No special sample preparation is required.

### A.11.3  TEM

Transmission Electron Microscopy (TEM) represents a heavily used microscopy for imaging down to the atomic scale. Indeed, state-of-the-art TEM instruments, using spherical correctors, can display sub-Angstrom resolution ($\sim$0.07 nm). TEM operates by directing a finely focused electron beam (source) of up to 300 keV at a specifically fabricated solid foil (lamella) of less than 100 nm in thickness. The electrons passing through this foil are then detected with the image representative of the atomic positions within the foil (the trajectories of these high energy electrons passing close to atomic nuclei are deflected according to the number of protons within the respective nuclei). Distribution maps can be generated using a static electron beam (TEM) by scanning the probe beam (Scanning-TEM or STEM). The images generated can then be useful in revealing structures and even the dispersion of different matrices. This arises from the fact that different materials have different secondary scattering characteristics. The spatial resolution is primarily limited by the probe beam diameter. As electrons are used, HV conditions or better are required. Like APT, however, highly specialized and extensive sample preparation is required. Furthermore, the effects of the ambient environment (noise, etc.) must be minimized.

### A.11.4  SPM (AFM and STM)-Based Techniques

Scanning Probe Microscopy (SPM) encompasses a group of techniques capable of deriving nanoscale information by scanning an atomically sharp tip over the outer surface of the solid of interest and measuring the appropriate response. This response may be in the form of current (I), voltage (V), capacitance (C), force (F), or the distance (D) by which the cantilever supporting the tip is deflected or a combination of the above. These techniques are not able to directly identify the composition of a material. That being said, these can in special cases derive the carrier concentration if a sufficient amount of information on the sample of interest already exists, i.e. can distinguish between an n-type dopant and a p-type dopant in Si. With the exception of STM, the advantages of these techniques are that no special sample preparation is required, and all analysis can be carried out under ambient conditions. The effects of the ambient environment (noise, etc.) must, however, be minimized.

Atomic Force Microscopy (AFM) is capable of routinely providing topographical maps to a spatial resolution of a few nanometers. Samples may be conductors, semiconductors, or insulators. Information is derived by scanning an atomically sharp tip over the surface of interest and measuring its deflection. Measurement of F versus D curves allows for surface tension, bond strength/elasticity/hardness data to be extracted. AFM allows for topography, mechanical, chemical, and electrical properties to be measured, sometimes simultaneously. Atomic-scale spatial resolution can be achieved, but only in highly specialized instrumentation in which a cooled sample is held under UHV.

Scanning Near-field Optical Microscopy (SNOM) also referred to as SONM is a technique capable of providing optical images to a spatial resolution of a few tens of nanometers. This is carried out by passing photons down a capillary tube that is some 20 nm in diameter situated within an AFM-like tip. Images are then derived by bringing the tip to within 10 nm and then scanning this tip/photon source. This near-field approach circumvents the diffraction limit by ensuring that all dimensions (capillary diameter and tip to surface distance) are much less than the wavelength of the photon source used. As in ATR, this exploits the unique properties of evanescent waves (a near-field standing wave formed at the boundary of different media).

Electrical Force Microscopy (EFM) can provide surface potential and charge distribution maps to a spatial resolution of a couple of nanometers. Information is derived by scanning an atomically sharp tip over the surface of interest and measuring the deflection resulting from the electrostatic force. This is sometimes termed voltage probing.

Magnetic Force Microscopy (MFM) is similar to EFM except for the fact that magnetic domains are mapped by the tip as opposed to electrostatic domains. This can be done to a spatial resolution of a couple of nanometers.

Kelvin Force Microscopy (KFM) is closely related to EFM in that it can also provide surface potential maps, but in the form of work–function variations. This is derived by measuring the Contact Potential Difference (CPD) between the surface and the tip. This can be done to a spatial resolution of a couple of nanometers.

Conductive Atomic Force Microscopy (CAFM) also known as CSAFM is capable of providing maps of the electrical conductivity ($dI/dV$) to a spatial resolution of a couple of nanometers. The sample may be a conductor or semiconductor. Information is derived by measuring the current ($100$ pA–$100$ μA) passing to/from the atomically sharp tip and the sample surface of interest. Measurement of I versus F or I versus V curves allows for localized data to be extracted.

TUNneling Atomic Force Microscopy (TUNA) is similar to CAFM in that it is also capable of providing electrical conductivity ($dI/dV$) maps to a spatial resolution of a couple of nanometers. Samples, however, must be less conductive than used in CAFM. Information is derived by measuring the current ($100$ fA–$100$ pA) passing to/from an atomically sharp tip and the sample surface. As conductivity is a function of film thickness, leakage paths, charge traps, and so on, these too can be mapped. Measurement of I versus F or I versus V curves allows for localized data to be extracted.

Scanning Capacitance Microscopy (SCM) is capable of providing carrier concentration maps to a spatial resolution of a few tens of nanometers. The sample may be a conductor or a semiconductor. Information is derived by measuring the capacitance variations noted on applying an AC bias between the atomically sharp tip and the sample surface of interest. This is useful for measuring concentrations of active $n$- and $p$-type dopants if present at concentrations greater than $\sim$1e16 atoms/cm$^3$. Measurement of the dC/dV versus V curves allows for differentiation between $n$- and $p$-type dopants. The polarization state in ferroelectrics can be derived from the slope of their C versus V curves.

Scanning Spreading Resistance Microscopy (SSRM), like SCM, can provide carrier concentration maps to a spatial resolution of around 2 nm. The sample may be a conductor or a semiconductor. Information is, however, derived by measuring the electrical conductivity or resistivity of a sample sandwiched between an atomically sharp tip and a conductive back plate. This is useful for measuring concentrations of active $n$- and $p$-type dopants if present at concentrations greater than $\sim$1e16 atoms/cm$^3$.

Scanning Tunneling Microscopy (STM) is capable of providing electron density maps (from which atomic distribution maps can be derived) to atomic scale spatial resolution ($<$0.3 nm). The sample may be a conductor or a semiconductor. Information is derived by scanning an atomically sharp tip over the surface of interest and measuring the current tunneling to, or from, this and the sample. Measurement of I-V or dI/dV-V curves allows for electronic properties such as work function to be derived.

## A.12  DIFFRACTION/REFLECTION TECHNIQUES OF INTEREST

Another means of extracting information is by studying the specular reflection or diffraction patterns produced on scattering of particles with solids. Specular reflection refers to the mirror-like reflection observed off of smooth surfaces, i.e., that in which deflection only occurs off of one plane (the surface plane). Diffraction, on the other hand, describes the coherent interaction of packages of particles off specific crystalline surface planes that may or may not be aligned with the surface plane.

More specifically, diffraction results from the interaction of particles with repeating lattice structure as described by the Bragg diffraction criteria. The diffraction is thus specific to the solid's lattice structure. This is particularly evident on crystalline substrates when the wavelength of the probe particles approaches atomic dimensions. For photons and electrons, this occurs at $\sim$5000 eV and $\sim$20 eV, respectively (defined through $E = hc/\lambda$). As this occurs well below 1 eV for ions, this diffraction is not observed in the ion scattering methods. Neutron diffraction relies on similar arguments. Note: Even though neutron and protons have essentially the same mass, neutrons have no charge.

The techniques of greatest interest (at least in the eyes of the Author) include those of:

1. XRD
2. GID

3. XRR
4. LEED
5. HREED
6. Neutron diffraction

XRD is included even though it is not a surface-specific technique as this is by far the most common of the diffraction-based techniques used, i.e., this is the standard method for solving crystal structures for both single crystal samples and poly crystalline samples inclusive of powders. Surface specificity is lost in XRD owing to the geometry used and the fact that photons have long path lengths within solids for both elastic and inelastic collisions. Neutron diffraction is included as it can be considered analogs to XRD, LEED, and related diffraction-based techniques using photons or electrons (Neutron diffraction uses neutrons). LEED and the remainder of those listed are surface-specific techniques. With the exception of LEED and RHEED, minimal sample preparation is required (LEED and RHEED require very clean surfaces), and analysis can be carried out under ambient conditions (LEED and RHEED require UHV).

### A.12.1   XRD and GID

X-Ray Diffraction (XRD) can be used to define the crystalline structure of any solid by directing an x-ray beam with a wavelength close to the separation of the atoms/ions making up the lattice and recording the diffraction pattern produced in the elastically scattered x-ray signal (those suffering Thompson scattering). At specific angles (that satisfying the Bragg diffraction criteria), signals are produced that are specific to the long-range lattice structure. This, therefore, limits this technique to crystalline structures. By rotating the sample/detector in 3D a full description of the lattice structure, distortion to the lattice structure (strain) and composition of the matrix can be attained (detection limits only allow major and minor components to be analyzed). This technique can also be applied to polycrystalline materials and powders, with average grain size data also accessible. Note: This is not a surface-specific technique.

### A.12.2   GID

Glancing Incidence Diffraction (GID) is a surface-specific variant of XRD. This is used to define the crystallinity within surface layers from 10 nm to 200 nm in thickness. This does so by directing x-rays at glancing angles ($\sim 2°$) with respect to the sample surface and measuring the diffraction pattern produced in the nonspecular beam (that deflected away from the plane normal to the surface that lies parallel to the incoming beam) when passed through a Soller slit arrangement. Analysis is carried out by holding the incident angle constant and rotating the detector through some angular range.

## A.12.3  XRR

X-Ray Reflectance (XRR), like GID, is a technique that directs an x-ray beam at the surface at glancing angles. However, instead of measuring the diffraction pattern as noted from crystalline samples, XRR records the specularly reflected beam. This is done as x-rays undergo total external reflection at small grazing angles, with the resulting signal providing insight into the film thickness, roughness, and density. As all surfaces exhibit some form of reflectivity, both crystalline and amorphous films can be examined.

## A.12.4  LEED

Low-Energy Electron Diffraction (LEED) is an electron analog of XRD that can provide the crystal structure over the outer 2–5 nm of a solid's surface. This surface specificity arises from the fact that unlike photons, low-energy electrons (20–200 eV) can only travel a very short distance within a solid before suffering elastic and inelastic collisions whereupon they lose energy. Electrons are directed normal to the surface and the backscattered signal (elastically scattered) is recorded. This displays an XRD-like diffraction pattern. As electrons are used, HV and clean surfaces are required.

## A.12.5  RHEED

Reflection High-Energy Electron Diffraction (RHEED) is an electron analog of GID that can provide the crystal structure of the outer 1–10 nm by directing a high-energy electron beam (~10–100 keV) at some glancing angle and measuring the forward scattered signal. The lattice order is revealed in the nonspecularly reflected beam as a result of the diffraction induced by the surface crystalline planes. Note: Although LEED provides better diffraction patterns, RHEED allows for improved sample access (needed in epitaxial growth studies). HV and clean surfaces are required.

## A.12.6  Neutron Diffraction

Neutron diffraction can be considered a neutron analog of XRD and LEED. As the name suggests, neutrons are directed at the solid and the backscattered signal (elastically scattered) is recorded. As this signal displays an XRD-like diffraction pattern specific to the long-range lattice structure of the sample, this technique can be used to derive the lattice structure, any distortion to the lattice structure (strain) and the composition (detection limits only allow major components to be analyzed). The primary difference among XRD, LEED, and Neutron diffraction is that in XRD and LEED, the incoming photons/electrons interact with the electron cloud of the atoms/ions making up the solid, whereas in neutron diffraction, the incoming neutrons scatter off of the nuclei of the atoms/ions making up the solid. Moreover, as neutrons posses a magnetic moment, the magnetism of the substrate can be examined. This makes this technique useful in the analysis of high-temperature superconductors. The primary drawback of this technique lies in the fact that nuclear reactors are needed (neutron source) along with additional radiation shielding. In addition, as neutrons travel significant distances (>many tens of micrometers) in solids, this is not generally considered a surface-specific technique. That being said, neutron scattering is able to provide detailed structural information not accessible by many of the other diffraction-based techniques. Like RHEED, analysis can be carried out *in situ*.

| | |
|---|---|
| AES | Auger Electron Spectroscopy |
| AFM | Atomic Force Microscopy |
| APT | Atom Probe Tomography |
| ATR | Attenuated Total Reflection |
| CAFM | Conductance Atomic Force Microscopy |
| CSAFM | Current Sensing Atomic Force Microscopy (also called CAFM) |
| CEEL | Core Electron Energy Loss spectroscopy |
| DRIFTS | Diffuse Reflectance Infra-red Fourier Transform Spectroscopy |
| EBSD | Electron Back-Scattered Diffraction |
| ESCA | Electron Spectroscopy for Chemical Analysis (also called XPS) |
| EELS | Electron Energy Loss Spectroscopy |
| EDS | Energy Dispersive x-ray Spectroscopy (also called EDX) |
| EDX | Energy Dispersive X-ray spectroscopy (also called EDS) |
| EFM | Electrical Force Microscopy |
| EPMA | Electron-Probe Micro-Analysis |
| ERDA | Elastic Recoil Detection Analysis (also called ERD, ERS, FRS, FReS) |
| ERD | Elastic Recoil Detection (also called ERDA, ERS, FRS, FReS, and HRS) |
| ERS | Elastic Recoil Spectroscopy (also called ERDA, ERD, FRS, FReS, and HRS) |
| ESI-MS | Electro Spray Ionization Mass Spectrometry |
| FIB | Focused Ion Beam. Note: This is a preparation technique as opposed to an analytical technique |
| FT-ICR | Fourier Transform Ion Cyclotron Resonance |
| FT-IR | Fourier Transform Infra-Red spectroscopy |
| FIM | Field Ionization Microscopy |
| FRS | Forward Recoil Scattering (also called ERD, ERDA, ERS, FReS and HRS) |
| FReS | Forward Recoil Spectrometry (also called ERD, ERDA, ERS, RRS, and HRS) |
| GD-MS | Glow Discharge Mass Spectrometry |

*Secondary Ion Mass Spectrometry: An Introduction to Principles and Practices*, First Edition.
Paul van der Heide.
© 2014 John Wiley & Sons, Inc. Published 2014 by John Wiley & Sons, Inc.

| | |
|---|---|
| GD-OES | Glow Discharge Optical Emission Spectroscopy |
| GID | Glancing Incidence Diffraction |
| HIM | Helium Ion Microscope |
| HREED | Reflection High-Energy Electron Diffraction |
| HREELS | High-Resolution Electron Energy Loss Spectroscopy |
| HRS | Hydrogen Forward Scattering (also called ERD, ERDA, ERS, FReS and HRS) |
| IR | Infra Red spectroscopy |
| ICISS | Impact Collision Ion Scattering Spectroscopy |
| ICP-MS | Inductively Coupled Plasma Mass Spectrometry |
| IMS | Ion Mobility Spectrometry (not be confused with Imaging Mass Spectrometry which is now referred to as Mass Spectrometry Imaging) |
| ISS | Ion Scattering Spectroscopy |
| KFM | Kelvin Force Microscopy |
| LA-TAP | Laser-Assisted TAP |
| LEED | Low-Energy Electron Diffraction |
| LEIS | Low-Energy Ion Scattering |
| LEXES | Low-Energy X-ray Emission Spectroscopy |
| MALDI | Matrix-Assisted Laser Desorption/Ionization |
| MEIS | Medium Energy Ion Scattering |
| MFM | Magnetic Force Microscopy |
| MSI | Mass Spectrometry Imaging |
| NRA | Nuclear Reaction Analysis |
| NSOM | Near field Scanning Optical Microscopy (also called SNOM) |
| PEEM | Photo-Electron Emission Spectroscopy |
| PIXE | Particle Induced X-ray Emission |
| RAIRS | Reflection-Adsorption IR Spectroscopy |
| RBS | Rutherford Back Scattering |
| REELS | Reflected Electron Energy Loss Spectroscopy |
| RHEED | Reflection High-Energy Electron Diffraction |
| RIMS | Resonance Ionization Mass Spectrometry |
| SAM | Scanning Auger Microscopy |
| SARIS | Scattering And Recoiling Imaging Spectroscopy |
| SERS | Surface Enhanced Raman Spectroscopy |
| SEM | Scanning Electron Microscopy |
| SEIRS | Surface Enhanced Infra-Red Spectroscopy |
| SCM | Scanning Capacitance Microscopy |
| SIMS | Secondary Ion Mass Spectrometry (Note: Although this acronym also refers to Stereoscopic Imaging and Measurement Systems this definition is not used in this text) |
| SNOM | Scanning Near-field Optical Microscopy |
| SNMS | Secondary Neutral Mass Spectrometry |
| SPELEEM | Spectroscopic Photo-Emission and Low-Energy Electron Microscopy |

| | |
|---|---|
| SPM | Scanning Probe Microscopy |
| SSMS | Spark Source Mass Spectrometry |
| SSRM | Scanning Spreading Resistance Microscopy |
| STM | Scanning Tunneling Microscopy |
| STEM | Scanning TEM |
| TAP | Tomographic Atom Probe |
| TEM | Transmission Electron Microscopy |
| TERS | Tip Enhanced Raman Spectroscopy |
| TOF-SARS | Time-Of-Flight Scattering And Recoiling Spectroscopy |
| TUNA | TUNneling Afm |
| UPS | Ultra-violet Photo-electron Spectroscopy |
| VASE | Variable Angle Spectroscopic Ellipsometry |
| VPD-ICP-MS | Vapor Phase Decomposition ICP-MS |
| XPS | X-ray Photo-electron Spectroscopy (also called ESCA) |
| TXRF | Total reflection X-Ray Fluorescence |
| UV vis | UV visible spectroscopy |
| VELS | Vibrational Energy Loss Spectroscopy |
| XPEEM | X-ray Photo-Electron Emission Microscopy |
| XPS | X-ray Photoelectron Spectroscopy |
| XRD | X-Ray Diffraction |
| XRR | X-Ray Reflectivity |
| XRF | X-Ray Fluorescence |
| WDS | Wavelength Dispersive x-ray Spectroscopy |

# ABBREVIATIONS COMMONLY USED IN SIMS

| | |
|---|---|
| BCA | Binary Collisions Approximation |
| DL | Detection Limit |
| DLD | Delay Line Detector |
| DD-EM | Discrete Dynode Electron Multiplier |
| DP | Depth Profile |
| Duo | Duoplasmatron |
| EDR | Extended Dynamic Range |
| EI | Electron Impact |
| EM | Electron Multiplier |
| ESA | Electro-Static Analyzer |
| EPS | Electrostatic Peak Switching |
| FC | Faraday Cup (FA is also used) |
| FPM | Fragment Pathway Mapping |
| FT-ICR | Fourier Transform Ion Cyclotron Resonance |
| FWHM | Full Width and Half Maximum peak intensity |
| G-SIMS | Gentle SIMS |
| GCIB | Gas Cluster Ion Beam |
| $HR^2$ | High mass resolution with high spatial resolution (term trade marked by Physical Electronics) |
| HV | High Vacuum |
| HMR | High Mass Resolution |
| ICR | Ion Cyclotron Resonance |
| IMF | Instrument Mass Fractionation |
| IMS | Imaging Mass Spectrometry (also known as: Ion Mobility Spectrometry) |
| KE | Kinetic Energy filtering |
| LMIG | Liquid Metal Ion Gun |
| LV | Low Vacuum |
| LTE | Local Thermal Equilibrium |
| $m/q$ | mass to charge ratio |
| MSI | Mass Spectrometry Imaging |
| MS-SIMS | Magnetic Sector Secondary Ion Mass Spectrometry |

*Secondary Ion Mass Spectrometry: An Introduction to Principles and Practices*, First Edition.
Paul van der Heide.
© 2014 John Wiley & Sons, Inc. Published 2014 by John Wiley & Sons, Inc.

| | |
|---|---|
| MCP-EM | Micro Channel Plate Electron Multiplier |
| MCs$^+$ method | Ceasium cluster ion method where M defined the element of interest |
| MD | Molecular Dynamics |
| MS | Magnetic Sector (mass filter) |
| MV | Medium Vacuum |
| PHD | Pulse Height Distribution |
| PCOR$^{TM}$ | Point by point CORection (name trademarked by the Evans Analytical Group (EAG)) |
| QCM | Quartz Crystal Micro balance |
| Quad | Quadrupole (mass filter) |
| RF | Radio Frequency (as in RF sources) |
| RG | Reverse Geometry |
| ROI | Region of Interest |
| Rp | Projected Range |
| RSF | Relative Sensitivity Factor |
| SF | Sensitivity Factor |
| SI | Specimen Isolation |
| SIM | Scanning Ion Microscope |
| SIMS | Secondary Ion Mass Spectrometry (also known as; Stereoscopic Imaging and Measurement and multiple SIMulationS) |
| SHRIMP | Sensitive High Resolution Ion Micro Probe |
| SMILES | Simplified Molecular Input Line Entry Specification |
| SRIM | Stopping Range of Ions in Matter |
| TOF | Time Of Flight (mass filter) |
| TRIFT | TRIple Focusing Time of flight |
| TRIM | TRansport of Ions in Matter |
| UHV | Ultra High Vacuum |
| ULE | Ultra Low Energy |
| VHV | Very High Vacuum |
| XHV | eXtreme High Vacuum |

**Aberrations** Distortions (to an image) resulting from incorrect transfer of information (optics)

**Ab-initio calculations** Calculations based on first principles (theory) and constants alone (c.f. semi-empirical calculations and empirical)

**Accuracy** Ability to derive the correct (actual) value

**Analytical cycle** A time interval over which a measurement/s is recorded

**Atom** This represents the smallest basic unit of matter which consists of a nucleus (protons and neutrons) around which electrons orbit

**Atomic mass** The actual mass of an isotope of an atom

**Atomic number (Z)** defined as the number of protons within an atom

**Atomic weight** The weighted average of the mass of all isotopes of the element

**Aufbau principle** From German meaning: *Building up*. This is a hypothetical process describing how the electron configuration of atoms is builtup starting with the filling of vacant levels of the lowest principle ($n$) and angular momentum ($l$) quantum numbers

**Auger charge transfer** An electron transfer process resulting in secondary ion formation or neutralization

**Bond breaking model** A model describing atomic secondary ion emission

**Centro-symmetric rotation** Rotation of the sample around the area being analyzed with the axis of rotation being perpendicular to surface

**Characteristic velocity** A term used to describe the likely hood of ion formation

**Charge compensation** A method used to remove localized electrical charge buildup

**Chromatic aberration** Type of optical distortion (see Appendix A.5)

**Cluster ions** Large molecular ions

---

*Secondary Ion Mass Spectrometry: An Introduction to Principles and Practices*, First Edition.
Paul van der Heide.
© 2014 John Wiley & Sons, Inc. Published 2014 by John Wiley & Sons, Inc.

**Cluster ion SIMS** A form of Dynamic SIMS in which large cluster primary ions are used

**Cluster model** A model describing molecular secondary ion emission (from MALDI)

**Collision cascade** Ion impact-induced collision sequences in which momentum transfer occurs

**Cohesive energy** Energy required to break a bond between a surface atom and the associated solid

**Core hole** The vacancy left behind when an electron is excited out of its core level stationary state within an atom

**Correlation diagram** Diagram displaying the formation and modification of Molecular orbitals with inter-atomic distance

**Coulomb explosion model** A model for describing sputtering

**Damage cross section** A parameter used in Static SIMS for describing the damage imparted during sputtering

**Dead time** A phenomena experienced by various detectors in which they remain insensitive for a short time following the detection of an intense signal

**Defect mediated sputtering model** A model for describing sputtering

**Depth profile** plot of intensity or concentration (abscissa) versus sputtering time or depth (ordinate)

**Depth resolution** Depth over which a signal from some abruptly appearing layer climbs from 16% of its maximum intensity to 84% (2 standard deviations)

**Desorption/Ionization model** A model describing molecular secondary ion emission

**Detection limit** Defines the smallest concentration that can be measured. This is also referred to as the Lower Limit of detection (LOD). This is not to be confused with *sensitivity*

**Diffraction** A process resulting from the interaction of waves with each other. These can interfere constructively (results in a signal) or destructively (no signal)

**Diffusion** Mixing of atoms in a solid through some physical forces

**Disappearance cross section** A parameter used in Static SIMS for describing the disappearance of a useful signal during sputtering

**Dynamic SIMS** SIMS carried out over more than one atomic layer per analytical cycle

**Electron Affinity (*E.A.*)** Minimum energy needed to add an electron to a free atom or molecule

**Electron tunneling model** A model describing atomic secondary ion emission

**Element** Atoms that display a specific set of properties. These are defined by the number of protons within the nucleus

**Empirical** Experimental results (no theory used)

**Energy** Product of ½ mass × velocity$^2$

**Enhanced transient effects** Additional transient effects noted during the onset of sputter depth profiling

**Excitons** Energy associated with electron hole pairs formed when an electron moves (is excited) across the band gap

*Ex-situ* Off site measurements

**Fermi edge ($E_F$)** Energy at which half the electronic orbitals in a solid are filled when at equilibrium. This is sometimes referred to as zero on the energy scale

**Gas flow model** A model describing molecular secondary ion emission

**Gaussian distribution** A type of statistical function (see Appendix A.4)

**Heat Spike model** A model describing secondary ion emission

**High Mass Resolution (HMR)** A means to separate isobaric interferences

**Holography** A method of extracting images from waves

**Image depth profile** A depth profile where images are also collected to reveal the full 3D structure

**Image field** Electrostatic field that extends a short distance from a solids surface. This is only experienced by charged particles in close proximity

**Infinite velocity method** A suggested SIMS quantification method

**Intense ultrafast electronic excitation model** A model for describing sputtering

*In-situ* On site measurements

**Ionization potential ($I$)** Minimum energy needed to remove an electron from a free atom/ion

**Isobaric interference** Form the Greek word meaning: *same* (iso) *weight* (baros). Such interferences arises when two or more ions have very similar mass

**Isotope** From the Greek work meaning; *same* (iso) *place* (topos)

**Kinetic energy ($K.E.$)** Energy contained within an electron or ion that pertains to its velocity, c.f. Potential energy

**Kinetic energy filtering** Measurement of higher energy atomic secondary ions for removing isobaric molecular signals

**Kinetic emission model** A model describing multiply charged atomic secondary ion emission

**Kinetic sputtering** Removal of atoms/molecules through kinematic processes

**Knock-on sputtering** Removal of atoms/molecules through momentum transfer

**Linear cascade model** A model for describing sputtering

**Local Thermal Equilibrium (LTE)** A model describing secondary ion emission

**Lorentzian distribution** A type of statistical function (see Appendix A.4)

**Mass Fractionation** Processes affecting isotopes in a different manner

**Mass spectra** plot of $m/q$ (abscissa) versus intensity (ordinate)

**Mathieu stability diagram** A diagram describing the operation of Quadrupole and ion trap mass filters

**Matrix effects (SIMS)** Effects induced by the substrates surface (matrix) that result in the variation of secondary ion yields

**Maximum Entropy** A deconvolution algorithm which uses Bayesian inference to satisfy the third law of thermodynamics (maximize entropy)

**Microprobe mode** An imaging mode in which points a projected on to a detector

**Microscope mode** An imaging mode in which the entire surface is projected on to a 2D detector

**Microscopy** Any technique providing images in 2D or 3D

**Momentum (momenta)** Product of mass × velocity

**Nascent emission model** A model describing molecular secondary ion emission

**Passivating oxide** Thin (~1 nm) stable protective oxide

**Pauli exclusion principle** A quantum mechanical law that states no two fermions (electrons) within the same atom can possess the same set of quantum numbers

**Peak stripping** A retrospective (mathematical) approach to remove isobaric interferences

**Pixel** The area over which a signal is summed (Voxel is 3d analog)

**Plasmons** Collective vibrations of electrons within a solid

**Phonons** Collective vibrations of atoms within a solid

**Photons** Electromagnetic radiation (a package of energy with zero rest mass)

**Poisson distribution** A type of statistical function (see Appendix A.4)

**Potential energy** Energy contained within an electron or ion that pertains to its excitation, c.f. Kinetic energy

**Potential sputtering** Removal of atoms/molecules through excitation processes

**Pooling mechanism** A model describing molecular secondary ion emission (from MALDI)

**Pre-cursor model** A model describing molecular secondary ion emission

**Precision** Variability associated with repeat measurements

**Principle component analysis** A mathematical procedure that attempts to recognize the minimum number of variables (principle components) in a data set

**Projected range (Rp)** The average depth incoming ions will be implanted into a solid

**Quantum numbers** A scheme based on quantum mechanics in which the energy, momenta, spatial distribution and spin of each electron bound to a specific element is specified

**Quantum mechanics** A mathematical interpretation of atomic structure

**Quasi-resonance charge transfer** An electron transfer process resulting in secondary ion formation or neutralization

**Recoil sputtering** Sputtering resulting from a single collision event

**Reduced cluster ion energy** Cluster ion energy divided by the number of atoms within the cluster

**Repeatability** Random uncertainty associated with data collection on the same instrument

**Reproducibility** Random uncertainty associated with data collection from different instruments

**Resonance charge transfer** An electron transfer process resulting in secondary ion formation or neutralization

**Ripple topography** Surface ripples that form under sputtering

**Selvedge model** A model describing secondary ion emission

**Secondary ion energy** Emission energy of secondary ions from a surface

**Secondary ion velocity** Emission velocity of secondary ions from a surface

**Secondary ion yield** The number of secondary ions produced per incoming primary ion

**Selvedge concept** A model describing molecular secondary ion emission

**Segregation** De-mixing of atoms within a solid as a result of chemical forces

**Semi-empirical** Calculations in which known parameters, i.e. structure, are input

**Sensitivity** Defines the smallest change in concentration that can be measured. This should not be confused with *detection limit*

**Space-charge affects** Electrostatic repulsion of electrons or ions in a charged particle beam

**Spatial resolution** Distance over which a signal from some abruptly appearing interface climbs from 16% of its maximum intensity to 84% (2 standard deviations)

**Specificity** In surface analysis this describes a means to examine a specific region, i.e. the surface region

**Spectroscopy** Any technique providing chemical information (different to a spectrometry as discussed in Appendix A.10)

**Spectrometry** Any technique providing chemical information (different to a spectroscopy as discussed in Appendix A.10)

**Spherical aberration** Type of optical distortion (see Appendix A.5)

**Specimen Isolation (SI)** A method used during the analysis of insulating samples

**Spectroscopic notation** A notation scheme used to describe stationary states

**Sputtering** The removal of atoms from a solid surface as a result of energetic ion impact

**Sputter rate** Rate in units of depth over time at which the surface is being removed through sputtering

**Sputter yield (absolute)** Number of surface atoms removed per incoming ion used to sputter the surface

**Sputter yield (useful)** Number of useful surface atoms (ions) removed per incoming ion. The term "useful": pertains to the number that can be measured

**Static limit** A parameter (in units of primary ion dose) that defines the upper limit under which Static SIMS can be carried out

**Static SIMS** Analysis of the top most monolayer with minimal molecular damage noted in subsequent ion impacts (primary ion dose is less than 1% of the surface density)

**Stationary state** An energy state (level or orbital) in which a bound electron can reside

**Surface** The outer most region of any solid or liquid substrate that dictates how the solid interacts with its environment

**Surface Binding energy** The energy binding of atoms/molecules to a surface

**Transient effects** A form of matrix effect prevalent during the initial stages of sputter depth profiling

**Transient width** The Width, or depth, over which transient effects are noted ($\sim 2.5 \times Rp$)

**Uncertainty Principle** Otherwise known as the Heisenberg Uncertainty Principle. This states that both the momentum and position of any particle can not be simultaneously defined to any degree of precision (only one or the other can be accurately defined)

**Vacuum** Any gaseous environment in which the density of particles exerts a lesser pressure than its surroundings

**Vacuum energy ($E_{vac}$)** The true zero on the energy scale. This is the energy that defines the border between free electrons in the gas phase and electrons bound to an atom, molecule or solid

**Valence electron** An atom or ion's outer electrons (c.f. core electrons)

**Voxel** Volume over which a signal is summed (Pixel is the 2d spatial analogue)

**Work function** Minimum energy required to remove an electron from a solid surface (the difference in energy between the Vacuum level and Fermi energy)

**X-ray notation** A notation scheme used to describe stationary states

**Zalar rotation** Trademarked name introduced by physical electronics to describe centro-symmetric rotation of the substrate

Question: Define the surface density of Silicon atoms in units of atoms/cm$^2$ for the Si(111) surface

Answer: Density $= 7.83 \times 10^{15}$ atoms/cm$^2$

Question: For a ground state ion that has 13 electrons, 14 protons, and 14 neutrons describe:

1) The isotope of the element along with its charge using the chemical symbol and the respective superscripts
2) The quantum numbers, $n$, and $l$ of inner two core electrons
3) The stationary state these electrons reside in using both spectroscopic and X-ray notation

Answers: 1) $^{28}$Si$^+$    2) $n = 1, l = 0$    3) 1 s, $K$

Question: At what mass-to-charge ratio will multiply charged atomic secondary ions of $^{28}$Si$^{2+}$ and $^{29}$Si$^{3+}$ be noted at

Answer: $^{28}$Si$^{2+}$ will be noted at 14 m/q and $^{29}$Si$^{2+}$ will be noted at 9.33 m/q

Question: Derive the theoretical surface Plasmon energy in units of electronvolts for electrons on a pure Silicon surface

Answer: Surface plasmon energy $= 11.73$ eV (Relation 2.6b)

Question: At atmospheric pressure, how quickly will a monolayer of Oxygen form on a clean Silicon surface assuming a sticking coefficient of unity.

Answer: $\sim$1 ns (as derived from Relation 4.4)

Question: What is the mass resolution required to separate the following signals

1) $^{12}$C$_2$$^1$H$_4$$^+$ from $^{12}$C$^1$H$_2$$^{14}$N$^+$
2) $^{16}$O$_2$$^-$ from $^{32}$S$^-$
3) $^{31}$P$^-$ from $^{30}$Si$^1$H$^-$

*Secondary Ion Mass Spectrometry: An Introduction to Principles and Practices*, First Edition.
Paul van der Heide.
© 2014 John Wiley & Sons, Inc. Published 2014 by John Wiley & Sons, Inc.

4) $^{40}Ca^+$ from $^{40}K^+$

5) $^{238}U^+$ from $^{238}Pu^+$ 193804

Answers: 1) ~2220,    2) ~1800,    3) ~3955,    4) ~28403,    5) ~193804

Question: Select options a–e which model best describes:

1) The sputtering of atomic ions from a Silicon surface under 10 keV Ar$^+$ primary ion impact at close to normal incidence.
2) The sputtering of large cluster ions from a dense substrate such as Iridium under 15 keV Au$^{3+}$ primary ion impact at close to normal impact.
3) The general trends noted in the ion yields from MALDI experiments.
   a) The Heat Spike model
   b) The Linear Cascade model
   c) The Electron Tunneling model
   d) The Kinetic Emission model
   e) The Cluster model

Answer: 1) b    2) a    3) e

Question: Select from options a–e which model best describes:

1) Si$^+$ secondary ion formation/survival from Silicon surface under 10 keV Cs$^+$ primary ion impact at close to normal incidence.
2) Si$^{2+}$ secondary ion formation/survival from Silicon surface under 10 keV Cs$^+$ primary ion impact at close to normal incidence.
3) The sputtering of large cluster ions from a dense substrate such as Iridium under 15 keV Au$^{3+}$ primary ion impact at close to normal impact.
   a) The Heat Spike model
   b) The Linear Cascade model
   c) The Electron Tunneling model
   d) The Kinetic Emission model
   e) The Cluster model

Answer: 1) c    2) d    3) a

Question: Select from options a–e which of the following primary ion sources should be used when:

1) Cs$^+$ primary ions of several tens of nA are required
2) O$_2$$^+$ primary ions of several tens of nA at ULE are required:
3) Bi$^+$ primary ions of 50 nm spot size is required:
   a) Electron impact source

b) Radio frequency source

c) Surface ionization source

d) Ion Cyclotron Resonance

e) Field ionization source also known as an LMIG

Answer: 1) c    2) b    3) e

Question: Select from options a–e which of the following secondary ion mass filters (not necessarily those used in commercially available SIMS instruments):

1) Provides the highest $m/\Delta m$ values

2) Provides the highest sensitivity to a specific isotope under Dynamic SIMS conditions

3) Has found greatest utilization in Static SIMS applications

a) Quadrupole mass filter

b) Ion trap mass filter

c) Magnetic Sector mass filter

d) Ion Cyclotron Resonance

e) Time of Flight

Answer: 1) d    2) c    3) e

Question: Select from options a–e which of the following modes of operation should be used when

1) The best possible spatial resolution is required in imaging studies:

2) Analyzing highly insulating samples in a pulsed primary ion Time of Flight instrument:

3) Examining dopant distributions in Silicon wafers:

a) Depth profiling mode

b) Microscope mode

c) Microprobe mode

d) Burst mode

e) Non-interlaced mode

Answer: 1) c    2) e    3) a

# ■■■■■■ REFERENCES[1,2]

Abramenko VA, Ledyankin DV, Urazgil'din IF, Yurasova VE. Nucl Instrum Meth Phys Res 1988;B33:547.

Achtnich T, Burri G, Ply MA, Ilegems M. Appl Phys Lett 1987;50:1730.

Alton DG. Rev Sci Instr 1988;59(7):1039.

AN 339. Available at: http://www.eag.com/documents/si-sio2-sims-detection-limits-AN339.pdf.

AN 434. Available at: http://www.eaglabs.com/documents/ule-b-implants-pcor-sims-AN434.pdf.

AN 435. Available at: http://www.eaglabs.com/documents/dosimetry-plasma-implanted-boron-pcor-sims-AN435.pdf.

Anderson CA. Int J Mass Spectrom Ion Phys 1970;3:413.

Anderson CA, Hinthorne JR. Anal Chem 1973;45:1421.

Andrã HJ. Electronic interactions of ions (atoms) with metal surfaces. In: Briggs JS, Kleinpoppen H, Lutz HO, editors. *Fundamental Processes of Atomic Dynamics*, NATA ASI series. New York: Plenum Press; 1987. p 631.

Arnot FL, Milligan JC. Proc R Soc London A 1936;156.

Aston FW. *Isotopes*. London: Arnold; 1922. p 152.

Aumayr F, Winter H. Nucl Instrum Meth Phys Res B 1994;90:523.

Aumayr F, Winter H. Phil Trans R Soc London A 2003;362:77.

Barat M, Lichten W. Phys Rev A 1972;6:211.

Barker GF. *Divisions of Matter. A Text-Book of Elementary Chemistry: Theoretical and Inorganic*. John F Morton & Co.; 1870.

Barth HJ, Muhling E, Eckstein W. Surf Sci 1986;166:458.

Behrisch R. In: Wittmaack K, editor. *Sputtering by Particle Bombardment. III*. Berlin: Springer; 1991.

Behrisch R, Eckstein W. *Sputtering by Particle Bonbardment; Experiments and Computer Simulations from Threshold to MeV Energies*. Berlin: Sprinler-Verlag; 2007.

Belu AM, Graham DJ, Castner DG. Biomaterials 2003;24:3635.

Benninghoven A. Physica Status Solidi 1969;34(2):K169–K171.

---

[1]Note I: This is not a comprehensive reference list, rather a condensed list.
[2]Note II: Full titles are only used for books or chapters of books.

---

*Secondary Ion Mass Spectrometry: An Introduction to Principles and Practices*, First Edition.
Paul van der Heide.
© 2014 John Wiley & Sons, Inc. Published 2014 by John Wiley & Sons, Inc.

Benninghoven A. Chem Phys Lett 1970;6:626.

Benninghoven A. Surf Sci 1973;35:427.

Benninghoven A. Surf Sci 1975;53:596.

Benninghoven, A. *Ion Formation from Organic Solids, Springer Series in Chemical Physics 25*. A. Benninghoven. Springer-Verlag: Berlin, (1983) p 77.

Benninghoven A. Angew Chem Int Ed Engl 1994;33:1023.

Benninghoven A, Loebach E. Rev Rev Sci Sci Instrum 1971;42:4949.

Benninghoven A, Rudenauer F.G., Werner, H.W. *Secondary Ion Mass Spectrometry: Basic Concepts, Instrumental Aspects, Applications and Trends*. Wiley & Sons: New York, (1987).

Berghmans B, van Daele B, Geenen L, Conrad T, Franquete A, Vandervorst W. Appl Surf Sci 2008;255:1316.

Bergland M, Weiser M. Pure and Appl Chem 2011;83(2):397.

Berthold W, Wucher A. Nucl Instrum Meth Phys Res B 1996;115:411.

Berthold W, Wucher A. Phys Rev B 1997;56:4251.

Betz G. Nucl Instrum Meth 1987;B27:104.

Biesack JP, Eckstein W. Appl Phys A 1984;34:73.

Boxer SG, Kraft ML, Weber PK. Annu Rev Biophys 2009;38:53.

Brene DA, Garrison BJ, Winograd N, Postawa Z, Wucher A, Blenkinsopp P. Phys Chem Lett 2011;2:2009.

Buyuklimanli TH, Marino JW, Novak SW. Appl Surf Sci 2004;231:636.

Castaing R, Slodzian GJ. J Microscopie 1962;1:395–399.

Castaing R, Slodzian GJ. J Phys 1981;E14:1119.

Chadwick JFRS. Proc R Soc A 1932;136:692.

Chandra S, Morrison GH. Biol Cell 1992;74:31.

Chait BT, Standing KG. Int J Mass Spectrom Ion Phys 1981;40:185.

Chryssoulis SL, Cabri LJ, Lennard W. Econ Geol 1989;84:1684.

Clegg BJ. Surf Interface Anal 1987;10:332.

Clegg BJ, Beal RB. Surf Interface Anal 1989;14:307.

Criegern R, Weitzel I. *SIMS V Proceedings*. Germany: Springer-Verlag Berlin; 1986. p 319.

Coplen TB et al. Pure Appl Chem 2002;74(10):1987.

Cuerno R, Barabasi A-L. Phys Rev Lett 1975;74:4746.

De De Bievre P, Peiser HS. Pure Appl Chem 1992;64(10):1535.

De Chambost E, Monsallut P, Rasser B, Schuhmacher M. Applied Surface Science, 2003;391:203–204.

Delcorte A, Wojciechowski I, Gonze X, Garrison BJ, Bertrand P. Int J Mass Spectrom 2002;214:213.

Deline VR, Katz W, Evans CA Jr, Williams P. Appl Phys Lett 1978;33:830.

Downard KM. Eur J Mass Spectrom 2007;13(3):177–190.

Dowsett MG. Appl Surf Sci 2003;203:5.

Dowsett MG, Barlow RD, Allen PN. J Vac Sci Technol B 1994;12(1):189.

Dowsett MG, Thompson SP, Corlett CA. *SIMS VIII proceedings*. New York: John Wiley & Sons; 1992. p 187.

Eckstein W. Surface and Interface Analysis 1989;14:799.

Fano U, Lichten W. Phys Rev Lett 1965;14:627.

Fayek M. *Seconday Ion Mass Spectrometry in the Earth Sciences*. Mineralogical Association of Canada Short Course, Vol 41; 2009.

Finzi-Hart JA et al. Proc Natl Acad Sci U S A 2009;106(15):6345.

Friedbacher, PhD Thesis, Technical University of Vienna, 1988.

Fletcher JS, Vickerman JC, Winograd N. Curr Opin Chem Biol 2011;15:1.

Franzreb K, van der Heide PAW. *SIMS XI Proceedings*. Vol. 883. New York: Wiley; 1998.

Garrison BJ, Postawa Z. Mass Spectrom Rev 2008;27:289.

Gehre D, et al. Characterization and Metrology for ULSI Technology: 2000 International Conference Proceedings, American Institute of Physics, 692 (2001).

Gilbert WDM. *Book 2, Chapter 2, London (1600)*, Translation by Mottelay, P.F. New York: Dover; 1958.

Gillen G, Roberson S. Rapid Commun Mass Spectrom 1998;12:1303.

Gilmore IS, Seah MP. Appl Surf Sci 2000;161:465.

Gilmore IS, Seah MP. Appl Surf Sci 2003;203:551.

Glish GL, Vachet W. Nat Rev Drug Discov 2003;2:140.

Gnaser H, Hutcheon ID. Phys Rev B 1987;35:877.

Gnaser H. Phys Rev B 1996a;54:16456.

Gnaser H. Phys Rev B 1996b;54:171411.

Gnaser H. *Low-Energy Ion Irradiation of Solid Surfaces*. Berlin: Springer; 1999.

Gnaser H, Ichiki K, Matsuo J. Rapid Common Mass Spectrom 2012;26:1.

Goldstien E. Erh Dtsch Phys Ges 1902;4:228.

Grasserbauer M, Stingeder G, Friedbacher G, Virag A. Surf Interface Anal 1989;14:623.

Green FM, Gilmore IS, Seah MP. Applied Surface Science 2008;255:852

Gries H. Surf Interface Anal 1985;7:29.

Groves WR. Phil Trans R Soc London 1852;142:87.

Guenther T, Jiang ZX, Kim K, Sieloff DDJ. Vac Sci Technol 2009;B27(2):677.

Harbich W. Chapter 4: Collisions of clusters with surfaces. In: Meiwes-Broer K-H, editor. *Metal Clusters at Surfaces, Structure, Quantum Properties, Physical Chemistry*. Berlin: Springer-Verlag; 2000.

Herzog RFK, Viehboeck F. Phys Rev 1949;76(6):855–856.

Higashi Y, Homma Y. Anal Sci 1998;14:281.

Hofer WO, Liebl H, Roos G, Straudemaier G. Int J Mass Spectrom Ion Phys 1976;19:327.

Hofmann S. Phil Trans R Soc London A 2004;362(1814):55–75.

Hofmann S, Sanz JM. In: Oechsner H, editor. *Topics in Current Physics*. Berlin: Springer, Vol 37; 1984. p 141.

Homma Y, Takano A, Higashi Y. Appl Surf Sci 2003;35:203–204.

Honda F, Lancaster GM, Fukuda Y, Rabalais JW. J Chem Phys 1978;69:4931.

Hsieh H, Averback RS, Sellers H, Flynn CP. Phys Rev B 1992;45:4417.

Hubbard J. Proc R Soc A 1963;176:328.

Hüfner S. *Photoelectron Spectroscopy: Principles and Applications*. 3rd ed, Springer: Berlin, (2003).

Hung KK et al. IEEE Transact Electron Device 1990;37(3):654.

Iltgen, K. PhD Thesis, Westfaelische Wilhelms-Universitaet: Munster, Germany, 1997.

Imada M, Fujimori A, Tokura Y. Rev Mod Phys 1998;70:1039.

Ireland T, Williams IS. Rev Mineral Geochem 2003;53(1):215.

Joyes P. Radiat Eff 1973;19:235.

Joyes P.J. de Phys 1969a;30:243.

Joyes P.J. de Phys 1969b;30:365.

Klemperer O. *Electron Optics*. 3rd ed. Cambridge University Press; 1971.

Klushin DV, Gusev MY, Lysenko SA, Urazgildin IF. Phys Rev B 1996;54:7062.

Knochenmuss R. In: Cole R, editor. *MALDI Ionization Mechanisms: An Overview (Chapter 5) in Electrospray and MALDI Mass Spectrometry: Fundamentals Instrumentation Practicalities and Biological Applications*. 2nd ed. New York: Wiley and Sons; 2010.

Knochenmuss R. J Mass Spectrom 2002;37:867.

Krauss AR, Gruen DM. Surface Science 1980;92:14.

Lancaster GM, Honda F, Fukuda Y, Rabalais JW. J Am Chem Soc 1979;101:1951.

Landau L. Z Physike Sowjetunion 1932;2:46.

Lawson JD. *The Physics of Charged Particle Beams*. New York: Oxford University Press; 1978.

Lechene C et al. J Biol 2006;5:20.

Lichten W. Phys Rev A 1967;164:131.

Lichten W. Phy Chem 1980;4:2102.

Liebl HJ. J Appl Phys 1967;38(13):5277–5280.

Liedke B, Heing K-H, Moller W. Nucl Instrum Meth Phys Res B 2013;316:56.

Losing R et al. *SIMS XI Proceedings*. Vol. 1017. John Wiley & Sons; 1998.

Lu C, Wucher A, Winograd N. Anal Chem 2011;83:381.

Ludermann HC, Redmond RW, Hillenkamp F. Rapid Comm Mass Spectrom 2002;16:1287.

Magee CA. Nucl Instrum Meth 1981;191:297.

Magee CW, Harrington W, Honig RE. Rev Scient Instrum 1978;49(4):477–485.

Magee CW, Honig RE. Surf Interface Anal, 4(2), 35, 1982.

Magee CW, Harrington WL, Honig RE. Rev Sci Instrum 1978;49:477.

Mahoney CM. *Cluster Secondary Ion Mass Spectrometry*: Wiley, Hoboken, New Jersey;2013.

Mahoney CM. Mass Spectrom Rev 2009;1.

Malherbe JB. Crit Rev Solid State Mater Sci 1994;19:55.

Mann K, Yu ML. Phys Rev B 1987;35:6043.

Marazov P, Samartsev AV, Wucher A. Appl Surf Sci 2006;252:6452.

March R. J Mass Spectrom 1997;32:351.

Mathieu. J Math Pure Appl (J Jiouville) 1868;13:137.

Metson JB et al. Surf Interface Anal 1983;5:181.

McGilvery, D.C. PhD Thesis, Latrobe University, Victoria, (1978).

McPhail D, Dowsett M. In: Vickerman JC, Gilmore IS, editors. *Surface Analysis: The Principle Techniques*. Wiley; 2009.

McPhail DS, Li L, Chater RJ, Yakovlev N, Seng H. Surf Interface Anal 2011;43:479.

Mochiji KJ. Anal Bioanal Technol 2011;S2:1.

Mott NF. Proc Phys Soc London A 1949;62:416.

Murray PT, Rabalais JW. J Am Chem Soc 1981;103:1007.

Newton G, Unsworth PJ. J Appl Phys 1976;47:70.

Nier A. Rev Sci Instrum 1947;18:398.

Nishi Y, Doering R. *Handbook of semiconductor manufacturing technology*, 2nd ed., CRC press Taylor & Francis Group, Florida; 2007.

Norskov JK, Lundqvist BI. Phys Rev B 1979;19:5661.

O'Connor DJ et al. Surf Sci 1988;197:277.

O'Connor DJ, Sexton BA. Smart R, St C. *Surface Analysis Methods in Materials Science*, 2nd ed.: Springer; 2003.

Oechsner H, Gerhard W. Surf Sci 1974;44:480.

Orloff J. Rev Sci Instrum 1993;64:1105.

Orphan VJ, House CH. Geobiology 2009;7:360.

Orth RG, Jonkman HT, Michl J. J Am Che Soc 1982;104:1834.

Pachuta SJ, Cooks RJ. Chem Rev 1987;87:647.

Palmblad M, Hakansson K, Hakansson P. et al. Eur J Mass Spectrom 2000;6:267.

Passarelli MK, Winograd N. Biochem Biophys Acta 2011;1811:976.

Paul W, Steinwedel H. Zeitschrift Fur Naturforschung 1953;8a:448.

PCOR-SIMS. Available at: http://www.eaglabs.com/documents/BRO16.pdf.

Plog C, Gerhard W. Zeitschrift fur Physik B Condensed Matter 1983a;54:59.

Plog C, Gerhard W. Zeitschrift fur Physik B Condensed Matter 1983b;54:71.

Redead PA, Hobson JP, Kornelsen EV. *The Physical Basis of Ultra High Vacuum*. London: Chapmann and Hall; 1968. Reprinted by the American Institute of Physics, New York, (1993).

Redhead PA, Hobson JP, Kornelsen EV. *The Physical Basis of Ultra high Vacuum, American Institute of Physics*, New York; 1968.

Rabalais JW, editor. *Low Energy Ion-Surface Interactions*. New York: Wiley; 1994.

Reed SJB. Mineralogical Magazine 1989;53:3.

Riviere JC, Myhra S. *Handbook of Surface and Interfacial Analysis: Methods for Problem-Solving*. 2nd ed. New York: CRC Press; 2009.

Robinson MA, Graham DJ, Castner DG. Anal Chem 2012;84(11):4880.

Rutherford E. Mineralogical Magazine 1911;21:669–688.

Rutherford E. Mineralogical Magazine 1919;37:537.

Rzeznik L et al. Phys Chem C 2008;112(2):521.

Schiott HE, Dan K. Vidensk Selsk Mat Fys Medd 1966;35:9.

Schiott HE. Radiat Eff 1970;6:107.

Schuetzle D et al. Rev Sci Instrum 1989;60:53.

Scymczak W, Wittmaack K. Nucl Instrum Meth Phys Res B 1993;82:220.

Seah MP, Nunney TS. J Phys D Appl Phys 2010;43:25.

Seah MP, Green FM, Gilmore ISJ. Phys Chem C 2010;114(12):5351.

Setz P, Knochenmuss R. J Phys Chem A 2005;106:4030.

Shimazu Y, Takano A, Itoh KM. Appl Surf Sci 2008;255:1345.

Sigmund P. Phys Rev 1969;184:383.

Sigmund P. Appl Phys Lett 1974;25:169.

Sigmund P. In: Behrisch R, editor. *Sputtering by Particle Bombardment. I. Physical Sputtering of Single Element Solids.* Vol. 47. Springer; 1981.

Smartsec AV, Duvendeck A, Wucher A. Phys Rev 2005;B72:115417.

Smith R. *Atomic and Ion Collisions in Solids and at Surfaces, Theory, Simulations and Applications.* Cambridge University Press; 2005.

Smith DF et al. Anal Bioanal Chem 2013;405(18):6069.

Smith DF et al. Anal Chem 2011;83:9552.

Smith NS, Tesch PP, Martin NP, Kinion DE. Appl Surf Sci 2008;255:1606.

Soddy F. Nature 1913;92:399.

Soldzian G, Henneguin JF. Compt Rend Acad Sci B 1966;263:1246.

Stuckleberg ECG. Helv Phys Acta 1932;5:369.

Szakal C et al. Phys Rev Lett 2006;96:216104.

Taylor S. Proceedings of the Royal Society A 280; 1964;1382–383.

Thomas GE. Radiat Eff 1977;31:185.

Thomson JJ. Mineralogical Magazine 1897;44:293.

Thomson JJ. Mineralogical Magazine 1910;20:752–767.

Thomson JJ. Proc R Society A 1913;89:1–20.

Thompson JS. Phys Rev 1931;38:1389.

Thompson MW. Mineralogical Magazine 1968;18:377.

Thompson SP et al. *SIMS VIII.* Vol. 183. New York: John Wiley & Sons; 1992.

Todd PJ, Schaaff TG. Journal of the American Society for Mass Spectrometry, 2002;13(9):1099–1107.

Tully JC, Tolk NH. In: Tolk NH et al., editors. *Inelastic Ion-Surface Collisions.* New York: Academic Press; 1977.

Urazgil'din IF, Borisov AG. Vacuum 1990;40:461.

Urbassek HM, Michl J. Nucl Intrum Meth Phys Res 1987;B22:480.

Valerie S. Surf Sci Rep 1993;17:85.

van der Heide PAW. Nucl Intrum Meth Phys Res 1994a;B93:421.

van der Heide PAW. Surf Sci Lett 1994b;302:L312.

van der Heide PAW. *SIMS XI Proceedings.* Vol. 821. New York: John Wiley & Sons; 1998.

van der Heide PAW. Nucl Instrum Meth Phys Res B 1999;157:126.

van der Heide PAW. Surf Sci 2004;555:193.

van der Heide PAW. Nucl Instrum Meth B 2005;229:35.

van der Heide PAW. Appl Surf Sci 2006;252:6433.

van der Heide PAW. J Electron Spectrosc Relat Phenom 2008;164:8.

van der Heide PAW. *X-ray Photoelectron Spectroscopy: An Introduction to Principles and Practices.* Hoboken, New Jersey: Wiley; 2012.

van der Heide PAW, Metson JB, Tui DL. Surf Sci 1993;280:208.

van der Heide PAW, Fichter DA. *SIMS XI Proceedings.* Vol. 747. New York: John Wiley & Sons; 1998.

van der Heide PAW et al. *SIMS XII Proceedings*. Vol. 485. Amsterdam: Elsevier; 2000.

van der Heide PAW, Karpusov D. *SIMS XI Proceedings*. Vol. 933. New York: Wiley; 1998.

van der Heide PAW, Karpusov D. *SIMS XII Proceedings*. Vol. 147. Amsterdam: Elsevier; 2000.

van der Heide PAW et al. Nucl Intrum Meth Phys Res B 2003;201:413.

van der Heide PAW, McIntyre NS. Surf Interface Anal 1993;20:1000.

van der Heide PAW, Warr BD, McIntyre NS. Scanning Microsc 1998;12(1):171.

van der Heide PAW et al. Surf Interface Anal 1994;21:747.

van der Weg WF, Bierman. J Physica 1969;44:708.

Vasile MJ. Phys Rev B 1984;29:3785.

Vickerman JC, Gilmore IS. *Surface analysis; The Principle Techniques*, 2nd ed.: Wiley; 2009.

Vickerman JC. Surf Sci 2009;603(10):1926.

Vickerman JC, Briggs D. *TOF-SIMS: Surface Analysis by Mass Spectrometry*. 2nd ed. Chichester, UK: IM Publications; 2012.

Wahl M, Wucher A. Nucl Instrum Meth B 1994;94:36.

Weast RC, Astle MJ, William HB, editors. *CRC Handbook of Chemistry and Physics*. 64th ed. Florida: Baton Rogue; 1985.

Wehner GK. J Appl Phys 1955:261056.

Wien W. Annal Phys 1902;8:260.

Wieser ME, Coplan TB. Pure Appl Chem 2011;83:359.

Williams P. Surf Sci 1979;90:588.

Williams P. Rev B 1981;23:6187.

Williams P, Evans CA. Surf Sci 1978;78:324.

Wilson RG, Stevie FA, Magee CW. *Secondary Ion Mass Spectrometry: A Practical Handbook for Depth Profiling and Bulk Impurity Analysis*. New York: Wiley-Interscience Publication; 1989.

Winograd N, Garrison BJ. Annu Rev Phys Chem 2010;5(61):305.

Wirtz T et al. J Appl Phys Lett 2012;101:041301.

Wittmaack K. Nucl Instrum Meth 1980;168:343.

Wittmaack K. Int J Mass Spectrom Ion Phys 1975;17:39–50.

Wittmaack K. Radiat Eff 1982;63:205.

Wittmaack K. Philos Trans R Soc London A 1996;354:2731.

Wittmaack KJ. Vac Sci Technol B 2000;18:1.

Wong K, Vongehr S, Kresin VV. Phys Rev B 2003;67:035406.

Woodcock KS. Phys Rev 1931;38:1696.

Wucher A, Oechsner H. Surf Sci 1988;199:567.

Yu ML. Phys Rev Lett 1978;40:574.

Yu ML. Phys Rev Lett 1981;47:1325.

Yu ML. Phys RevB 1982;26:4731.

Yu MLJ. Vac Sci Technol A 1983;1(2):500.

Yu M. Charged and Excited states of sputterd atoms (Chapter 3). In: Behrisch R, Wittmaack K, editors. *Sputtering by Particle Bombardment III: Characteristics of Sputtered Particles, Technical Applications.* Berlin: Springer Verlag; 1991.

Yu ML, Lang ND. Phys Rev Lett 1983;50:127.

Yu ML, Lang ND. Nucl Instrum Methods Phys Res B 1986;14(403).

Yu ML, Mann K. Phys Rev Lett 1986;57:1476.

Zalar A. Thin Solid Films 1985;124:223.

Zalm PC. Rep Prog Phys 1995;58:1321.

Zannan J, Sawastsky GA, Allan JW. Phys Rev Lett 1985;55:418.

Zener C. Proc R Soc London A 1932;137:696.

*Secondary Ion Mass Spectrometry: An Introduction to Principles and Practices*, First Edition.
Paul van der Heide.
© 2014 John Wiley & Sons, Inc. Published 2014 by John Wiley & Sons, Inc.

Printed and bound by CPI Group (UK) Ltd, Croydon, CR0 4YY

16/04/2025

14658530-0003